Lecture Notes in Artificial Intelligence   3863

Edited by J. G. Carbonell and J. Siekmann

Subseries of Lecture Notes in Computer Science

T0216269

Michael Kohlhase (Ed.)

Mathematical Knowledge Management

4th International Conference, MKM 2005
Bremen, Germany, July 15-17, 2005
Revised Selected Papers

 Springer

Series Editors

Jaime G. Carbonell, Carnegie Mellon University, Pittsburgh, PA, USA
Jörg Siekmann, University of Saarland, Saarbrücken, Germany

Volume Editor

Michael Kohlhase
International University Bremen
School of Engineering and Science
Campus Ring 1, 28758 Bremen, Germany
E-mail: m.kohlhase@iu-bremen.de

Library of Congress Control Number: 2006920149

CR Subject Classification (1998): I.2, H.3, H.2.8, I.7.2, F.4.1, H.4, C.2.4, G.4, I.1

LNCS Sublibrary: SL 7 – Artificial Intelligence

ISSN   0302-9743
ISBN-10   3-540-31430-X Springer Berlin Heidelberg New York
ISBN-13   978-3-540-31430-1 Springer Berlin Heidelberg New York

Springer is a part of Springer Science+Business Media

springer.com

© Springer-Verlag Berlin Heidelberg 2006
Printed in Germany

Typesetting: Camera-ready by author, data conversion by Scientific Publishing Services, Chennai, India
Printed on acid-free paper   SPIN: 11618027   06/3142   5 4 3 2 1 0

Preface

This volume contains the proceedings of the Fourth International Conference on Mathematical Knowledge Management MKM 2005 held July 15–17, 2005 at International University Bremen, Germany. Previous conferences have been at the Research Institute for Symbolic Computation (RISC) Linz, Austria (September 2001), at Bertinoro, Italy (March 2003), and Bialowiecze, Poland (September 2004).

Mathematical knowledge management (MKM) is a field in the intersection of mathematics and computer science, providing new techniques for managing the enormous volume of mathematical knowledge available in current mathematical sources and making it available through the new developments in information technology.

The annual MKM Conference brings together mathematicians, software developers, publishing companies, math organizations, math users, and educators to exchange their views and approaches, current activities and new initiatives.

For the first time, MKM 2005 chose to have post-conference proceedings, as otherwise the submission deadline would have collided with other conferences and crimped time since MKM 2004 in September 2004. The decision also facilitated keeping the conference open to new ideas as well as keeping up the maturity of the papers necessary for inclusion into archival proceedings. With a May 15 deadline, MKM 2005 received 38 submissions. Each submission was reviewed by at least three programme committee members. The committee decided to accept 27 papers for presentation at the conference. Out of these, 26 papers were accepted for publication in the conference proceedings after re-evaluation by the Programme Committee since they included significant improvements triggered by the referee reports and the discussions at the conference.

As MKM is a small conference with a tightly knit community of authors, submissions by Programme Committee members were allowed: six submissions included committee members, but the review process was kept inaccessible to them. One submission was co-authored by the Program Chair; its review process was organized independently by Bill Farmer.

The papers in this volume cover the whole area of mathematical knowledge management. Topics range from foundations and the representational and document-structure aspects of mathematical knowledge, over process questions like authoring, migration, and consistency management by automated theorem proving to applications in eLearning and case studies.

I am grateful to Tom Hales for agreeing to give an invited talk at MKM 2005, to the Programme Committee, and the external reviewers for their excellent work and dedication to the MKM 2005 program. The work of the Programme Committee and the preparation of the proceedings were greatly simplified by Andrei Voronkov's excellent `EasyChair` system.

October 2005   Michael Kohlhase

Conference Organization

Programme Chair

Michael Kohlhase, International University Bremen, Germany

Programme Committee

Andrew A. Adams, The University of Reading, UK
Andrea Asperti, University of Bologna, Italy
Richard Baraniuk, Rice University, USA
Christoph Benzmüller, Saarland University, Germany
Olga Caprotti, University of Helsinki, Finland
Mike Dewar, NAG, Ltd., UK
William Farmer, McMaster University, Canada
Tetsuo Ida, University of Tsukuba, Japan
Fairouz Kamareddine, Heriott Watt University, UK
Robert Miner, Design Science, USA
Till Mossakowski, University of Bremen, Germany
Andrzej Trybulec, University of Bialystok, Poland
Stephen Watt, University Western Ontario, Canada

Local Organization

The local organization was carried out by the Conference Chair in collaboration
with Events4, a conference organization company founded and operated by IUB
students.

External Reviewers

Morten Andersen
Grzegorz Bancerek
Rebhi Baraka
Marc Buckley
Chad Brown
Paul Cairns
David Carlisle

Arjeh Cohen
Armin Fiedler
Yukiyoshi Kameyama
Artur Kornilowicz
Klaus Luettich
Christoph Lüth
Mircea Marin

Andreas Meier
Yasuhiko Minamide
Immanuel Normann
Martijn Oostdijk
Matti Pauna
Martin Pollet

Krzysztof Prazmowski
Christoph Schwarzweller
Andreas Strotmann
Diedrich Wolter
Jürgen Zimmer

Table of Contents

Session V: MKManagement Tools

Session VI: Documents

Session VII: MKM Case Studies

Session VIII: Course Materials

Session IX: Migration

A Proof-Theoretic Approach to Hierarchical Math Library Organization

Kamal Aboul-Hosn and Terese Damhøj Andersen

Department of Computer Science, Cornell University, Ithaca, New York, USA
kamal@cs.cornell.edu, katten@kattens.dk

Abstract. The relationship between theorems and lemmas in mathematical reasoning is often vague. No system exists that formalizes the structure of theorems in a mathematical library. Nevertheless, the decisions we make in creating lemmas provide an inherent hierarchical structure to the statements we prove. In this paper, we develop a formal system that organizes theorems based on *scope*. Lemmas are simply theorems with a local scope. We develop a representation of proofs that captures scope and present a set of proof rules to create and reorganize the scopes of theorems and lemmas. The representation and rules allow systems for formalized mathematics to more accurately reflect the natural structure of mathematical knowledge.

1 Introduction

The relationship between theorems and lemmas in mathematical reasoning is often vague. What makes a statement a lemma, but not a theorem? One might say that a theorem is "more important," but what does it mean for one statement to be "more important" than another? When writing a proof for a theorem, we often create lemmas as a way to break down the complex proof, so perhaps we expect the proofs of lemmas to be shorter than the proofs of theorems. We also create lemmas when we have a statement that we do not expect to last in readers' minds, i.e., it is not the primary result of our work. The way we make these decisions while reasoning provides an inherent hierarchical structure to the set of statements we prove. However, no formal system exists that explicitly organizes proofs into this hierarchy.

Theorem provers such as NuPRL, Coq, and Isabelle provide the ability to create lemmas. But their library structures are flat, and no formal distinction exists between lemmas and theorems [1, 2, 3]. The reasons to distinguish lemmas from theorems in these systems is the same as the reasons in papers: to ascribe various levels of importance and to introduce dependency or scoping relationships.

We seek to formalize these notions and provide a proof-theoretic means by which to organize a set of proofs in a hierarchical fashion that reflects this natural structure. Our thesis is that the qualitative difference between theorems and lemmas is in their *scope*. Scope already applies to mathematical notation. Never in a paper would one need to define the representation of a set ($\{\ldots\}$) nor operators such as union and intersection. Set notation is standard, thus has a

M. Kohlhase (Ed.): MKM 2005, LNAI 3863, pp. 1–16, 2006.

global scope that applies to any proof. However, one often defines operators that are only used for a single paper; the author does not intend for the notation to exist in other papers with the same meaning without being defined again. Similarly, a *theorem* is a statement that can be used in any other proof. Its scope is global, just as set notation. A *lemma* is a statement with a local scope limited to a particular set of proofs. We want a system that represents and manipulates scope formally through the structure of the library of proofs.

In this paper, we propose such a system. First, we propose a formal definition of scoping for proof libraries. Next, we describe a representation of proofs that is able to capture this definition of scope based on work by Kozen and Rama-narayanan [4]. We provide a set of formal rules to create and reorganize the scopes of theorems and lemmas.

We believe that the ability to create and manage complex scoping and dependency relationships among proofs will allow systems for formalized mathematics to more accurately reflect the natural structure of mathematical knowledge.

2   A Motivating Example

Consider reasoning about a Boolean algebra $(B, \vee, \wedge, \neg, 0, 1)$. Boolean algebra is an equational theory, thus contains the axioms of equality:

$$\text{ref} : x = x \tag{1}$$

$$\text{sym} : x = y \rightarrow y = x \tag{2}$$

$$\text{trans} : x = y \rightarrow y = z \rightarrow x = z \tag{3}$$

$$\text{cong}_\wedge : x = y \rightarrow (z \wedge x) = (z \wedge y) \tag{4}$$

$$\text{cong}_\vee : x = y \rightarrow (z \vee x) = (z \vee y) \tag{5}$$

$$\text{cong}_\neg : x = y \rightarrow \neg x = \neg y \tag{6}$$

All variables are implicitly universally quantified in these axioms. Suppose we wanted to prove the following elementary fact:

Theorem 1.

$$\forall a, b, c, z.a = b \rightarrow a = c \rightarrow z \vee (a \wedge b) = z \vee (a \wedge c) \tag{7}$$

Here is how a proof might go. First, we could prove a lemma.

Lemma 1.

$$\forall x, y, z.x = y \rightarrow z \vee (x \wedge x) = z \vee (x \wedge y) \tag{8}$$

Using $a = b$ and $a = c$ from the statement of our theorem, we could apply the lemma under the substitutions $[x/a, y/b, z/z]$ and $[x/a, y/c, z/z]$ to deduce

$$z \vee (a \wedge a) = z \vee (a \wedge b) \tag{9}$$

$$z \vee (a \wedge a) = z \vee (a \wedge c) \tag{10}$$

Next, we know from applying symmetry to (9) that

$$z \vee (a \wedge b) = z \vee (a \wedge a) \tag{11}$$

Finally we conclude from transitivity, (9), and (11) that

$$z \vee (a \wedge b) = z \vee (a \wedge c)$$

which is what our theorem states.

We may decide that (8) does not apply to theorems other than (7), and consequently, should only have a scope limited to the proof of (7). Our representation of proofs makes explicit the limited scope of (8).

Another important observation is that in all places we use (8), the variable z from (7) is always used for the variable z in the lemma. We may wish not to universally quantify z for both (7) and (8) individually, but instead universally quantify z once and for all so that it can be used by both proofs:

$$\forall z, \ \forall a, b, c, a = b \rightarrow a = c \rightarrow z \vee (a \wedge b) = z \vee (a \wedge c)$$
$$\text{and } \forall x, y, x = y \rightarrow z \vee (x \wedge x) = z \vee (x \wedge y) \tag{12}$$

Moving the quantifier for z looks like a simple task, applying the first order logic rule

$$(\forall z. \varphi) \wedge (\forall z. \psi) \equiv \forall z. (\varphi \wedge \psi)$$

However, the proof of the lemma itself must also change, as must any proof that is dependent on this lemma.

Although either version of the lemma can be used to prove the theorem, note that their meanings are subtly different because of the placement of the quantification. Placing a separate quantification of z as in (8) makes the lemma read: "Lemma 1: For all x, y, and z,..." In this case, z is a variable in the lemma for which we expect there to be a substitution whenever the lemma is used in a proof. Using one quantification for both the theorem and the lemma as in (12) makes the lemma read: "Let z be an arbitrary, but fixed boolean value. Lemma 1: For all x and y..." In this case, z is a fixed constant for the lemma.

In this simple example, using (8) or (12) does not matter. However, in other cases, the choices made for quantification may reflect a general style in one's proofs. One may like lemmas to be as general as possible, universally quantifying any variables that appear in the lemma and relying on no constants. On the other hand, one may want to make lemmas as specific as possible, applying only in a select few proofs in order to minimize the number of quantifications. We want to capture this subtle difference formally in our representation of proofs in order to allow the user to choose the representation that best fits the intended meaning.

3   Proof Representation

For representing theorems and lemmas like those in Section 2, we use proof terms similar to those defined in a paper by Kozen and Ramanarayanan [4]. Their paper presents a *publish-cite* system, which uses proof rules with an explicit library to formalize the representation and reuse of theorems. The system of [4] uses universal Horn equational logic, and we do as well, since it is a good vehicle for illustrating the organization and reuse of theorems. There is no inherent limitation in the system that requires the use of this logic; it could be extended to work with more complex deductive systems.

We use the word "theorem" to mean a theorem, lemma, or axiom. We build theorems from terms and equations. Consider a set of *individual variables* $X = \{x, y, \ldots\}$ and a first-order signature $\Sigma = \{f, g, \ldots\}$. An *individual term* $s, t, \ldots$ is either a variable $x \in X$ or an expression $f t_1 \ldots t_n$, where f is an n-ary function symbol in Σ and $t_1 \ldots t_n$ are individual terms. An equation $d, e, \ldots$ is between two individual terms, such as $s = t$.

A *theorem* is a universally quantified Horn formula of the form

$$\forall x_1, \ldots x_m . \varphi_1 \rightarrow \varphi_2 \rightarrow \cdots \rightarrow \varphi_n \rightarrow \psi \tag{13}$$

where the φ_is are equations representing *premises*, ψ is an equation representing the conclusion, and $x_1 \ldots x_m$ are the variables that occur in the equations $\varphi_1, \ldots, \varphi_n, \psi$. A formula may have zero or more premises. These universally quantified formulas allow arbitrary specialization through term substitution. An example of this is the use of (8) with substitutions to get (9) and (10).

Let $\mathcal{P}$ be a set of *proof variables* $p, q, \ldots$. A proof of a theorem is a λ-term abstracted over both the proof variables for each premise of a theorem proven by the proof and the individual terms that appear in the proof. A *proof term* is:

- a variable $p \in \mathcal{P}$
- a constant, referring to the name of a theorem
- an application $\pi\tau$, where π and τ are proof terms
- an application πt, where π is a proof term and t is an individual term
- an abstraction $\lambda p.\tau$, where p is proof variable and τ is a proof term
- an abstraction $\lambda x.\tau$, where x is an individual variable and τ is a proof term

When creating proof terms, we have the typing rules seen in Table 1. These typing rules are what one would expect for a simply-typed λ-calculus. The typing environment Γ maps variables and constants to types. According to the Curry-Howard Isomorphism, the type of a well-typed λ-term corresponds to a theorem in constructive logic and the λ-term itself is the proof of that theorem [5]. For example, a theorem such as (13) viewed as a type would be realized by a proof term representing a function that takes an arbitrary substitution for the variables x_i and proofs of the premises φ_i and returns a proof of the conclusion ψ.

In [4], a library of theorems is represented as a flat list of proof terms. All of the theorems have global scope, i.e., they are able to be cited in any other proof in the library.

The goal of this paper is to provide a scoping discipline so that naming and use of variables can be localized. The proof term itself should tell us in which proofs we can use a lemma. We use a construct similar to the SML let expression, which limits the scope of variables in the same way we wish to limit the scope of lemmas.

Table 1. Typing rules for proof terms

$$\overline{\Gamma, p : e \vdash p : e}$$

$$\frac{\Gamma \vdash \pi : e \to \varphi \quad \Gamma \vdash \tau : e}{\Gamma \vdash \pi\tau : \varphi}$$

$$\frac{\Gamma, p : e \vdash \tau : \varphi}{\Gamma \vdash \lambda p.\tau : e \to \varphi}$$

$$\overline{\Gamma, c : \varphi \vdash c : \varphi}$$

$$\frac{\Gamma \vdash \pi : \forall x.\varphi}{\Gamma \vdash \pi t : \varphi[x/t]}$$

$$\frac{\Gamma \vdash \tau : \varphi}{\Gamma \vdash \lambda x.\tau : \forall x.\varphi}$$

In order to represent theorems in a hierarchical fashion, we add two kinds of proof terms:

- a sequence $\tau_1; \ldots; \tau_n$, where $\tau_1, \ldots, \tau_n$ are proof terms. This allows several proofs to use the same lemmas. Sequences cannot occur inside applications.
- an expression let $L_1 = \tau_1 \ldots L_n = \tau_n$ in τ end. This term is meant to express the definition of a set of lemmas for use in a proof term τ. The τ_is are proof terms, each bound to an identifier L_i. With the existence of the sequences, each τ_i may define the proof for more than one lemma. The identifiers L_i are arrays, where the jth element, denoted $L_i[j]$, is the name of the lemma corresponding to the jth proof in τ_i not bound to a name in τ_i, denoted $\tau_i[j]$. The let expression binds names to the proofs and limits their scope to proof terms that appear later in the let expression. In other words, a lemma $L_i[j]$ can appear in any proof $\tau_k, k > i$, or in τ. The name of a lemma has the same type as the proof to which it corresponds. This scoping discipline for lemmas corresponds exactly to the variable scoping used in SML let expressions.

These new rules have corresponding typing rules, in Table 2.

Table 2. Typing rules for proof terms

$$\frac{\Gamma \vdash \tau_1 : \varphi_1 \quad \ldots \quad \Gamma \vdash \tau_n : \varphi_n}{\Gamma \vdash \tau_1; \ldots; \tau_n : \varphi_1 \wedge \ldots \wedge \varphi_n}$$

$$\frac{\begin{array}{l}\Gamma \vdash \tau_1 : \varphi_1 \\ \Gamma, L_1 : \varphi_1 \vdash \tau_2 : \varphi_2 \\ \ldots \\ \Gamma, L_1 : \varphi_1, \ldots, L_{n-1} : \varphi_{n-1} \vdash \tau_n : \varphi_n \\ \Gamma, L_1 : \varphi_1, \ldots, L_n : \varphi_n \vdash \tau : \varphi\end{array}}{\Gamma \vdash \text{let } L_1 = \tau_1 \ldots L_n = \tau_n \text{ in } \tau \text{ end} : \varphi_1 \to \ldots \to \varphi_n \to \varphi}$$

The rule for a sequence of proof terms is relatively straightforward; the type of a sequence is the conjunction of the types of the proof terms in the sequence. The typing rule for the let expression is based on the scoping of the proofs. We must be able to prove that each proof τ_k has type φ_k under the assumption that all variables $L_i, i < k$ have the type φ_i, where τ_i is assigned to L_i. Finally, we must be able to prove that τ has the type φ under the assumption that every L_i has type φ_i.

As an example, we represent the proofs of (7) and (8) as

$$
\begin{aligned}
&\mathsf{thm} = \\
&\quad \mathsf{let\ lem} = \lambda x \lambda y \lambda z \lambda P.(\text{Proof of lemma}) \\
&\quad \mathsf{in} \\
&\qquad \lambda a \lambda b \lambda c \lambda z \lambda Q \lambda R.\mathsf{trans\ (sym\ (lem\ } Q))\ (\mathsf{lem}\ R) \\
&\quad \mathsf{end}
\end{aligned}
$$

where thm is the name assigned to (7) and lem is the name assigned to (8). For ease of reading, we have omitted the applications of proof terms to individual terms, which represent the substitution for individual variables. P, Q, and R are proofs of type $x = y$, $a = b$, and $a = c$, respectively.

If we choose to universally quantify z only once as in (12), we represent the proof as

$$
\begin{aligned}
&\mathsf{thm} = \\
&\quad \lambda z.\mathsf{let\ lem} = \lambda x \lambda y \lambda P.(\text{Proof of lemma}) \\
&\qquad \mathsf{in} \\
&\qquad\quad \lambda a \lambda b \lambda c \lambda Q \lambda R.\mathsf{trans\ (sym\ (lem\ } Q))\ (\mathsf{lem}\ R) \\
&\qquad \mathsf{end}
\end{aligned}
$$

As we can see, there is a one-to-one correspondence between the positions of λ-abstractions and where individual variables are universally quantified. We formally develop the proof terms for thm and lem in Section 5.

4   Proof Rules

We provide several rules for creating and manipulating proofs. The rules allow one to build proofs constructively. They manipulate a structure of the form $\mathcal{L};\mathcal{C};\mathcal{T}$, where

- $\mathcal{L}$ is the library of theorems, $T_1 = \pi_1, \ldots, T_n = \pi_n$, where T_i is an array of identifiers with the jth element denoted $T_i[j]$, naming the jth proof in π_i, denoted $\pi_i[j]$,
- $\mathcal{C}$ is the list of lemmas currently in scope, $L_1 = \tau_1, \ldots, L_m = \tau_m$, with components defined as they are for $\mathcal{L}$, and
- $\mathcal{T}$ is a list of annotated *proof tasks* of the form $A \vdash \pi : \varphi$, where A is a list of assumptions, π is a proof term, and φ is an unquantified Horn formula.

In these rules, we use the following notational conventions:

- α and β are proof variables or individual variables.
- $\overline{X}$ is a set of elements $\{X_1, \ldots, X_n\}$, where X_i can be an individual variable or a proof variable.
- $T = \pi$ binds a proof term π to an identifier T. The term π may define the proof for more than one theorem. Therefore, the identifier T is an array,

where the jth element, denoted $T[j]$, is the name of the theorem corresponding to the jth proof in π not bound to a name in π, denoted $\pi[j]$.

- $\overline{T} = \overline{\pi}$ is a sequence of bindings $T_1 = \pi_1, \ldots, T_n = \pi_n$.
- $\overline{T} : \overline{\varphi}$ is a sequence of type bindings $T_1 : \varphi_1, \ldots, T_n : \varphi_n$, where $\varphi = \varphi_1 \to \ldots \to \varphi_n$.
- $\pi[\overline{x}/\overline{t}]$ means for all i, replace element $x_i \in \overline{x}$ in π with $t_i \in \overline{t}$.
- Given a binding $T = \pi$, $X[T/\pi]$ means for all i, replace $T[i]$ with $\pi[i]$ in X, where X is a proof term, a list of theorems, or a list of proof tasks.
- For a proof term π, a sequence of identifiers $\overline{T} = T_1 \ldots T_n$, and a variable α, $\pi[\overline{T}/\overline{T}\,\alpha]$ means for all i and j, replace $T_i[j]$ with $T_i[j]\,\alpha$, where juxtaposition represents functional application.
- Given a binding $T = \ldots \lambda\alpha_i\lambda\alpha_j \ldots \pi$, $\mathcal{C}[T(i,j)/T(j,i)]$ means for all k, swap the ith and j^{th} term or proof to which $T[k]$ is applied in $\mathcal{C}$.
- $FV(\varphi)$ is the set of free individual variables in the Horn formula φ.

The structure $\mathcal{L};\mathcal{C};\mathcal{T}$ must also be well typed, according to the rules in
Table 3. The typing rules enforce an order on the list of theorems and lemmas.
The rules look very similar to the rules for the let expression.

Table 3. Typing rules for proof library

$$\frac{\begin{array}{c} \Gamma \vdash \pi_1 : \varphi_1 \\ \Gamma, T_1 : \varphi_1 \vdash \pi_2 : \varphi_2 \\ \ldots \\ \Gamma, T_1 : \varphi_1, \ldots, T_{n-1} : \varphi_{n-1} \vdash \pi_n : \varphi_n \end{array}}{\Gamma \vdash \overline{T} = \overline{\pi} : \varphi_1 \to \ldots \to \varphi_n}$$

$$\frac{\begin{array}{c} \Gamma \vdash \overline{T} = \overline{\pi} : \varphi_{T1} \to \cdots \to \varphi_{Tn} \\ \Gamma, \overline{T} : \overline{\varphi_T} \vdash \overline{L} = \overline{\tau} : \varphi_{L1} \to \cdots \to \varphi_{Lm} \\ \Gamma, \overline{T} : \overline{\varphi_T}, \overline{L} : \overline{\varphi_L} \vdash \mathcal{T} : \psi \end{array}}{\Gamma \vdash \overline{T} = \overline{\pi}; \overline{L} = \overline{\tau}; \mathcal{T} : \varphi_{T1} \to \cdots \to \varphi_{Tn} \to \varphi_{L1} \to \cdots \to \varphi_{Lm} \to \psi}$$

We must also have a typing rule for the proof tasks $\mathcal{T}$. The rule is a meta-
typing rule on deductions of the form $A \vdash \pi : \varphi$, which we omit for brevity.

The proof rules fit into two categories: rules that manipulate the proof tasks
and rules that manipulate the structure of proof terms that appear in $\mathcal{C}$.

4.1 Rules for Manipulating Proof Tasks

The first set of rules is in Table 4. Note that the **(reorder)** rule has a side condition $(*)$ explained below. The first four rules are the same as the rules in [4].

The **(collect)** rule works on a set of tasks with no further assumptions, i.e.,
tasks with completed proofs. The rule

1. gives the collection of the tasks a new name L that does not appear in the
   library or the current list of lemmas,

Table 4. Rules for manipulating proof tasks

(assume)
$$\frac{\mathcal{L} \, ; \, \mathcal{C} \, ; \, \mathcal{T}, \quad A \vdash \tau : e}{\mathcal{L} \, ; \, \mathcal{C} \, ; \, \mathcal{T}, \quad A, p : d \vdash \tau : e}$$

(ident)
$$\frac{\mathcal{L} \, ; \, \mathcal{C} \, ; \, \mathcal{T}}{\mathcal{L} \, ; \, \mathcal{C} \, ; \, \mathcal{T}, \quad p : e \vdash p : e}$$

(mp)
$$\frac{\mathcal{L} \, ; \, \mathcal{C} \, ; \, \mathcal{T}, \quad A \vdash \pi : e \to \varphi \quad A \vdash \tau : e}{\mathcal{L} \, ; \, \mathcal{C} \, ; \, \mathcal{T}, \quad A \vdash \pi \, \tau : \varphi}$$

(discharge)
$$\frac{\mathcal{L} \, ; \, \mathcal{C} \, ; \, \mathcal{T}, \quad A, p : e \vdash \tau : \varphi}{\mathcal{L} \, ; \, \mathcal{C} \, ; \, \mathcal{T}, \quad A \vdash \lambda p.\tau : e \to \varphi}$$

(collect)
$$\frac{\mathcal{L} \, ; \, \overline{M} = \overline{\pi} \, ; \quad \vdash \tau_1 : \varphi_1 \ldots \vdash \tau_n : \varphi_n}{\mathcal{L} \, ; \quad \begin{array}{l} L = \text{let } \overline{M} = \overline{\pi} \\ \quad \text{in } \lambda \overline{x}_1.\tau_1; \ldots ; \lambda \overline{x}_n.\tau_n \text{ end} \end{array}} \quad ; \qquad \overline{x}_i = FV(\varphi_i)$$

(publish)
$$\frac{\mathcal{L} \qquad ; \, \overline{L} = \overline{\tau} \, ; \qquad}{\mathcal{L}, \overline{L} = \overline{\tau} \, ; \qquad \qquad ;}$$

(tcite)
$$\frac{\mathcal{L}_1, T = \pi, \mathcal{L}_2 \, ; \, \mathcal{C} \, ; \, \mathcal{T}}{\mathcal{L}_1, T = \pi, \mathcal{L}_2 \, ; \, \mathcal{C} \, ; \, \mathcal{T}, \quad \vdash T[j] \, \overline{t} : \varphi[\overline{x}/\overline{t}]} \quad T[j] : \forall \overline{x}.\varphi$$

(lcite)
$$\frac{\mathcal{L} \, ; \, \mathcal{C}_1, L = \pi, \mathcal{C}_2 \, ; \, \mathcal{T}}{\mathcal{L} \, ; \, \mathcal{C}_1, L = \pi, \mathcal{C}_2 \, ; \, \mathcal{T}, \quad \vdash L[j] \, \overline{t} : \varphi[\overline{x}/\overline{t}]} \quad L[j] : \forall \overline{x}.\varphi$$

(tforget)
$$\frac{\mathcal{L}_1, T = \pi, \mathcal{L}_2 \, ; \, \mathcal{C} \qquad ; \, \mathcal{T}}{\mathcal{L}_1, \mathcal{L}_2[T/\pi] \quad ; \, \mathcal{C}[T/\pi] \, ; \, \mathcal{T}[T/\pi]}$$

(lforget)
$$\frac{\mathcal{L} \, ; \, \mathcal{C}_1, L = \pi, \mathcal{C}_2 \, ; \, \mathcal{T}}{\mathcal{L} \, ; \, \mathcal{C}_1, \mathcal{C}_2[L/\pi] \quad ; \, \mathcal{T}[L/\pi]}$$

(promote)
$$\frac{\mathcal{L} \, ; \, \mathcal{L}_1, L = \text{let } \overline{M} = \overline{\tau} \text{ in } \pi \text{ end}, \mathcal{L}_2 \, ;}{\mathcal{L} \, ; \, \mathcal{L}_1, \overline{M} = \overline{\tau}, L = \pi, \mathcal{L}_2 \qquad ;}$$

(reorder)
$$\frac{\mathcal{L} \, ; \, \mathcal{C}_1, L = \lambda \alpha_1 \ldots \lambda \alpha_i \lambda \alpha_j \ldots \lambda_n.\pi, \mathcal{C}_2 \qquad \qquad ;}{\mathcal{L} \, ; \, \mathcal{C}_1, L = \lambda \alpha_1 \ldots \lambda \alpha_j \lambda \alpha_i \ldots \lambda_n.\pi, \mathcal{C}_2[L(i,j)/L(j,i)] \, ;} \, (*)$$

2. forms the universal closures of the φ_is and the corresponding λ-closures of the τ_is, and
3. moves the proofs to the list of lemmas currently in scope.

Any lemmas that were in scope for the proof tasks are explicitly made lemmas with the let statement. These lemmas are no longer immediately available to proof tasks. However, one can access a lemma moved into a let by using the **(promote)** rule. If no lemmas currently exist, a let expression is not created and instead the name L is bound to the λ-closures of the τ_is.

The **(publish)** rule moves the current lemmas to the library, at which point they become theorems.

The **(tcite)** rule is the elimination rule for the universal quantifier for theorems in the library. This rule specializes the theorem with a given substitution $[\overline{x}/\overline{t}]$. It is important to note that the proof $\pi_i[j]$ of $T_i[j]$ is not copied into the proof tasks. If this were the case, then β-reduction on the proof could make it impossible to distinguish between a proof that cited $T_i[j]$ and a proof that developed $\pi_i[j][\overline{x}/\overline{t}]$ explicitly. Instead, the name of the theorem serves as a citation token, with the same type as the proof itself. The **(lcite)** rule does the same for lemmas from $\mathcal{C}$.

The **(tforget)** rule removes all citations of the forgotten theorems and replaces them with the proofs of the theorems. With the proof instead of the citation token, β-reduction on citations of a theorem can take place during proof normalization, creating the specialized version of the proof we did not create in the **tcite** rule. All citations of the theorems T are replaced with a specialized version of the proof π. The **(lforget)** rule does the same for lemmas in $\mathcal{C}$.

The **(promote)** rule moves a set of lemmas from inside a let expression to the list of lemmas currently in scope. This makes these lemmas again available to be cited.

The **(reorder)** rule changes the order of abstractions in a proof term. Correspondingly, citations of any lemmas defined by that proof term must be changed to have the order of their applications changed. The condition $(*)$ is that if α_i is an individual variable and α_j is a proof variable with type φ, then α_i does not occur anywhere in φ. If α_i did occur in φ and we performed **(reorder)**, φ would contain an unbound variable.

4.2   Rules for Manipulating Proof Terms

The set of rules for manipulating proof terms that appear in $\mathcal{C}$ is in Table 5. These rules do not change any proofs of theorems currently in scope for the proof tasks, so we know that any changes in proofs do not have to be reflected in the current tasks. Some of these rules have side conditions, which are marked with a symbol in $(\cdot)$ and explained below.

The **(push)** rule moves an abstraction from the front of a sequence to each proof in the sequence. This rule does not change the types of the proofs; it only duplicates $\lambda\alpha$. One would anticipate using this rule after performing a **(generalize)**.

The **(pull)** rule is the inverse of the **(push)** rule. It moves an abstraction from the front of every proof in a sequence to the front of the entire sequence. This rule would most likely be used before a **(specialize)**.

The **(generalize)** rule moves an abstraction from the outside of a let statement to each proof term in the list of defined lemmas and to the proof term τ. This does not change any theorem whose proof is in τ. The proofs and types of the lemmas $\overline{L}$ do change, because they are now abstracted over another variable.

Correspondingly, we have to change any citations of the lemmas. From the scoping discipline, we know exactly where these citations can be: in the proofs of the lemmas, $\overline{\pi}$, or in the proof τ. Before performing **(generalize)**, all the lemmas and τ referred to the same α. Now, the first abstraction for any of the

Table 5. Rules for manipulating proof terms in $\mathcal{C}$

(push)
$$\frac{\lambda\alpha.(\pi_1;\ldots;\pi_n)}{\lambda\alpha.\pi_1;\ldots;\lambda\alpha.\pi_n}$$

(pull)
$$\frac{\lambda\alpha.\pi_1;\ldots;\lambda\alpha.\pi_n}{\lambda\alpha.(\pi_1;\ldots;\pi_n)}$$

(generalize)
$$\frac{\lambda\alpha.\text{let } \overline{L} = \overline{\pi} \text{ in } \tau \text{ end}}{\text{let } \overline{L} = \lambda\alpha.\pi[\overline{L/L}\ \alpha] \text{ in } \lambda\alpha.\tau[\overline{L/L}\ \alpha] \text{ end}}$$

(specialize)
$$\frac{\text{let } \overline{L} = \overline{\lambda\alpha.\pi} \text{ in } \lambda\alpha.\tau \text{ end}}{\lambda\alpha.\text{let } \overline{L} = \pi[\overline{L\ \alpha/L}] \text{ in } \tau[\overline{L\ \alpha/L}] \text{ end}} \quad (**)$$

(split)
$$\frac{\text{let } \overline{L} = \overline{\pi_L}, \overline{M} = \overline{\pi_M} \text{ in } \tau \text{ end}}{\text{let } \overline{L} = \overline{\pi_L} \text{ in let } \overline{M} = \overline{\pi_M} \text{ in } \tau \text{ end end}}$$

(merge)
$$\frac{\text{let } \overline{L} = \overline{\pi_L} \text{ in let } \overline{M} = \overline{\pi_M} \text{ in } \tau \text{ end end}}{\text{let } \overline{L} = \overline{\pi_L}, \overline{M} = \overline{\pi_M} \text{ in } \tau \text{ end}}$$

(rename)
$$\frac{\lambda\alpha.\pi}{\lambda\beta.\pi[\alpha/\beta]} \quad (\#)$$

lemmas is over α. Consequently, any citation of the lemmas must be changed to have the first application be to a term that matches α explicitly. Since all of the proofs referred to the same α before the operation, we can simply use the α in the applications and replace all occurrences of $L_i[j]$ with $L_i[j]\ \alpha$.

The types of the L_is and π_is also change. If α is an individual variable, we add another universal quantification to the front of the type. If α is a proof variable, we add another implication, corresponding to a premise.

The **(specialize)** rule does the opposite of **(generalize)**. A variable that was universally quantified for the lemmas L now becomes a constant for them when we move α to the outside of the let. As stated, the rule requires $\lambda\alpha$ to precede every proof π. This is not actually a requirement for correctness, but it makes stating the side condition easier. The side condition $(**)$ is that any citation of a lemma $L_i[j]$ is of the form $L_i[j]\ \alpha$. In other words, the same variable used in the λ-abstraction for the lemma must be the first variable to which the lemma is applied. Otherwise, the proof may no longer be correct, since another term used in the place of α may have different assumptions than those of α. Given this condition and the scoping discipline, we know exactly which citations need to change: those of the form $L_i[j]\ \alpha$ that appear in the π_is or in τ.

The **(split)** rule takes a list of lemma definitions and separates them into two sets of definitions, one in the same place and one nested in a new let expression within the in part of the original let. The proofs of the lemmas do not change at all, so no citations need to change. The **(merge)** rule is the inverse of the **(split)** rule.

The **(rename)** rule changes the name of a single variable. The side condition (#) is that the new name β must not occur anywhere in π. This corresponds to α-conversion.

Soundness for the proof system requires that a sequence of applications of the rules transforms a proof term of a type φ into a new proof term of a type ψ that is equivalent modulo first-order equivalence. Let $\pi \Rightarrow \tau$ mean that the proof term τ is derivable from π using our proof rules in one step.

Theorem 2. *If $\pi \Rightarrow \tau$ and $\Gamma \vdash \pi : \varphi$, then $\Gamma \vdash \tau : \psi$, where φ and ψ are equivalent modulo first-order equivalence.*

Proof. The proof is by induction on the proof terms.

In order to prove the cases for **(generalize)**, we need a couple lemmas about substitution. We state the lemmas as meta-typing rules.

Lemma 2.
$$\frac{\Gamma, p : \varphi_p, L : \varphi \vdash \tau : \psi}{\Gamma, p : \varphi_p, L : \varphi_p \to \varphi \vdash \tau[L/L\ p] : \psi}$$

where $L = \pi$ does not appear in τ.

Lemma 3.
$$\frac{\Gamma, L : \varphi \vdash \tau : \psi}{\Gamma, L : \forall x.\varphi \vdash \tau[L/L\ x] : \psi}$$

where $L = \pi$ does not appear in τ.

Proof. The proof for both lemmas is by induction on proof terms.

We need similar lemmas for the **(specialize)** rule as well. The details of the proof are omitted due to space constraints. It is interesting to note, however, that the proof of soundness demonstrates that the types for let expressions and our environment $\mathcal{L}, \mathcal{C}, \mathcal{T}$ are correct.

5   A Constructive Example

To demonstrate the use of the proof rules, we develop the proofs of (8) and (7). Recall, we wish to prove

$$\forall a, b, c, z.a = b \to a = c \to z \vee (a \wedge b) = z \vee (a \wedge c) \tag{14}$$

using the lemma

$$\forall x, y, z.x = y \to z \vee (x \wedge x) = z \vee (x \wedge y) \tag{15}$$

We use the following axioms

$$\mathsf{sym} : \forall x, y.x = y \to y = x \tag{16}$$

$$\mathsf{trans} : \forall x, y, z.x = y \to y = z \to x = z \tag{17}$$

$$\mathsf{cong}_\wedge : \forall x, y, z.x = y \to (z \wedge x) = (z \wedge y) \tag{18}$$

$$\mathsf{cong}_\vee : \forall x, y, z.x = y \to (z \vee x) = (z \vee y) \tag{19}$$

The library $\mathcal{L}$ initially contains all of our axioms. Until we need them, we omit both $\mathcal{L}$ and $\mathcal{C}$ for readability. We also omit term substitutions when performing cites.

First, we prove the lemma. By (**ident**), we have

$$P : x = y \vdash P : x = y \tag{20}$$

We use (**tcite**) with the substitutions $[x/x, y/y, z/x]$ and (**assume**) to add

$$P : x = y \vdash \mathsf{cong}_\wedge : x = y \rightarrow (x \wedge x) = (x \wedge y) \tag{21}$$

Applying (**mp**) to (20) and (21) gives

$$P : x = y \vdash \mathsf{cong}_\wedge P : (x \wedge x) = (x \wedge y) \tag{22}$$

We use (**tcite**) with the substitutions $[x/x \wedge x, y/x \wedge y, z/z]$ and (**assume**) to add

$$P : x = y \vdash \mathsf{cong}_\vee : (x \wedge x) = (x \wedge y) \rightarrow z \vee (x \wedge x) = z \vee (x \wedge y) \tag{23}$$

Applying (**mp**) to (22) and (23) gives

$$P : x = y \vdash \mathsf{cong}_\vee \mathsf{cong}_\wedge P : z \vee (x \wedge x) = z \vee (x \wedge y) \tag{24}$$

Now we apply (**discharge**) to (24) to get

$$\vdash \lambda P.\mathsf{cong}_\vee \mathsf{cong}_\wedge P : x = y \rightarrow z \vee (x \wedge x) = z \vee (x \wedge y) \tag{25}$$

We can use the (**collect**) rule to add (25) to our current term, given it the name lem. Our entire state is

$$\mathcal{L}; \mathsf{lem} = \lambda x \lambda y \lambda z \lambda P.\mathsf{cong}_\vee \mathsf{cong}_\wedge P : \forall x, y, z.x = y \rightarrow z \vee (x \wedge x) = z \vee (x \wedge y);$$

Now we start on the proof of the theorem. First we use (**ident**) to add the task

$$Q : a = b \vdash Q : a = b \tag{26}$$

Next, we use (**lcite**) with the substitutions $[x/a, y/b, z/z]$ and (**assume**) to get our lemma from the current term

$$Q : a = b \vdash \mathsf{lem} : a = b \rightarrow z \vee (a \wedge a) = z \vee (a \wedge b) \tag{27}$$

Applying (**mp**) to (26) and (27) gives

$$Q : a = b \vdash \mathsf{lem}\, Q : z \vee (a \wedge a) = z \vee (a \wedge b) \tag{28}$$

We now use (**cite**) with the substitutions $[x/z \vee (a \wedge a), y/z \vee (a \wedge b)]$ and (**assume**) to introduce

$$Q : a = b \vdash \mathsf{sym} : z \vee (a \wedge a) = z \vee (a \wedge b) \rightarrow z \vee (a \wedge b) = z \vee (a \wedge a) \tag{29}$$

Applying **(mp)** to (28) and (29) gives

$$Q : a = b \vdash \mathsf{sym}\ (\mathsf{lem}\ Q) : z \vee (a \wedge b) = z \vee (a \wedge a) \qquad (30)$$

Next, we use **(ident)** to introduce

$$R : a = c \vdash R : a = c \qquad (31)$$

Next, we use **(lcite)** with the substitutions $[x/a, y/c, z/z]$ and **(assume)** to get our lemma from the current term again

$$R : a = c \vdash \mathsf{lem} : a = c \to z \vee (a \wedge a) = z \vee (a \wedge c) \qquad (32)$$

Applying **(mp)** to (31) and (32) gives

$$R : a = c \vdash \mathsf{lem}\ R : z \vee (a \wedge a) = z \vee (a \wedge c) \qquad (33)$$

Applying **(tcite)** with the substitutions $[x/z \vee (a \wedge b), y/z \vee (a \wedge a), z/z \vee (a \wedge c)]$ allows us to add

$$\vdash \mathsf{trans} : z\vee(a\wedge b) = z\vee(a\wedge a) \to z\vee(a\wedge a) = z\vee(a\wedge c) \to z\vee(a\wedge b) = z\vee(a\wedge c)$$
$$(34)$$

Applying **(assume)** to (30), (33), and (34) gives

$$Q : a = b, R : a = c \vdash \mathsf{sym}\ (\mathsf{lem}\ Q) : z \vee (a \wedge b) = z \vee (a \wedge a) \qquad (35)$$
$$Q : a = b, R : a = c \vdash \mathsf{lem}\ R : z \vee (a \wedge a) = z \vee (a \wedge c) \qquad (36)$$
$$Q : a = b, R : a = c \vdash \mathsf{trans} : (a \wedge b) = z \vee (a \wedge a) \qquad (37)$$
$$\to z \vee (a \wedge a) = z \vee (a \wedge c) \to z \vee (a \wedge b) = z \vee (a \wedge c)$$

Two applications of **(mp)** using (35), (36), and (37) gives

$$Q : a = b, R : a = c \vdash \mathsf{trans}\ (\mathsf{sym}\ (\mathsf{lem}\ Q))\ (\mathsf{lem}\ R) : z \vee (a\wedge b) = z\vee(a\wedge c) \quad (38)$$

We use **(discharge)** on each assumption in (38) to get

$$\vdash \lambda Q.\lambda R.\mathsf{trans}\ (\mathsf{sym}\ (\mathsf{lem}\ Q))\ (\mathsf{lem}\ R) : a = b \to a = c \to z \vee (a\wedge b) = z\vee(a\wedge c) \quad (39)$$

We can use the **(collect)** rule to add (39) to our current term, give it the name thm, and make lem a lemma by introducing a let expression. Our new $\mathcal{C}$ term is

thm =
> let lem = $\lambda x \lambda y \lambda z \lambda P.\mathsf{cong}_\vee \mathsf{cong}_\wedge P : \forall x, y, z.x = y \to z \vee (x \wedge x) = z \vee (x \wedge y)$
> in
>
> $\lambda a \lambda b \lambda c \lambda z \lambda Q.\lambda R.\mathsf{trans}\ (\mathsf{sym}\ (\mathsf{lem}\ Q))\ (\mathsf{lem}\ R) : \forall a, b, c, z.a = b \to a = c$
> $\to z \vee (a \wedge b) = z \vee (a \wedge c)$
>
> end

At this point, we could apply **(publish)** to add thm to the library. However, we may first wish to make thm and lem use the same z. To do this, we apply **(reorder)** to the term several times to get

thm =
let lem $= \lambda z \lambda x \lambda y \lambda P.\text{cong}_\vee \text{cong}_\wedge P : \forall z, x, y.x = y \rightarrow z \vee (x \wedge x) = z \vee (x \wedge y)$
in

$\lambda z \lambda a \lambda b \lambda c \lambda Q.\lambda R.\text{trans (sym (lem } Q))$ (lem R) $: \forall z, a, b, c.a = b \rightarrow a = c$
$\rightarrow z \vee (a \wedge b) = z \vee (a \wedge c)$

end

We now apply (**specialize**) to move λz to the front of the **let** expression

thm =
$\lambda z.$let lem $= \lambda x \lambda y \lambda P.\text{cong}_\vee \text{cong}_\wedge P : \forall x, y.x = y \rightarrow z \vee (x \wedge x) = z \vee (x \wedge y)$
in

$\lambda a \lambda b \lambda c \lambda Q.\lambda R.\text{trans (sym (lem } Q))$ (lem R) $: \forall z, a, b, c.a = b \rightarrow a = c$
$\rightarrow z \vee (a \wedge b) = z \vee (a \wedge c)$

end

6   Related Work

Several people have looked at the problem of proof reuse and library organization. Limiting the scope of variables and assumptions is handled by Isabelle's
locales, which limit the use of a set of local variables and assumptions to a current
theory [6, 7]. In fact, the system allows one to create nested locales and move
them outward in the nesting, corresponding to our (**specialize**) rule. However,
theorems themselves are not a part of these locales and cannot be moved in the
same way; the library of theorems is still a flat structure, without a complete
notion of scope for theorems.

Melis and Schairer have looked at proof reuse in formal software verification
[8]. In their proofs, subgoals are often very similar, so the reuse of completed
proofs is instrumental in reducing the time required to verify programs. They
have a notion of a lemma, where a proof used in an earlier subgoal can be
reused within later subgoals of the same proof. The system can attempt to detect
these similar proofs automatically or the user can specify them. However, the
relation between these subgoals is never stored in the proof, so a later analysis
of the proof would not reflect the fact that similar subgoals were found and
reused. Moreover, lemmas are not stored or reusable in different theorems. Given
the similarities within proofs, one can imagine that there would also be several
similarities between proofs for which storage of some of the more fundamental
lemmas could be justified.

Lorigo et al. have worked on applying WWW search techniques to obtain
information about the structure of libraries of proofs and theorems. In [9],
they describe how this can be used to find the structure of mathematical topics and categories of theorems in libraries depending on inter theorem usage.

The approach is meant to be used with already existing libraries of formal mathematics, and work one way, in the sense that it gathers information from the library and presents it to the user, but does not re-order the theorems in the library itself into the discovered relationships. In contrast, our approach intrinsically groups related theorems and lemmas already during their proof and keeps them together unless specifically moved by the user.

7 Future Work

We see many benefits to an automated theorem prover using a library with such a formal hierarchical structure. First of all, we would expect the structure of the library to indicate which theorems are more closely related–theorems that use the same variables, assumptions, or lemmas would be grouped together in let expressions and share abstractions. Large mathematical libraries could naturally be broken down into smaller parts based on these groupings.

One can imagine several heuristics that could be improved by the structure of the library. A system could first look at citing lemmas currently in scope before searching the entire library. The number of lemmas in scope is likely to be smaller than the number of theorems. Heuristics that automatically detect similar subproofs and create lemmas from them should also be possible. Given the formal structure of proofs, finding shared lemmas is a form of common subexpression elimination. In discovering these lemmas automatically, the library takes on the structure natural to the theorems proven. It could also provide guidance to a user proving a new theorem, knowing that the current proof being worked on and other theorems already proven share a few lemmas.

Currently, we have a basic implementation of all of the operations in a system that works on Kleene algebra with tests [10]. The system, written in Java, has a command line interface that allows one to create, manipulate, and save proofs in a tree structure, which corresponds naturally to the let expressions and local scoping. We hope to add to the system the ability to view and manipulate the library as a figure, given that the tree structure lends itself well to direct graphical depiction. One would easily be able to see and to alter the relationship between theorems while their manipulations would be guided by a strong underlying formalism.

Acknowledgements

We are indebted to Dexter Kozen, Ganesh Ramanarayanan, and Stephen Chong for valuable ideas and comments. This work was supported in part by NSF Grant CCR-0105586 and ONR Grant N00014-01-1-0968. The views and conclusions contained herein are those of the authors and should not be interpreted as necessarily representing the official policies or endorsements, either expressed or implied, of these organizations or the US Government.

References

1. Kreitz, C.: The Nuprl Proof Development System, Version 5: Reference Manual and User's Guide. Department of Computer Science, Cornell University. (2002)
2. The Coq Development Team: The Coq Proof Assistant Reference Manual – Version V7.3. (2002) http://coq.inria.fr.
3. Wenzel, M., Berghofer, S.: The Isabelle System Manual. (2003)
4. Kozen, D., Ramanarayanan, G.: A proof-theoretic approach to knowledge acquisition. Technical Report 2005-1985, Computer Science Department, Cornell University (2005)
5. Sørensen, M.H., Urzyczyn, P.: Lectures on the Curry–Howard isomorphism. Available as DIKU Rapport 98/14 (1998)
6. Kammüller, F.: Modular reasoning in isabelle. In McAllester, D.A., ed.: CADE. Volume 1831 of Lecture Notes in Computer Science., Springer (2000) 99–114
7. Ballarin, C.: Locales and locale expressions in isabelle/isar. In Berardi, S., Coppo, M., Damiani, F., eds.: TYPES. Volume 3085 of Lecture Notes in Computer Science., Springer (2003) 34–50
8. Melis, E., Schairer, A.: Similarities and reuse of proofs in formal software verification. In: EWCBR. (1998) 76–87
9. Lorigo, L., Kleinberg, J.M., Eaton, R., Constable, R.L.: A graph-based approach towards discerning inherent structures in a digital library of formal mathematics. In: MKM. (2004) 220–235
10. Kozen, D.: Kleene algebra with tests. Transactions on Programming Languages and Systems **19** (1997) 427–443

An Exploration in the Space of Mathematical Knowledge

Andrea Kohlhase[1] and Michael Kohlhase[2]

[1] DiMeB, Dept. of Computer Science and Mathematics, University Bremen
kohlhase@informatik.uni-bremen.de
[2] School of Engineering and Science, International University Bremen
m.kohlhase@iu-bremen.de

Abstract. Although knowledge is a central topic for MKM there is little explicit discussion on what 'knowledge' might actually be. There are specific intuitions about form and content of knowledge, about its structure, and epistemological nature that shape the MKM systems, but a conceptual model is missing.

In this paper we try to rationalize this discussion to give MKM a firmer footing, to start a discussion among MKM researchers and help relate the MKM intuitions and discourses to other communities.

Starting from the observation that many concrete realizations of mathematical knowledge objects are considered equivalent, we propose a conceptual model of the space of (mathematical) knowledge objects graded by levels of abstraction and presentational explicitness and draw conclusions for MKM markup formats.

1 Handles on (Mathematical) Knowledge

The concept of 'knowledge' is investigated by many scientific disciplines, some take a microscopic, ontological view, some a macroscopic, epistemological view and still others a pragmatic view. The latter seems to be the dominant one in the field of Mathematical Knowledge Management (MKM), but ever so often we find pragmatic limits and have to cross the border. There are multiple ways of looking at mathematical knowledge; for instance there is much discussion about whether we should focus on the essence or the visual appearance of mathematical objects and where to determine the borders between these as they seem to be fluctuating.

In this paper, we start an exploration into the world of mathematical knowledge. Reflections on this mathematical space were inspired in part by an article by SEYMOUR PAPERT, called "An Exploration in the Space of Mathematics Educations" [Pap96]. There, he investigates different math educational approaches, but instead of contrasting them he relates them by interpreting them as axes in an n-dimensional space. Here, we investigate essence/appearance approaches concerning knowledge objects and are interested in the resulting *knowledge space*, hoping that this perspective yields new and unexpected dependencies and relations.

M. Kohlhase (Ed.): MKM 2005, LNAI 3863, pp. 17–32, 2006.

1.1 Knowledge and Context

Information theory assumes that the fundamental concepts of data, information, and knowledge are not interchangeable concepts. In particular, the transitive combination of "Lots of available data" and "Information are good data" and "Knowledge is created with information" readily accepted in the Internet Bubble cannot be held.

As data are visually accessible, we need to consider yet another concept: a 'glyph' is an arrangement of pixels on a screen (or dots of ink on a sheet of paper) into a recognizable shape. In contrast to the usage of data (which contain something even if we don't know what), the usage of 'glyphs' emphasizes the pure presentation of a single character without any underlying semantics. In order to close in on 'knowledge', we want to take a closer look at the meaning of glyphs, data, and information and their relationships and differences based on an established knowledge management model. PROBST ET AL. (see [PRR97]) posit that glyphs, data, information, and last but not least knowledge can be seen as stages of a pipeline that is shown in Figure 1.

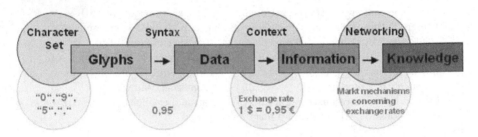

Fig. 1. From Mere Glyphs To Valuable Knowledge

In particular, **glyphs** are just a set of characters or symbols like {0;9;5;,} without any structure. A first set of rules imposed on the glyphs — the syntax — yields **data** which can be handled by machines. For obtaining meaning from such data we still need another component: the context. Usually, we discern data from **information** by viewing information as data with a message or data with an intention. DAVENPORT and PRUSAK think of information "as data that makes a difference" [DP98]. Data becomes information if a user can interpret the data in regard to a specific goal (or a local context) e.g. using the string '0,95' as number in an equation concerning exchange rates in our example. In contrast, information becomes **knowledge**, if a user can interpret the information in regard to a global context like understanding the exchange rate equation in the area of specific market behavior with respect to change of exchange rates.

1.2 Communities of Practice as Knowledge Context

We described in Section 1.1 how knowledge in the field of Knowledge Management is dealt with. In 1991, BROWN and DUGUID investigated more closely the global context which transforms information into knowledge (see "The Social

Life of Information" [BD00]). In [BD91] they identify LAVE and WENGER's influential concept of "Community of Practice" [LW91–p. 98] as the *social life of information*, i.e. they link Communities of Practice with organizational learning and hence with knowledge. A **Community of Practice (CoP)** is

> "a group of people (e.g. professionals) informally bound to one another through exposure to a common class of problems, common pursuit of solutions, and thereby themselves embodying a store of knowledge."   [WMS02]

In mathematical terms, scientific groups can only build a CoP if their members agree on the validity of certain equivalence relations (which we will call "substance equivalences" in 2.1).

CoPs are considered as the locus of knowledge as opposed to the learner's mind. The process of obtaining knowledge (learning) is described as "a process that takes place in a participation framework" [LW91–p. 14] where "participation is always based on situated negotiation and renegotiation of meaning in the world" [LW91–p. 51]. So far, the assignment of meaning is done in MKM via semantic annotations, but the necessary agreement on CoP-dependent substance equivalences are not yet paid attention to, even though they seem to play a decisive role in the Mathematical Knowledge Space.

1.3   The Pragmatics of Mathematical Knowledge

In order to make knowledge amenable to management, it has to be 'captured'. More specifically, it has to be reified, so that it can be stored, transfered, or retrieved as **knowledge object**. But even if we set aside for the moment the problem of explicit and implicit (tacit) knowledge that is well-studied in learning theories, we have to look at the relationship between a knowledge object and the represented knowledge itself. Therefore, we need to look at the question what knowledge is made of and whether or what part of it 'exists'. We focus on the philosophy of Mathematics that is concerned with this question and find that it has occupied many famous people like CARNAP, BERNAYS, BENACERRAF, or PUTNAM (see for example "The existence of mathematical objects" in [BP64–pp. 241-311]). An important distinction in this consideration is the one of **substance** and **accidence**[1]. Substance is the unchanging essence of an object, whereas accidence is the object's appearance. These terms form a *dialectic pair*: even though an object's substance and appearance can be differentiated, they are inseparable, they form a unity, so that one cannot think of one without the other. The question whether a knowledge object (especially a mathematical object) exists in "'being' or 'thinking' " [Isr79–7] is mostly irrelevant to mathematicians as long as it can be described. Its answer depends on a person's underlying ontological belief (see for example BENACERRAF's essay "What numbers could not be" [BP64–pp. 272-295]). But in real life, mathematicians are pragmatists, they use abstract objects independent from their existence. Analogously, epistemological issues are pragmatically ignored by (most) MKM systems.

---

[1] This distinction was used by Kant, there are many similar ones, including: essence/appearance (Hegel), matter/form (Aristoteles), content/form (MKM).

However, the differentiation between content and form found its way into the general MKM discourse. It is consensus in MKM that for a mathematical object we can distinguish its form from its content and express both aspects in markup systems. For instance, the MATHML format [ABC$^+$03] has two sub-languages: *presentation*-MATHML describes the two-dimensional layout in an expression like $\sqrt[3]{x+2}$, and *content*-MATHML, which can express its functional structure as the application of the cubic root function to a sum with the variable x and the number 2.

In general, it could be argued, that it makes no difference whether we take the symbol '$\mathbb{R}$' for the real numbers or maybe simply 'R'. We could just call it "different notation". But do you really believe that your personal selection of symbols is a matter of accident? Especially mathematicians do take great care in this selection out of coherence and consistence reasons [Hei00], but also because they know that different presentations and conceptualizations do have different associations and they make pragmatic use of it. In philosophic terms, we might call this the dialectic character of the substance/accidence aspects of a knowledge object. In many cases, the choice of conceptualization and presentation can make the difference whether a problem is solvable at all; see e.g. [Rob91] for a collection of striking examples.

In the following, if we use the pair **substance/accidence** we want to stress the different perspectives one can take looking at objects. This view is concerned with the relevance and the timeliness of the respective objects. In contrast, if we look at concrete objects, i.e. manifestations of knowledge, we can speak of their **content** and **form**. Here, we can think of content and form as the object's constitutive elements. They give rise to a knowledge space spanned by substance and accidence, inhabited by knowledge objects with certain "content and form coordinates".

2 A Conceptual Model for Knowledge Spaces

The fundamental observation is that knowledge can only be observed or communicated, if it is in a concrete form, e.g. written down in a book or uttered by a colleague or teacher. For this realization — which we can consider as knowledge *object* — a lot of conceptual and presentational aspects have to be fixed. Some seem to contribute to the meaning of the object, while others are thought of as rather personal choices like the page size of the book that contains the knowledge. In this section, we have a closer look at what the mathematical community deems substantial, yielding substance of knowledge as the totality of traits (which can be modeled as equivalence classes) that constitute the meaning.

We will use the following group definitions as a running example in this paper. It is well-known that groups can alternatively be described in two ways:

Definition 1 [KM79]. A **group**$_1$ is a set G together with an associative binary operation $\circ\colon G \times G \to G$, such that there is a unit element e for $\circ$ in G, and all elements have inverses.

Definition 2 [Hal59]. A **group₂** is a set G, together with a (not necessarily associative) binary operation $/: G \times G \to G$, such that $a/a = b/b$, $a/(b/b) = a$, $(a/a)/(b/c) = c/b$, and $(a/c)/(b/c) = a/b$ for all $a, b, c \in G$.

For any group₁ $(G, \circ)$, we can define a binary operation $/_\circ$ by $a/_\circ b := a \circ b^{-1}$ that shows that $(G, /_\circ)$ is a group₂, and vice versa (using $a \circ_/ b := a/b_/^{-1}$ with $b_/^{-1} := ((b/b)/b))$. So we see that the two definitions are isomorphic (which we could capture as a structure $\mathcal{G} := (G, \circ, e, \cdot^{-1}, /_\circ) = (G, \circ_/, a/a, \cdot_/^{-1}, /)$; see [CS98] for a formal account). In Mathematics it is usual to represent a structure like a group simply as the pair $\mathcal{G}_1 := (G, \circ)$ or a pair $\mathcal{G}_2 := (G, /)$, since in a group₁ the unit e and the inverse operation $\cdot^{-1}$ are uniquely determined by G and $\circ$ (and similarly for a group₂). So, we can view $\mathcal{G}$ as the substance of group and $\mathcal{G}_1$ and $\mathcal{G}_2$ as its accidences.

Mathematicians frequently speak of $\mathcal{G}_1$ and $\mathcal{G}_2$ as different *re*presentations. Note, that there often is a mix-up between the terms 'presentation' and '*re*presentation'. Principally, '**presentation**' is used to describe an explicit realization whereas '*re*presentation' is used to describe an implicit formalization[2]. In the example, instead of the usage of the symbols $\{(G, \circ); e; \cdot^{-1}\}$ just as well the symbols $\{\langle S, + \rangle; 1; - \}$ respectively could have been used in the presentation. In order to avoid confusions and for the purposes of the discussion in this paper we prefer to phrase these *re*presentations as "**conceptualizations**" to mark them off their "presentations". The term '**representation**' is therefore freed and serves as superordinated expression for conceptualizations as well as presentations.

2.1 Substance Equivalence

In Mathematics and in the natural sciences it is customary to consider presentational aspects like the (natural) language to be irrelevant for the meaning of a mathematical text. In this view, any document can be translated to any natural language without loss of meaning. As we have seen in the groups example above, Mathematics knows an even stronger equality notion — isomorphism. These distinct notions of equality of representations give rise to equivalence relations like $=_{lang}$ or $=_{log}$ which we will call **substance equivalences**.

In particular, we can consider the relationship model of $\mathcal{G}$ as knowledge reification along equivalence relations (Figure 2). The oval nodes are knowledge objects, whereas the various edges in the triangular graph signify the relations between the objects. Starting the description in a bottom-up way from right to left, the nodes $\mathcal{G}^{i,*}$ where $* \in \{e, g\}$ and $1 \leq i \leq 2$ stand for concrete variants of Definition group$_i$ in English (e) and German (g). As Mathematics considers translations between natural languages to be meaning-preserving, these are

---

[2] In German, this connection is exemplified in the language itself: presentation translates to 'Darstellung' whereas *re*presentation translates to 'Darstellungsweise', i.e. the mode of presentation. This corresponds to the frequent usage of the term 'presentation' in combination with the preposition 'for' in contrast to '*re*presentation'-usage concentrating on the 'of'-object. In other words, presentation is targeted with respect to the potential audience, whereas *re*presentation is focused on the content and its structure.

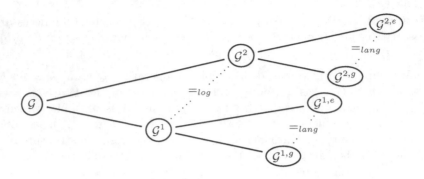

Fig. 2. Knowledge Reification in Mathematics

considered to be "conceptually equal" (see the dotted line between them). This allows us to obtain the knowledge objects $\mathcal{G}^i$ for the two conceptualizations of groups discussed above, which we model as the equivalence classes of $\mathcal{G}^{i,*}$: $\mathcal{G}^i := [\mathcal{G}^{i,e}]_{=_{lang}} = [\mathcal{G}^{i,g}]_{=_{lang}}$. These are logically equivalent, so they give rise to another knowledge object given as the equivalence class of all logically equivalent conceptualizations of groups, which we denote with $\mathcal{G} := [\mathcal{G}^1]_{=_{log}}{}^3$ (that corresponds to the structure $\mathcal{G}$ above).

We can see the diagram in Figure 2 as a visualization of the passage from implicitly represented knowledge objects on the left to explicitly presented ones on the right, making representation choices along the way.

2.2 Substance Equivalences and MKMarkup Formats

The substance equivalences we discussed above are generally accepted in Mathematics. If we look at other disciplines, e.g. in the Arts or Humanities, the assumption that representations can be translated without loss of substance would be highly controversial; there are literary texts (e.g. poems) that are considered "un-translatable". Hence, the diagram in Figure 2 would look completely different for e.g. literary science[4]. As Communities of Practice are marked by their collective value judgments about knowledge, we argue that substance equivalences are the defining characteristics of CoPs. Moreover, we have to assume that at least some are (implicitly) inscribed into the representation formats used by a CoP. But exactly how are the mathematical substance equivalences inscribed into the MKM formats? We differentiate between the well-known formula, statement, and theory level of mathematical knowledge objects and give examples in several MKMarkup formats.

We will start our analysis with the simplest case: the *formula level* and have a look at various MKM formats. In TeX/LaTeX, we can specify the exact sizes,

---

[3] Note, that the equivalence class construction of $\mathcal{G}$ is independent of the order of intermediate layers, i.e. the equivalence relations commute and therefore it is done modulo the transitive closure of $=_{log} \cup =_{lang}$.

[4] We suspect that it would be a left-right-inverted (dual) version of the triangle for Mathematics, but leave the investigation of this to further research.

colors, or fonts of the glyphs that make up the two-dimensional layout of a mathematical formula. Since we are given handles how to specify all these, we have to assume that these parameters matter, and therefore that the format does not inscribe equivalence of formulae where the glyphs differ in size, color, or font. In presentation MATHML, the specification of these traits is not possible in the prime vocabulary, but relegated to a CSS style system (which allows the specifications to be overridden in the client by standard means) which we take as a hint that stronger substance equivalences are in effect than in TEX/LATEX. OPENMATH [BCC+04] is of course the most radical in the substance equalities it assumes. It is impossible to specify the presentation of an OPENMATH object, as this format is geared towards communication of mathematical objects between systems. Communication with humans will be done via OPENMATH editors and presentation systems; which are free to choose any presentation suitable. Obviously, any two presentational variants e.g. $\binom{n}{k}$, $_nC^k$, C_k^n, and C_n^k are substance equivalent, since they all mean the same: $\frac{n!}{k!(n-k)!}$ (see [Koh05b] for a discussion).

At the *level of mathematical statements*, where e.g. our groups example is located, things are more complicated. We have already seen that, here, issues like the (natural) language employed in a definition, or the conceptualization play a major role. This leads us to another way, in which substance equivalences can be inscribed into MKMarkup formats. For instance, our own OMDoc format [Koh05a] has an explicit concept of language variants e.g. in the `definition` element (which represents a definition such as the one for group₁): it can incorporate a multilingual collection of `CMP` elements that contain definitional text fragments that are explicitly considered language variants of each other. So, the substance equivalence $=_{lang}$ from our example in Figure 2 is inscribed into OMDoc. We can see that the substance equivalence $=_{log}$ is inscribed into OMDoc as well. It is provided by the `alternative` element, which in our example would allow to phrase the definition of group₂ as an alternative definition to group₁ as long as we have proofs for the equivalence. In this situation, OMDoc only provides one concept for a group, a clear sign that $=_{log}$ is assumed in OMDoc.

At the *level of theories*, OMDoc has still another way of inscribing substance equalities into the format as it supports *theory morphisms*, i.e. structures that allow to prove that one theory is included in (or even isomorphic to) another modulo a variety of translations. In particular, isomorphic theories are considered as logically interchangeable (even if they are pragmatically different), another materialization of $=_{log}$.

2.3   The Conceptual Model of MKS

We will now take a look at how the reified knowledge (text fragments marked up in an MKM format) fit into a conceptual model of the **Mathematical Knowledge Space (MKS)**. We develop the intuition for MKS by constructing the MKS for groups. Its generalization we leave to the gentle reader.

It is a central observation, that — even though we may actually *want* to write down an abstract object like $\mathcal{G}^1$ or even $\mathcal{G}$ in Figure 2 — we only *can* write down a leaf. Given the discussion in the last section, we have to assume that when we express mathematical knowledge in an MKM format, we actually write down a **markup pair** consisting of a concrete realization and the assumed substance equivalence relation. For instance, to formulate the conceptualization group$_1$, we can either type the markup pair $\langle \mathcal{G}^{1,e}, =_{lang}\rangle$ or $\langle \mathcal{G}^{1,g}, =_{lang}\rangle$. Note, that these markup pairs contain enough information to reconstruct $\mathcal{G}^1$ as $[\mathcal{G}^{1,e}]_{=_{lang}}$ or $[\mathcal{G}^{1,g}]_{=_{lang}}$. We can consider $\mathcal{G}^1$ as their substance and $\mathcal{G}^{1,*}$ as their accidences.

We can lift $=_{lang}$ to an **markup equivalence relation** $\hat{=}_{lang}$ by setting $\langle x, R\rangle \hat{=}_{lang}\langle y, R\rangle$ iff $x =_{lang} y$. As $\mathcal{G}^{1,g} =_{lang} \mathcal{G}^{1,e}$ the pairs $\langle \mathcal{G}^{1,e}, =_{lang}\rangle$ and $\langle \mathcal{G}^{1,g}, =_{lang}\rangle$ are $\hat{=}_{lang}$-equivalent, giving rise to an equivalence class $\mathcal{G}^1_{lang}$, which we consider to be the *"language-independent markup object for a group"*. For example, $\mathcal{G}^1_{lang}$ is naturally *represented* by the multilingual `definition` element in OMDoc.

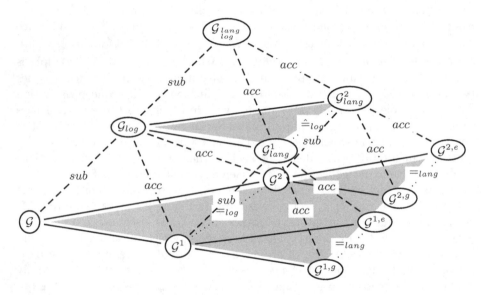

Fig. 3. The Mathematical Knowledge Space for Groups

In Figure 3 we picture the substance and accidence relations **sub** and **acc** resp. with dashed lines, interpreting the triangle from Figure 2 as the base face of a tetrahedral graph and positioning $\mathcal{G}^1_{lang}$ in the first level above it. An analogous construction yields $\mathcal{G}^2_{lang}$. This gives us license to construct a knowledge object $\mathcal{G}_{log}$ as the equivalence class of the $\mathcal{G}^i_{lang}$ modulo $\hat{=}_{log}$, on the first level just as we did for the lower level in Section 2.1. Note, that

$$\mathcal{G}_{log} := \left[\mathcal{G}^1_{lang}\right]_{\hat{=}_{log}} = \left[\left[\langle \mathcal{G}^{1,e}, =_{lang}\rangle\right]_{\hat{=}_{lang}}\right]_{\hat{=}_{log}} \equiv \left[\left\langle [\mathcal{G}^{1,e}]_{\hat{=}_{lang}}, =_{lang}\right\rangle\right]_{\hat{=}_{log}}$$

The right-hand side of this is again an equivalence class of markup pairs, so we can consider $\mathcal{G}$ as the substance of $\mathcal{G}_{log}$ and the $\mathcal{G}^i$ as its accidences, making $\mathcal{G}_{log}$ the *"conceptualization-independent conceptualizations of group"*. In particular, all the relations in the lower part of Figure 3 commute. Note, that just as in the lower level, the objects become more explicit from left to right. Finally, we complete the picture by iterating the construction to obtain a knowledge object $\mathcal{G}_{\substack{lang \\ log}}$ for the group that is *"independent of everything 'relevant'"* , where the relevance is determined by the knowledge object's author's CoP, in our example the conceptualizations with respect to *log* and *lang*.

In particular, *we obtain a knowledge object that no longer contains anything that the given Community of Practice deems substance-irrelevant.*

3   Interpretations of MKS

In Section 2.1 we have presented a model of the reification of knowledge based on substance equivalences. We can interpret Figure 2 — i.e. the *base face of the MKS tetrahedron* — as the perspective of an *author* who writes down her knowledge with an audience in mind. Naturally, her membership in a Community of Practice (see 2.2) determines the employed *implicit* substance equivalences.

In 2.3, we completed this picture by extending the analysis with an account of markup processes resp. *markup formats*, yielding the mathematical knowledge space in Figure 3. Here, we can interpret the *right face of the MKS tetrahedron* (i.e. the triangle area between $\mathcal{G}_{\substack{lang \\ log}}$, $\mathcal{G}^{1,g}$, and $\mathcal{G}^{2,e}$) as the markup process, starting out with concrete materialization of knowledge, ending with a knowledge object in a markup format with *explicit* or *inscribed* substance equivalences.

3.1   MKS and the Content/Form Distinction

Let us now consider the *front face of the MKS tetrahedron* (i.e. the triangle area between $\mathcal{G}_{\substack{lang \\ log}}$, $\mathcal{G}$, and $\mathcal{G}^{1,g}$)[5]. Starting at the top with $\mathcal{G}_{\substack{lang \\ log}}$ which we call the **Knowledge Object**, we can distinguish its content from its form arriving at what we call the "**Form Object**" and the "**Content Object**" — which can be recurrently subjected to the same analysis (see Figure 4 for the resulting view of the front face of the MKS). With the substance perspective on the Content Object we arrive at what we call the "**Platonic Object**"[6]. Successively looking down the substance branch of the tree, we arrive at more and more fundamental, abstract objects. In particular, these are increasingly liberated from their conceptualization as well as presentation. In contrast, looking down the accidence branch we arrive at more and more concrete and tangible objects. In detail, the accidence view on the Content Object leads to its conceptualization level (the "**Conceptualized Object**"), where we have a representation of the content in which certain decisions of how to think about it have been taken (e.g. do we

---

[5] Note, that the front face of the MKS tetrahedron is the only surface conceptually left as the back face's interpretation is analogous as it is just a variant.

[6] The existence of such an object is not discussed, since either ontological assumption has no consequences for the conceptual model. As soon as we start reifying implicit knowledge (independent from the underlying ontology) we have to choose a form which in turn materializes the object.

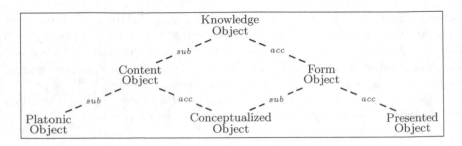

Fig. 4. The Analysis Triangle of a Knowledge Object

want to think about a group as an object, where the associative operation ○ is the primary concept or not?).

Now, let us look at the accidence aspect of the Form Object. As it becomes more and more concrete, we are lead to a presentation level and therefore to the concrete "**Presented Object**". The substance perspective on the Form Object reveals again a conceptualization level, which by our analysis above is the Conceptualized Object. Let us clarify this with the group example: if we want to talk about what 'the group' really is (i.e. the Platonic Object) we have to decide on a representation (otherwise communication is impossible). This selection determines which of the above definitions will be applied. In other words, the choice of the definition fixes the conceptualization of a group. The MKM community seems to concentrate on conceptualizations as *semantic* representations (accidence of the content = substance of the form).

Interestingly, so far capturing knowledge has always aimed at those knowledge objects that are "independent of everything" and *not* at the Platonic Objects themselves (possibly because we mistook them for the same).

3.2   MKS and the MKM User

Now we want to look at the MKS from the perspective of the recipient of knowledge, i.e. *the user* or learner who starts with the concrete materialization of knowledge like a certain document. The user heads for the knowledge itself — the Platonic Object — which is a Knowledge Object's author's point of departure. A reader has to differentiate between the potential content and the concrete form of a document. Depending on her personal choice what content and what form is, she understands and builds up her own knowledge. In contrast to the sender of knowledge, who knows the used equivalence relations (and more) and actively chooses the representation of content, the recipient of knowledge has to infer the applicable equivalence relations.

We claim that the user perspective is already present in the analysis triangle that we have studied in the last section: let us look at a student confronted with a book. It contains the knowledge in its final presented representation (*Presented Object*), but the student is aiming at an understanding of the underlying substance (*Platonic Object*). In order to decide what the content or the form is in the Presented Object, the student has to envisage a *Knowledge Object*, i.e. a potential model of the real knowledge to be learned. From this hypothetical Knowledge Object she can infer the *Content Object* and the *Form Object*. This dramatically reduces the search space of possible interpretations of the Presented Object to the presentations of the Form Object. Here, "understanding" means that the student is able to distinguish between the content of the Form Object (*Conceptualized Object*) and the Presented Object as its form.

Again, interestingly, the user generally is thought of as either modeling the Platonic Object (e.g. in case of a lecture) or the Knowledge Object (e.g. in case of an MKM system), whereas we conjecture that the user is building a Conceptualized Object as approximation of the Platonic Object. Taking this seriously might help to understand how MKM systems need to be positioned in a learning cycle.

3.3   MKS and Narratives

In the discussion of knowledge/document markup formats on the level of theories, it is always difficult to decide what to mark up; the underlying knowledge or the structure of the document that conveys it. Note, that the underlying structures depend on the choice of conceptualization and therefore can be quite different.

Take for instance a didactically enhanced document that introduces a new concept by first presenting a naive, reduced approximation $\mathcal{N}$ of the real theory $\mathcal{F}$, only to show an example $\mathcal{E}_\mathcal{N}$ of where this is insufficient. Then the document proposes a first (straw-man) solution $\mathcal{S}$, and shows an example $\mathcal{E}_\mathcal{S}$ of why this does not work. Based on the information gleaned from this failed attempt, the document builds the eventual version $\mathcal{F}$ of the concept and demonstrates that

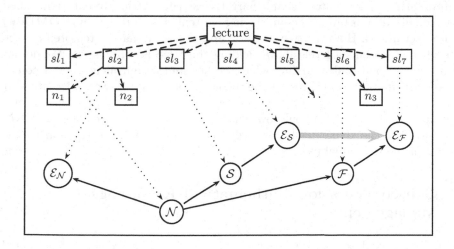

Fig. 5. An Introduction of a Concept via Straw-Man

this works on $\mathcal{E}_{\mathcal{F}}$. Let us visualize the narrative- and content structure in Figure 5. The structure with the solid lines and boxes at the bottom of the diagram represents the content structure, where the boxes $\mathcal{N}$, $\mathcal{E}_{\mathcal{N}}$, $\mathcal{S}$, $\mathcal{E}_{\mathcal{S}}$, $\mathcal{F}$, and $\mathcal{E}_{\mathcal{F}}$ signify theories for the content of the respective concepts and examples. The arrows mark the conceptual dependency structure, e.g. theory $\mathcal{F}$ imports theory $\mathcal{N}$.

The top part of the diagram with the dashed lines stands for the narrative structure, where the arrows mark up the document structure. For instance, the slides sl_i are grouped into a lecture. The dashed lines between the two documents are pointers into the content structure. In the example in Figure 5, the second slide of "lecture" presents the first example: the text fragment n_1 links the content $\mathcal{E}_{\mathcal{N}}$, which is referenced from the content structure to slide 1. The fragment n_2 might say something like "this did not work in the current situation, so we have to extend the conceptualization...".

If we look carefully, we can see that the lower level of the diagram represents the content of the knowledge (structured by the inherent semantic relations of the objects involved), and the upper part the form (structured, so that humans are motivated to concern themselves with the material, understand why some definitions are stated in just this way, and get the new information in easily digestible portions). For instance, the OMDOC format [Koh05a] contains theory-level content- and presentation markup infrastructure for these aspects. The *theory-level content markup* contains the constitutive representations structured by OMDOC theories and their semantic relations (e.g. inheritance), and the *narrative markup* contains the document structure (e.g. that of a course divided into lectures and further into slides), motivating narrative, course-specific information ("When is the final exam?"), etc.

Just as for content-based systems on the formula level, there are now MKM systems that generate presentation markup from content markup, based on general presentation principles, also on this level. For instance, the ACTIVEMATH system [SBF^{+}00] generates a simple narrative structure (the presentation; called a personalized book) from the underlying content structure (given in OMDOC) and a user model. However, a systematic analysis as we have attempted for the formula and statement levels above yielding the MKS is still missing. We do not even have a good understanding what the substance equivalences (and consequently the markup primitives) at the theory level might be. We conjecture that a thorough understanding of the substance/accidence aspects of the theory level (and a theory-level MKS) could eventually lead to a new generation of MKM systems, that can dynamically play with the content/form distinction to the benefit of the individual user.

4   Consequences for Mathematical Knowledge Management

Let us now speculate about the consequences of the suggested conceptual model for the field of mathematical knowledge management and the knowledge representation formalisms employed there.

The first consequence is that we have to extend the MKM representation formats with an explicit representation of substance equivalence relations. First steps into this direction have already been taken in [MKB04, KBHL+03], where a mathematical knowledge base was extended by a concept of *"variants"* to model language, verbosity, formalism, and partially also versioning variants. Our model here suggests that this approach needs to be systematized and explicit representations for 'higher-level' objects need to be introduced. Furthermore, we have to take stock of the various (sub)-relations of the substance equivalences. In [KA03], we have studied how (rather low-level, technical) substance equivalences interact with distribution and versioning of mathematical knowledge and documents. We will have to extend this to the semantic substance equivalences discussed here; [Hut04] goes first steps into this direction.

Another obvious consequence is that we will have to model *Communities of Practice* together with the mathematical knowledge in order to make the CoP-dependencies explicit. However, it is not directly obvious how to model CoPs and their relations to each other. There are large CoPs, like the CoP shared by all the **STEM** fields[7], and small ones whose members agree on particular mathematical objects and differ on others. For instance, it is a matter of CoP in Mathematics whether you accept the law of excluded middle or the axiom of choice. Such CoP differences can already be modeled in MKM formalisms that have a notion of theories that are ordered by inclusion or inheritance. But inclusion of accepted theories is not the only relation between CoPs. For instance, there are two CoPs in theoretical physics, one standardizing the Ricci tensor to twice the other. To model the equivalence (modulo renormalization), we need rich theory structure with theory morphisms like the ones assumed in [Far00, Koh05a]. The differing group conceptualizations can be handled in the same way.

But as we have seen, the influence of the CoP reaches much farther: Even *the set of substance equivalences is determined by the CoP*. Currently, the assumptions about these seem to be hardwired into the representation formats discussed and utilized in MKM. Depending on how extensive such influences turn out to be, we may have to make representation formats parametric to be able to model such influences explicitly. This also begs the question, whether future MKM knowledge bases will be CoP-specific (severely limiting their usefulness) or whether we will be able to have CoP-spanning knowledge bases. In the latter case, we could annotate documents with new kind of metadata, e.g. the CoP (or CoPs) of the author, or of the intended audience. This has been studied informally by projects that attempt large-scale multi-disciplinary knowledge collections like the Connexions project (see [CNX05, HBK03]) and turned out to be of central relevance for integration, navigation, and quality assurance.

5 Conclusion and Further Work

We have explored the space of (mathematical) knowledge MKS spanned by the substance/accidence distinction (with some philosophical excursions) that

---

[7] STEM — Sciences, Technology, Engineering and Mathematics.

triggers content/form coordinates of a knowledge object which are discussed in various communities (like Artificial Intelligence, Logic & Foundations of Mathematics, document markup, or MKM). We have extended or contrasted the discussion there with cognitions in other scientific disciplines, specifically Knowledge Management and Social Sciences. In the former, the relevance of the context of knowledge is understood, whereas in the latter, the context of knowledge is studied and identified as Community of Practice. Based on this broadened view, we propose a more fine-grained, multi-layered model of the content/form distinction, which explains the apparent multi-faceted nature of the content/form debate. This model is based on two assumptions:

(i) the existence of a CoP-determined set of substance equivalences that identify the substance of a knowledge object by equating accidental representation commitments, and

(ii) the dialectic property of the substance and accidence aspects of a knowledge object

In our view, assumption (i) is very natural in the field of Mathematics and neighboring disciplines, and the CoP-dependency is often neglected, since it seems to hold for the large CoP shared by the STEM disciplines. This CoP also delineates the applicability of MKM techniques (which currently seem to hard-wire the substance equivalences into the representation formats) to the STEM fields. Turned positively, we conjecture that

MKM techniques can go wherever the substance equivalences of Mathematics hold!

On the technological side, our fine-grained knowledge space and its CoP-dependence open whole areas of applications. CoP-information would allow to personalize presentations that are generated from content without assuming total knowledge about the user's preferences. Knowledge about the substance equivalences will (in principle) allow automatic translation (generation of variants tailored to the user and situation).

We hope that the musings presented in this paper will be taken up by others and contribute to a consensus about the foundations of mathematical knowledge, so that we can better manage it. In particular, we have to leave to further investigations the tasks of coming up with a content-oriented model of CoPs and their interrelations (we have only been able to motivate the necessity of this and identify some guiding questions in Section 4) and that of fully exploring the consequences for Mathematical Knowledge Management.

Unfortunately, the model of the knowledge space we have presented opens up as many questions as it helps answer, e.g. what does the knowledge space look like where other forms of substance equivalences are involved. For instance, in the Arts and Humanities, a similar model might be applicable, only that form of an artifact or representation is considered its substance, e.g. in a poem, whereas the meaning is considered its accidence. After all, we 'interpret a poem' giving it multiple possible meanings. This suggests the existence of a tetrahedral knowledge space that is 'dual' to the one depicted in Figure 3. This could help solve the riddle that knowledge is considered "objective" in some communities and "subjective" by others.

References

[ABC⁺03]   Ron Ausbrooks, Stephen Buswell, David Carlisle, Stphane Dalmas, Stan Devitt, Angel Diaz, Max Froumentin, Roger Hunter, Patrick Ion, Michael Kohlhase, Robert Miner, Nico Poppelier, Bruce Smith, Neil Soiffer, Robert Sutor, and Stephen Watt. Mathematical Markup Language (MathML) version 2.0 (second edition). W3c recommendation, World Wide Web Consortium, 2003. Available at http://www.w3.org/TR/MathML2.

[BCC⁺04]   Stephen Buswell, Olga Caprotti, David P. Carlisle, Michael C. Dewar, Marc Gaetano, and Michael Kohlhase. The Open Math standard, version 2.0. Technical report, The Open Math Society, 2004. http://www.openmath.org/standard/om20.

[BD91]   John Seely Brown and Paul Duguid. Organizational Learning and Communities of Practice:Toward a Unified View of working, Learning and Innovation. *Organization Science*, 2(1):40–57, 1991.

[BD00]   John Seely Brown and Paul Duguid. *The Social Life of Information*. Harvard Business School Press, 2000.

[BP64]   Paul Benacerraf and Hilary Putnam, editors. *Philosophy of Mathematics: Selected Readings*. Cambridge University Press, 2nd edition 1983 edition, 1964.

[CNX05]   CONNEXIONS. Project home page at http://cnx.rice.edu/, seen January 2005.

[CS98]   Lassaad Cheikhrouhou and Volker Sorge. Planning equivalence proofs. In *Workshop on Using AI Methods in Deduction at CADE-15*, 1998.

[DP98]   Thomas H. Davenport and Laurence Prusak. *Working Knowledge*. Harvard Business School Press, 2000 edition, 1998.

[Far00]   William Farmer. An infrastructure for intertheory reasoning. In David McAllester, editor, *Automated Deduction – CADE-17*, number 1831 in LNAI, pages 115–131. Springer Verlag, 2000.

[Hal59]   Marshall Hall. *The Theory of Groups*. The Macmillan Company, New York, 1959.

[HBK03]   Geneva Henry, Richard G. Baraniuk, and Christopher Kelty. The connexions project: Promoting open sharing of knowledge for education. In *Syllabus, Technology for Higher Education*, 2003.

[Hei00]   Bettina Heintz. *die innenwelt der mathematik. zur kultur und praxis einer beweisenden disziplin*. Springer-Verlag Wien, 2000.

[Hut04]   Dieter Hutter. Towards a generic management of change. In Christoph Benzmüller and Wolfgang Windsteiger, editors, *Computer-Supported Mathematical Theory Development*, number 04-14 in RISC Report Series, pages 7–18. RISC Institute, University of Linz, 2004. Proceedings of the first "Workshop on Computer-Supported Mathematical Theory Development" held in the frame of IJCAR'04 in Cork, Ireland, July 5, 2004. ISBN 3-902276-04-5. Available at http://www.risc.uni-linz.ac.at/about/conferences/IJCAR-WS7/.

[Isr79]   Joachim Israel. *The Language of Dialectics and the Dialectics of Language*. Munksgaard, Humanities Press, USA, 1 edition, 1979. (American) ISBN 0 391 01000X.

[KA03]   Michael Kohlhase and Romeo Anghelache. Towards collaborative content management and version control for structured mathematical knowledge. In Andrea Asperti, Bruno Buchberger, and James Harold Davenport, editors, *Mathematical Knowledge Management, MKM'03*, number 2594 in LNCS, pages 147–161. Springer Verlag, 2003.

[KBHL⁺03]   B. Krieg-Brückner, D. Hutter, A. Lindow, C. Lüth, A. Mahnke, E. Melis, P. Meier, A. Poetzsch-Heffter, M. Roggenbach, G. Russell, J.-G. Smaus, and M. Wirsing. Multimedia instruction in safe and secure systems. In *Recent Trends in Algebraic Development Techniques*, volume 2755 of *LNCS*, pages 82–117. Springer Verlag, 2003.

[KM79]   M. I. Kargapolov and J. I. Merzljakov. *Fundamentals of the Theory of Groups*. Graduate Texts in Mathematics. Springer Verlag, 1979.

[Koh05a]   Michael Kohlhase. OMDoc an open markup format for mathematical documents (version 1.2), 2005. Manuscript, http://www.mathweb.org/omdoc/omdoc1.2.ps to appear in Springer LNAI.

[Koh05b]   Michael Kohlhase. Semantic markup for TeX/LaTeX. Manuscript, available at http://kwarc.eecs.iu-bremen.de/software/stex, 2005.

[LW91]   Jean Lave and Etienne Wenger. *Situated Learning: Legitimate Peripheral Participation (Learning in Doing: Social, Cognitive and Computational Perspectives S.)*. Cambridge University Press, 1991.

[MKB04]   Armin Mahnke and Bernd Krieg-Brückner. Literate ontology development. In Robert Meersman, Zahir Tari, and Angelo Corsaro et al., editors, *On the Move to Meaningful Internet Systems 2004: OTM 2004 Workshops*, number 3292 in LNCS, pages 753–757. Springer Verlag, 2004.

[Pap96]   Seymour Papert. An Exploration in the Space of Mathematics Educations. *International Journal of Computers for Mathematical Learning*, 1(1):95–123, 1996.

[PRR97]   G. Probst, St. Raub, and Kai Romhardt. *Wissen managen*. Gabler Verlag, 4 (2003) edition, 1997.

[Rob91]   J. A. Robinson. Formal and informal proofs. In R. S. Boyer, editor, *Automated Reasoning: Essays in Honor of Woody Bledsoe*, pages 267–282. Kluwer, London, 1991.

[SBF⁺00]   Jörg Siekmann, Christoph Benzmüller, Armin Fiedler, Andreas Franke, George Goguadze, Helmut Horacek, Michael Kohlhase, Paul Libbrecht, Andreas Meier, Erica Melis, Martin Pollet, Volker Sorge, Carsten Ullrich, and Jürgen Zimmer. Adaptive course generation and presentation. In P. Brusilovski, editor, *Proceedings of the Fifth International Conference on Intelligent Tutoring Systems—Workshop W2: Adaptive and Intelligent Web-Based Education Systems*, pages 54–61, Montreal, 2000.

[WMS02]   E. Wenger, R.A. McDermott, and W. Snyder. *Cultivating of Communities of Practice*. Harvard Business School Press, 2002.

Authoring Presentation for OpenMath

Shahid Manzoor, Paul Libbrecht, Carsten Ullrich, and Erica Melis

Deutsches Forschungszentrum für Künstliche Intelligenz
{manzoor, paul, cullrich, melis}@activemath.org

Abstract. Some mathematical objects can have more than one notation. When a system compiles mathematical material from multiple sources, a management effort to maintain uniform and appropriate notations becomes necessary. Additionally, the need arises to facilitate the notations editing of the mathematical objects with authoring tools. In this paper, we present our work towards those needs. We have designed a framework that defines an authoring cycle supported by series of tools, which eases the creation of notations for the symbols in the process of publishing mathematics for the web.

1 Introduction

ACTIVEMATH [MAB+01] is an adaptive and interactive web-based learning environment for mathematics. It dynamically generates content adapted to the students profile i.e. goals, preferences (e.g. personalized presentation and field of interest etc), capabilities, and knowledge. The mathematical content in ACTIVE-MATH is represented in OMDoc [Koh04] and are stored in a knowledge base. The motivation for our work comes from a number of problems we experienced with authors when they write mathematical course material in the ACTIVEMATH environment. For instance, in different languages, different mathematical notations for a symbol are used: the slope symbol in English is written as $slope(F, p)$, or $steigung(F, p)$ in German. Moreover, different authors want to use different notations for the same mathematical objects such as $\frac{1}{2}$ or $1 : 2$, $a * b$ or ab, $\frac{d}{dx}f$ or $f'x$.

However, authors are challenged when writing the presentation of the mathematical expression. The current approaches provide a *meta stylesheet* (an XML encoding to represent the notations for the symbols) as an authoring support, which is converted into XSLT-templates. That is, the authors lack authoring tools which ease their notations editing and tools to ease the publishing tasks.

We propose a framework that defines an authoring cycle for the editing of symbols' notations which involves a series of tools that support the process from editing to previewing the notations. This framework simplifies the symbols' notation authoring process.

We start with a description of the current approaches for authoring OPEN-MATH symbols' notations. Thereafter, we describe problems in these authoring processes. In Sec. 4, we explain the XML encoding for the notation and annotations. We discuss our authoring tools and environment for the notations editing.

M. Kohlhase (Ed.): MKM 2005, LNAI 3863, pp. 33–48, 2006.

Finally we describe how the notations are processed in the presentation architecture of ActiveMath.

2   Previously Existing Approaches to Presentation Generation for OpenMath

MathML is a W3C standard for representing 2-dimensional mathematical formulæ on the web. It has been embedded inside XHTML. It comes with two languages, Presentation MathML (PMML) and ContentMathML-content. The PMML concentrates on the presentation of the expression. On the other hand, MathML-content organizes mathematical formulæ in trees of operators, variables, and numbers with well defined semantics. Its set of possible symbols is, however, fixed.

OpenMath's primary goal is to serve as a communication of mathematical objects between applications. Similarly to MathML-content, it organizes formulæ in trees of mathematical symbols. Contrary to MathML-content, OpenMath has a well defined extensibility mechanism: one can write content-dictionaries (CD) to provide a description of new symbols.

Generally, XSLT is used to transform the OpenMath objects into the output formats, such as HTML or TEX. In [Car00] and [Koh04], an XSLT based algorithm is described to generate the presentation. There, an template rule is required for each symbol. Each template generates the notation in output format recursively. The match rule for each template is built with **name** and **cd** attributes. Below is an example of an XSLT template for the **divide** symbol.

```
<xsl:template match="om:OMA[om:OMS[position()=1 and
@name='divide'\break and @cd='arith1']]">
  <mfrac>
    <xsl:apply-templates select="*[2]" />
    <xsl:apply-templates select="*[3]" />
  </mfrac>
</xsl:template>
```

OMDoc [Koh04] provides a `<presentation>` element to write notations aiming to facilitate the authoring support, as hand written XSLT templates are tedious and error-prone. The `<presentation>` element points to the symbol for which the presentation is being written. It uses the `<style>`, `<xslt>` and `<use>` elements to generate a particular presentation of a symbol. For complex notations such authoring requires is defining the body of an XSLT template. For simple notation binary operators, subscripts, list, and fraction, the easy syntax of the `<use>` element is sufficient. This approach supports the XSLT templates generation for multiple output formats. But, authors have to specify the notation for each output format.

In Fig. 1, the example of `<presentation>` element represents the notation for the **divide** OpenMath symbol in three output formats. Each `<use>` element represents the notation, specified as character data for HTML (/) and LATEX (\frac) or element attribute for MathML (mfrac).

```
<presentation role="applied" for="divide" theory="arith2">
  <use format="html">/</use>
  <use format="mathml" element="mfrac" />
  <use format="latex">\frac</use>
</presentation>
```

Fig. 1. Example of presentation tag in OMDoc

Another approach is discussed in [NW01], in which Naylor and Watt have introduced the concept *meta stylehseet* that generates the XSLT stylesheet automatically for PMML and MATHML-content formats. They proposed the *meta stylesheet* as an extension to Content Dictionaries, in which the notation of the symbol in PMML and TEX along with the OPENMATH expression is represented together. An example for the **divide** symbol is shown below. The notation is represented in **<version>** element and the OPENMATH expression in **<semantic_template>**.

```
<Notation>
  <version precedence="200" style="1" >
    <math>
      <mfrac>
        <mi xref="arg1">a</mi>
        <mi xref="arg2">b</mi>
      </mfrac>
    </math>
  </version>
  <semantic_template>
  <OMOBJ>
    <OMA>
      <OMS cd="arith1" name="divide" />
      <OMV id="arg1" name="a" />
      <OMV id="arg2" name="b" />
    </OMA>
  </OMOBJ>
  </semantic_template>
</Notation>
```

Fig. 2. An example of the notation of the *divide* symbol following [NW01]

In this approach, the XPath of the arguments in the generated XSLT templates is determined by linking the symbols and the notations with their attributes as in Fig. 2: the elements with **id** attribute represent arguments in the OPENMATH expression) and **xref** indicates the argument in PMML.

3   Symbols Presentation Authoring Problems

Although *meta stylesheet* approaches discussed in the previous section provide a basic authoring support, authors may experience difficulty in writing notations

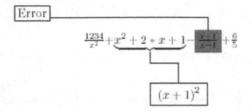

Fig. 3. An example of a mathematical expression with annotations

for the OPENMATH symbols. The problems are as follows: first, some mathematical symbols can be presented in many ways. The automatic adaptation of a notation (from many notations for a symbol) in different contexts such as class, book and student is an issue, especially when different materials are presented or merged. For instance, in ACTIVEMATH, materials may be compiled by merging the contents from different authors or collections. Secondly, authors, who have little or no background in programming, are required to write presentation in complex XML or XSLT structures. And last, but not least, authors may wish to highlight mathematical expressions by (a) making a box around the contents, (b) changing the background color, or, (c) labeling etc (See Fig. 3).

Most authors need an easy to use authoring environment, where they can edit the notations for the OPENMATH symbols as well as preview it in the target presentation.

4   Solving Authors' Problems

Generally, authors use specialized editors for content authoring, which provide facilities that are best suited to their task. The editor selection depends on the content type or format. For instance, for XML documents editing there are specialized editors that understand the structural behavior and grammar of the documents in general. If an editor is aware of the usage of the XML document, for example when editing OMDoc documents for ACTIVEMATH, it is possible to make the editing easier and more productive.

jEditOQMath [Lib04] is such an authoring environment: it is a package distributed to the authors of content for the ACTIVEMATH learning environment. It is based on the open-source text-editor jEdit and its rich support for the editing of XML documents allows the easy creation and maintenance of valid XML documents. It adds several search facilities specialized for OMDoc and is integrated with the ACTIVEMATH server of the author: it hosts publishing routines which, report consistency errors; it provides quick-open and quick-link facilities using the drag-and-drop paradigm. jEditOQMath is freely available.[1]

Editing mathematical formulæ is done in the OQMath language[2] which is converted to OPENMATH formulae. It converts a highly readable, linear, syntax

---

[1] Please see http://www.activemath.org/projects/jEditOQMath/.
[2] Please see http://www.activemath.org/projects/OQMath/.

to OPENMATH expressions (thanks to the usage of Unicode characters such as
∃ or Ω) in a way that authors can extend with new (input-level) notations. The
experience with jEditOQMath thus far has proven that the OQMath verbosity
is acceptable for mathematical documents. It has also proven that error feedback
is important when using an environment where such a rich presentation process
is applied.

Our solution provides a framework that defines an authoring process for notation editing to the publishing in ACTIVEMATH. The framework consists of 2
phases. The Phase I deals with OPENMATH and notation editing. The notations are edited by our tool called OPENMATH Presentation Editor (OMPE).
OMPE is run as a plug-in of jEditOQMath providing the notation authoring
facility within the same environment where the authors edit their mathematical
documents. Phase II is initiated passively, when the contents are previewed in
ACTIVEMATH. In this phase the notations can be adapted according to contextual information such as language, class/group, student and book. The whole
process utilizes a `<symbolpresentation>` element, which we discuss in the next
section.

4.1   Knowledge Representation of the Symbols Notations

In order to define different notations of a symbol, we have introduced the element `<symbolpresentation>`. It encapsulates the data to generate the XSLT
templates for multiple output formats, currently HTML, MATHML and LATEX.
The grammar of the `<symbolpresentation>` element is shown in the table
below.

tag	attributes	children
symbolpresentation	xref, id	(notation)+
notation	notation, precedence, format, language, lbrack, rbrack, style	(math,OMOBJ)

symbolpresentation: represents notations for a symbol. Its **xref** attribute
points to the symbol for which notation is being written. It can contain
multiple notations of a symbol, and each notation is represented by the
`<notation>` element.

notation: Each notation element represents one notation of the symbol. For
this purpose, it contains two structures, ie. PMML and OPENMATH. Its
format attribute contains a list of output formats for the symbol. Three
attributes, **precedence**, **lbrack** and **rbrack**, control the bracket printing
around the symbol. The left bracket is only printed, if the symbol **precedence**
is less than the parent symbol. The same algorithm works with the right
bracket with the corresponding attributes. Its **style** and **language** attributes
are used to define selectors for a symbol that will be used at rendering time
to choose the appropriate notation.

OMOBJ: The OMOBJ element represents the OPENMATH expression
containing the prototype of the symbol for which a presentation is being
written. The OPENMATH symbols in this expression are used in three possible ways: 1) as function, applied to its arguments represented inside the

OMA element, 2) standalone, or 3) having an attribute inside the attribution object (OMATTR). This expression allows the production of XSLT-templates applied to symbols or more complex expressions. The arguments are represented by <OMV> elements.

math: This element contains Presentation MATHML (PMML) expression representing the notation for a symbol. The arguments in the notation (PMML) is represented by an <mi> element. The arguments in both PMML and OPEN-MATH expression are mapped by pairing the **name** attribute of the <OMV> element and content of the <mi>. The XPath expression in the XSLT template is calculated by matching the arguments in both the expressions.

An example of the **plus** notation in <symbolpresentation> element is shown in Fig. 4.

```
<symbolpresentation xmlns="http://www.activemath.org/namespaces/am_content"
    id="arith1plus_77_24" xref="mbase://openmath−cds/arith1/plus">
  <notation format="html|pmathml|TeX" precedence="110" lbrack="("rbrack=")">
    <math xmlns="http://www.w3.org/1998/Math/MathML">
      <mrow>
        <mi>a</mi>
        <mo>+</mo>
        <mi>b</mi>
      </mrow>
    </math>
    <OMOBJ xmlns="http://www.openmath.org/OpenMath">
      <OMA>
        <OMS cd="arith1" name="plus" />
        <OMV name="a" />
        <OMV name="b" />
      </OMA>
    </OMOBJ>
  </notation>
</symbolpresentation>
```

Fig. 4. Example of <symbolpresentation> element for the **plus** symbol

4.2   Dealing with Different Notations for the Same Symbol

Our structure allows the authors to define different notations for a symbol. For each notation, a new definition of <notation> element is required. Below are contexts we have identified to influence the adaptation in symbol notation.

Language: This deals with internationalization for the symbols. Some symbols have different notations depending on the languages, e.g. the greatest common divisor symbol is written $\gcd(a, b)$ in English but is written $\mathrm{kgV}(a, b)$ in German. Using our notation infrastructure allows to solve the internationalized symbol presentation in LEACTIVEMATH as discussed in the report [LW05]. To associate a notation with a language, we are using the **language** attribute of the <notation> element.

Different patterns of the arguments: One of the variations in the symbol rendering occurs when it has a different organization or number of the arguments. Consider the two notations, $\sum_{x=1}^{n} x$ and $\sum_{x \in m} x$, of the sum symbol. In the example, both sum symbols differ in their first arguments, i.e. be it the interval 1 to n or the set m. For each case like this, an author has to define a `<notation>` element.

Authors Styles: There are situations in which the symbol notation differs depending on the authors' styles even though the symbols are alike in all respects. Example of such presentations are: $\frac{1}{2}$ or $1:2$, $a*b$ or $a \cdot b$ or ab, $\frac{df}{dx}$ or $f'x$. For this, we provide a style attribute of the notation which allows the authors to define their own specific style notation for a symbol.

This multiple styles notations raise the question of the selectivity of a particular notation especially, when a material is compiled by merging the contents from different sources. We defined the following priorization for a consistent material presentation in ACTIVEMATH from least to highest priority:

- **System Defaults** has the least priority in the system. All collections XSLT templates are merged into one, the import order of XSLT rules is used to manage this priority.
- **Author/Collection** This is the authors' default notation defined for a particular collection. It is automatically selected, if no other priority level is defined.
- **Book** A book can be generated by selecting a list of learning concepts from different content collections. At this priority level and other following levels, priority should be assigned by the notation selector tool which should store the priority information in the book configuration.
- **Group** The group represents a group of learners, for example a class, studying a common course. We expect, for example, teachers responsible of a course to adapt notations for their classes.
- **Individual** The students himself may be able to assign the priority to a notation. His choices, which we expect to be rare, have the highest priority level.

Within the Same Collection: In this context, the authors want to define different notations of the same symbol within the same collection. Take, for example, the associativity law of the **plus** symbol, i.e $a+(b+c) = (a+b)+c$. The default notation for the **plus** symbol will not print the bracket, but explicit brackets are required to explain the law. This kind of cases is handled with OPENMATH attribution object. We have introduced a **type** attribute for the application OPENMATH object to define for multiple notations of the same symbol. The **type** attribute abstractly defines a class of the symbols for which special treatment is made during the presentation generation. So, whenever an author wants a special notation other than the default in his book, he can assign this attribute to the instance of the application object containing the particular symbol. In Fig. 5, an example of the **plus** notation for a specific occasion is shown.

```
<symbolpresentation id="arith1plus_3_88" xref="mbase://openmath-cds
/arith1/plus">
  <notation format="html|pmathml|TeX" precedence="110" lbrack="
(" rbrack=")">
    <math xmlns="http://www.w3.org/1998/Math/MathML">
      <mrow><mo>(</mo>
        <mi>a</mi>
        <mi>+</mi>
        <mi>b</mi>
        <mo>)</mo></mo></mrow>
    </math>
    <OMOBJ xmlns="http://www.openmath.org/OpenMath">
      <OMATTR>
        <OMATP>
          <OMS cd="am.presentation" name="type"/>
          <OMV name="associative"/>
        </OMATP>
        <OMA>
          <OMS cd="arith1" name="plus" />
          <OMV name="a" />
          <OMV name="b" />
        </OMA>
      </OMATTR>
    </OMOBJ>
  </notation>
</symbolpresentation>
```

Fig. 5. Example of the `plus` symbol notation with `type` attribute

4.3   Knowledge Representation for the OPENMATH Objects Annotations

Sometimes the authors wish to emphasize a sub-expression to elaborate a concept by annotating it with styles such as colors, borders, fonts etc or by adding a label. We have built a mechanism for attaching mathematical expressions to styles of a *stylesheet* using the attribution elements. The *stylesheet* contains the style definitions like color, background color, border, etc. Additional information can be stored in an attribution object. For instance, an author can assign an attribute *error* to OPENMATH expression (as in Fig. 6) to highlight a mistake in an exercise step by rendering the background in the `red` color as is described Fig. 9.

We use the attribution object to define the additional information for the presentation systems. Adding attributes to the OPENMATH objects does not change the meaning of the object. Moreover, OPENMATH applications can ignore the attributes, if they do not know its meaning. Therefore attributes are the ideal place to store the data about the specialized presentation.

Presentation Attributes. In the following we propose a list of attributes to be used for storing the information about annotations:

```
<OMOBJ xmlns="http://www.openmath.org/OpenMath">
  <OMA>
    <OMS name="plus" cd="arith1" />
    <OMV name="a"/>
    <OMATTR>
      <OMATP>
        <OMS name="type" cd="am.presentation">
          <OMV name="error" />
        </OMATP>
        <OMV name="b" />
    </OMATTR>
  </OMA>
</OMOBJ>
```

Fig. 6. An example of stylistic annotation using an OPENMATH attribution

type: This attribute assigns a type to an OPENMATH object. The value of this variable can be any string in the `name` attribute of `<OMV>`. In our approach, we use the `type` attribute for defining multiple notations of a symbol (see Fig. 5).

The `type` attribute is also used to invoke the `stylesheet` for the presentation of a symbol; as in Fig. 6, `error` identifier is attributed to an OPENMATH object.

label: Labels attach media information (images, text and math expression) with the mathematical sub-expressions to illustrate its underlying meaning. This attribute arranges labels into eight logical positions around the mathematical expressions as shown in the Fig. 7.

The logical position is assigned via the `orientation` attribute assigned to the `value` object of the `label` attribute. In the Fig. 8, the `label` attribute is assigned to the variable object x. The position of the text *Variable* is set **down**.

Stylesheet: A **stylesheet** stores style attributes, i.e color, font-size, font-style, background, border and border-color. These attributes are common for each output formats (HTML, MATHML and LᴬTEX). The style attributes are grouped under `<styleset>` element with a logical name represented in `id` attribute, e.g. The `<styleset id=''error''>` style contains border and red background in the

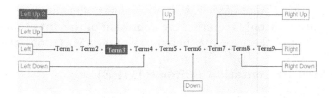

Fig. 7. Various positions for the presentation of labels

$label(orientation(down),"Variable",x)$

```
<OMOBJ xmlns="http://www.openmath.org/OpenMath">
  <OMATTR>
    <OMATP>
      <OMS name="label" cd="am.presentation">
      <OMATTR>
        <OMATP>
          <OMS name="orientation" cd="am.presentation">
          <OMV name="down" />
        </OMATP>
        <OMSTR>Variable</OMSTR>
      </OMATTR>
    </OMAT>
    <OMV name="x" />
  </OMATTR>
</OMOBJ>
```

Fig. 8. An example OPENMATH formula with attribution to denote a label. Above it is the OQMath formula that produces it: this is what we expect the author to input.

```
<Styles>
  <styleset id="error">
    <style name="border" value="solid" />
    <style name="background" value="red" />
  </styleset>
</Styles>
```

Fig. 9. Definition of Error Style

stylesheet (an example is shown in Fig. 9). The XSLT presentation system of each output format applies these logical styles to the mathematical expression by translating the attributes into their native implementations. For example, the `background` attribute , MATHML has the `mathbackground` attribute of the `<style>` element, in HTML, it has `background-color` in CSS and in LaTeX it is implemented via the `colorbox` macro. The style definition (`<styleset>`) is only instantiated by matching `id` of the style definition with the `type` attribute of the OPENMATH objects.

From the authoring point of view, the `stylesheet` offers an easy way to write stylistic information by only requiring the definition of the `<styleset>` element and the assignment of the `type` attribute to the OPENMATH objects.

4.4   OPENMATH **Presentation Editor (OMPE)**

OMPE is an authoring tool for the symbol notation (`<symbolpresentation>`) editing. It accpets a linear syntax for both OPENMATH and PMML expressions, and, reduces the burden of writing long complex XML expressions: the syntax

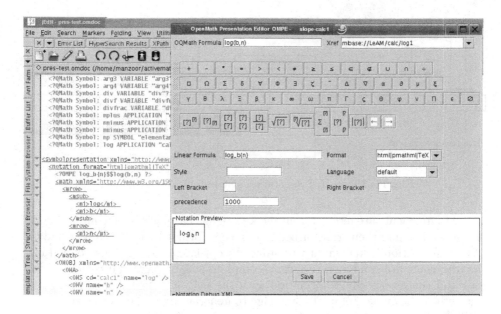

Fig. 10. Screenshot of OMPE editor

for the OPENMATH input is based on OQMath and the PMML input is similar to LaTeX.

The editor also provides a tool bar, containing buttons for the maths notations, operators and symbols to help in editing the linear expression for PMML.

OMPE is developed on top of the Java OPENMATH Editor (JOME).[3] Originally, JOME was capable of producing a limited set of OPENMATH symbols and content MATHML from a linear syntax. We have extended it to produce PMML, and the notation of the symbol in the `<symbolpresentation>` tag. OMPE runs as a plug-in of jEditOQMath as shown in Fig. 10.

To build a new notation, an author has to invoke the editor from an OQMath document where mathematical input notations are already defined. There, he provides the required information in the input boxes and then saves it. The new notation is pasted at the location of the cursor in the document. Alternatively, the author has to move his cursor inside the `<symbolpresentation>` element in the document and then invoke the OMPE editor from the plug-in menu. The editor loads the notation in its environment and gives the author the opportunity to edit it. After editing, the author saves the notation, OMPE replaces the old notation with the edited one.

5   Automatic Presentation Generation

The content presentation process in ACTIVEMATH consists of series of steps making a presentation pipeline [ULWM04]. It is a 2-stage process. In the first stage,

---

[3] Please see `http://jome.sourceforge.net/`.

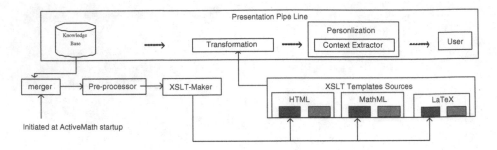

Fig. 11. Phase II: Presentation generation for the symbols

contents fragments are fetched from the knowledge base, with some intermediate preprocessing, and then transformed into output format by using XSLT; no adaptivity information is known, at this stage, except language. In the second stage, the fragments are combined to form a complete output page and enriched with user and context-specific information with the help of velocity code (Velocity[4] is a high performance template language to generate dynamic web pages).

The presentation generation comes in the phase II of our authoring process. It is started automatically in the ACTIVEMATH environment. This phase deals with the adaptation that consists in selecting a notation among the several ones available using contextual information. It is made of the following two processes:

5.1   XSLT **Generation for the Notations**

The generation of the XSLT code from `<symbolpresentation>` element is a three step process (illustrated in Fig. 11). In the first step, the **merger** extracts all the `<symbolpresenttion>` elements from the knowledge base and groups the notations in one `<symbolpresentation>` element for each symbol. In the second step, the merged file is passed to the **pre-processor**, which decorates the notations code with styles, XPath and conditions if required. In the last step, the decorated notations are passed to the XSLT-maker, which finally generates XSLT templates for the three output formats. The **blue** box in the figure represents, the generated XSLT code for the symbols.

Below are examples of XSLT code generated depending on the `<notation>` elements for each symbol available in the system.

Example 1. The XSLT code for the **plus** symbol for two cases: 1) having **associative** attribute for explicit brackets printing, 2) or default.

```
<xsl:template match="om:OMA[om:OMS[@name='plus' and @cd='arith1']]">
   <xsl:choose>
     <xsl:when test="om:OMATTR[om:OMATP[om:OMS[@name='type' and
     cd='am.presentation'and following−sibling::om:OMV[position() =1
        and @name='associative']]]]">
```

---

[4] Please see `http://jakarta.apache.org/velocity/`.

```
    <mrow>
      <mo>(</mo>
      <xsl:apply-templates select="*[2]" />
      <mo>+</mo>
      <xsl:apply-templates select="*[3]" />
      <mo>)</mo>
    </mrow>
  </xsl:when>
  <xsl:otherwise>
    <mrow>
      <xsl:apply-templates select="*[2]" />
      <mo>+</mo>
      <xsl:apply-templates select="*[3]" />
    </mrow>
  </xsl:otherwise>
  </xsl:choose>
</xsl:template>
```

Example 2. In this example, the language adaptivity (German, i.e. **de**, on line 3) is checked at XSLT level. Further adaptation is made for the author style (**m-notation**) in the default language case (on line 11). This XSLT template produces Velocity code (e.g. on lines 12, 19 and 27) where branching between the notations is done by an `#if` condition.

```
1   <xsl:template match="om:OMA[om:OMS[@name='slope' and @cd='calc1']]">
2     <xsl:choose>
3       <xsl:when test="$language='de'">
4         <mrow><msub><mo>steigung</mo>
5           <xsl:apply-templates select="*[2]" /></msub>
6           <mrow><mo>(</mo>
7           <xsl:apply-templates select="*[3]" />
8           <mo>)</mo></mrow>
9         </mrow>
10      </xsl:when>
11      <xsl:otherwise>
12        <xsl:text>#if ($style.contains("m-notation")) then </xsl:text>
13          <mrow><msub><mo>m</mo>
14            <xsl:apply-templates select="*[2]" /></msub>
15            <mrow><mo>(</mo>
16            <xsl:apply-templates select="*[3]" />
17            <mo>)</mo></mrow>
18          </mrow>
19        <xsl:text> #else</xsl:text>
20          <mrow><mo>slope</mo>
21            <mo>(</mo>
22            <xsl:apply-templates select="*[2]" />
23            <mo>,</mo>
24            <xsl:apply-templates select="*[3]" />
25            <mo>)</mo>
26          </mrow>
```

```
27          <xsl:text> #end</xsl:text>
28        </xsl:otherwise>
29      </xsl:choose>
30    </xsl:template>
```

Example 3. The example for the sum symbol with different arguments patterns, i.e set and default.

```
<xsl:template match="om:OMA[om:OMS[@name='sum' and @cd='arith1']]">
  <xsl:choose>
    <xsl:when test="om:OMA[position()=2 and om:OMS[@name='integer_interval'
    and @cd='interval1']]">
    −
    </xsl:when>
    <xsl:otherwise>
    −
    </xsl:otherwise>
  </xsl:choose>
</xsl:template>
```

5.2 Notation Generation in Output Formats

Current approaches for the notation generation that we reviewed in Sec. 2 only deal with XSLT generation. In order to adaptively present mathematical notations, information from the context is needed which can only be acheived by server side scripting languages such as Servelts, PHP and Velocity. In ACTIVE-MATH, this context information is available at Velocity interpretation time, i.e. in the personalization phase.

The presentation generation pipeline (shown in Fig. 11) is initiated with a user request from the browser. At the stage of transformation, the annotation interpreter (a set of XSLT templates, shown in green box in the Fig. 11) checks whether the OPENMATH objects (in the contents) are attributed. In that case, it applies the <styleset> from the stylesheet document or generate supporting code for the labels. The interpreter implementation is specific to each output format. The labeling annotation can not be fully handled at the XSLT level, because XSLT does not know about the dimension and position of the expression. For this, we have written output format-specific routines that executes at final rendering level: for HTML & MATHML, they are executed in the client browser, for LATEX-based output-formats (currently PDF), a LATEX package was written. The labeling, in MATHML, requires a DOM facility and thus is restricted to the Mozilla browser currently. At the stage of Personalization in the pipeline, the Velocity code for multiple notations (discussed in Sec. 5.1) is executed by matching the *styles list* returned from our ContextExtractor. The ContextExtractor should retrieve the styles list based on the priority of the context. Currently, it only extracts style at the collection level.

6   Conclusion and Future Work

In this paper, we have described a solution to the authoring problems that we have identified in our work with the ACTIVEMATH learning environment. We believe that other mathematical web presentation systems could benefit from our work as well.

We are polishing the tools so that authors adopt it as soon as possible, in particular in the LEACTIVEMATH EU project.[5] Further work is also being done in order to make notations user-visible aside of the description of a symbol.

Currently, adaptivity of a notation is only acheived at the collection level. Using a notation selection tool similar to[6] we intend to offer the management of notations to a system: it will be of use to authors and editors to edit their books' notations, to teachers to edit the classes' notations, and, potentially, to individual users. This will allow the adaptation of a notation as discussed in Sec.4.2.

Moreover, the notations selected for a presentation have to be synchronized with presentation of mathematical tools. For instance, an input editor will be integerated into LEACTIVEMATH; it will enable learners to input mathematical formulæ into, e.g, the search interface or exercise interactions. The Wiris input editor[7] chosen for this task allows the definition of *domain files* which describe the notations for each symbol: an export from the `<symbolpresentation>` elements to domain files is under work. The integeration of the input editor within ACTIVEMATH will take advantage of the notations available for the context and will, thus, enable learners to input mathematical formulæ using notations consistent with the mathematical content that is presented to them.

References

[Car00]   David Carlisle. Openmath, MathML, and XSL. In *ACM SIGSAM Bulletin*, volume 34, number 2, pages 6–11, June 2000. ISSN:0163-5824.

[Koh04]   Michael   Kohlhase.   OMDOC   an   open   markup   format   for mathematical   documents   (version   1.2),   2004.   Manuscript, http://www.mathweb.org/omdoc/.

[Lib04]   P. Libbrecht. Authoring web content in activemath: From developer tools and further. In Alexandra Christea and Franca Garzotto, editors, *Proceedings of the Second International Workshop on Authoring Adaptive and Adaptable Educational Hypermedia, AH-2004: Workshop Proceedings, Part II, CS-Report 04-19*, pages 455–460. Technische Universiteit Eindhoven, 2004.

[LW05]   "Paul   Libbrecht   and   Stefan   Winterstein".   "internationalizing leactivemath".   Technical report, LeActiveMath consortium, "http://www.leactivemath.org", "2005".

---

[5] Please see http://www.leactivemath.org/.

[6] Please   see   Notation   Selection   Tool   at   http://www.orcca.on.ca/MathML/NotationSelectionTool/.

[7] See http://www.wiris.com/.

48   S. Manzoor et al.

[MAB+01]   E. Melis, E. Andrès, J. Büdenbender, A. Frischauf, G. Goguadze, P. Lib-
           brecht, M. Pollet, and C. Ullrich. Activemath: A generic and adaptive
           web-based learning environment. *International Journal of Artificial In-
           telligence in Education*, 12(4):385–407, 2001.

[NW01]   Bill Naylor and Stephen Watt. Meta style sheets for the conver-
           sion of mathematical documents into multiple forms. In *International
           Workshop on Mathematical Knowledge Management*, September 2001.
           http://www.emis.de/proceedings/MKM2001.

[ULWM04]   C. Ullrich, P. Libbrecht, S. Winterstein, and M. Mühlenbrock. A flexible
           and efficient presentation-architecture for adaptive hypermedia: Descrip-
           tion and technical evaluation. In Kinshuk, C. Looi, E. Sutinen, D. Samp-
           son, I. Aedo, L. Uden, and E. Kähkönen, editors, *Proceedings of the
           4th IEEE International Conference on Advanced Learning Technologies
           (ICALT 2004)*, pages 21–25, 2004.

Translating Mathematical Vernacular into Knowledge Repositories

Adam Grabowski[1] and Christoph Schwarzweller[2]

[1] Institute of Mathematics, University of Białystok,
ul. Akademicka 2, 15-267 Białystok, Poland
adam@math.uwb.edu.pl
[2] Department of Computer Science, University of Gdańsk,
ul. Wita Stwosza 57, 80-952 Gdańsk, Poland
schwarzw@math.univ.gda.pl

Abstract. Defining functions is a major topic when building mathematical repositories. Though relatively easy in mathematical vernacular, function definitions rise a number of questions and problems in fully formal languages (see [4]). This becomes even more important for repositories in which properties of the defined functions are not only stated, but also proved correct. In this paper we investigate function definitions in the Mizar system. Though most of them are straightforward and follow the intuition, we also found a number of examples differing from mathematical vernacular or where different solutions seem equally reasonable. Sometimes there even do not seem to exist solutions not somehow "ignoring mathematical vernacular". So the question is: Should we seek for some kind of standard, that is a "formal mathematical vernacular", or should we accept that different authors prefer different styles?

1 Introduction

Mathematical knowledge management aims at providing both tools and infrastructure supporting the organization, development, and teaching of mathematics on computers. Large repositories of mathematical knowledge are of major concern since they provide the user with a knowledge base of verified mathematical facts. However, this knowledge is often not easy to access due to the formal language in which it is presented and verified. On the other hand the acceptance of repositories and hence of mathematical knowledge management systems heavily relies on the way mathematics is presented to the user; thus the closer to "everyday" mathematics the used language is, the more likely users of the system will be found.

The language actually used by mathematicians, however, is rather vague and imprecise: working mathematicians use what is called the "mathematical vernacular" [3, 9], a language rather to communicate than to be completely formal. As stated by Davenport [4] "It turns out to be remarkably hard to write 'correct' mathematics in the mathematical vernacular." The reason is that the knowledge implicitly used in the vernacular must be made explicit for 'correct' mathematics. The same holds for knowledge repositories, especially if such a repository

M. Kohlhase (Ed.): MKM 2005, LNAI 3863, pp. 49–64, 2006.

is connected with a theorem prover or checker and is not just a collection of (textual) definitions and theorems. Here, for example, we do not have "obvious" special cases that "need not to be taken into accounct".

On the other hand existing theorem provers and checkers provide languages successfully used to formalize and prove numerous advanced theorems. The languages to do so, however, are usually highly specialized and hard to understand from the viewpoint of working mathematicians. The reason is that here the language has to be not only formal but also semantically exact in order to produce reliable proofs of theorems. As a consequence, there is a clash between what mathematicians and computers – that is computer scientists who design and implement theorem provers and checkers – consider comfortable. For theorem proving it might be reasonable to use languages "bizar" to a mathematician, as the goal is "simply" to find a (representation and) a computer proof for a specially chosen theorem.

In mathematical repositories the situation is somewhat different: here we look for general methods describing (and proving) knowledge from different – if not all – areas of mathematics. In addition this knowledge is to be accessed and used by non-specialists also, so that the knowledge should not be hidden by the formal language of the system. Nevertheless the language used has to semantically exact to produce reliable results. So the question is: Should we develop mathematical knowledge management systems as closely as possible to the vernacular of working mathematicians in order to please them as potential users? Or should we include other language elements or slightly different definitions in case they are more convenient from the theorem proving point of view?

In this paper we discuss this question by inspecting the Mizar language and the Mizar Mathematical Library. We focus on definitions, in particular function definitions, which are often given partially or by case distinctions (see [4]). This "impreciseness" is not further considered by mathematicians: theorems are stated without really worrying about the "easy special cases". In mathematical repositories, however, this is not possible and therefore Mizar provides language constructs to cope with such situations. However, as we will see, these do not allow for a one-to-one translation of the "mathematical vernacular", some decisions remain up to the author. In addition we also present example situations which a) do not strictly follow the "mathematical vernacular" and b) provide more elegant proving and reuse in a repository.

The plan of the paper is as follows. After a brief review how functions can be defined in Mizar in the next section, we start with an investigation of the empty set and its elements in section 3. This easy example already indicates, that there exist different possibilities to realize mathematical vernacular in repositories. That this is no accident is shown in section 4 and 5 where a number of examples from different areas such as trigonometric functions and arithmetics are presented. Problems concerning more involved topics such as modularity of repositories and ambiguities are discussed in section 6. These observations imply that maintaining and revising of repositories will stay an important topic in the future. Section 7 discusses software built for the Mizar Mathematical Library to support this task.

2   Defining Functions in Mizar

Mathematical functions often cannot be defined uniformly on their domains; there are defined by case distinctions such as for example the signum function or even partially by giving additional conditions for the arguments as in the case of inverse trigonometric functions. Of course one can introduce new domains on which such functions are then totally defined; this, however, seems to be rather artificial and in addition would lead to an inflation of domains not acceptable in a mathematical repository.

The Mizar system basically provides two language constructs to cope with such situations: the `assume`-clause to express restrictions of arguments and the `if`-clause for defining case distinctions. In this section we give some introductory examples for using (and abusing) these constructs before we discuss their implications for mathematical knowledge repositories.

A standard example for restricting domains is the square root functions which is defined for non-negative real numbers only. The straightforward Mizar definition is as follows.

```
definition let a be real number;
  assume 0 <= a;
  func sqrt a -> real number means
    0 <= it & it^2 = a;
end;
```

Note, however, that this definition implies that for each application of `sqrt` a non-negative argument `a` is necessary, that is one has to show or state as an assumption that `0 <= a`. Things become more puzzled when considering for example trigonometric functions: `tan a` is defined only if `cos a` is not zero, we thus get

```
definition let a be real number;
  assume not ex k being Integer st a = Pi / 2 + k * Pi;
  func tan a -> real number equals
    sin(a) / cos(a);
end;
```

and, given `a`, to get the value `tan a` the assumption is evident and has to be shown explicitly. The situation looks different when it comes to case distinctions using the `if`-clause. Though defining functions this way requires proving consistency – the cases need not be distinct, so one has to show that the corresponding values do not contradict each other – most examples are straightforward and intuitive such as

```
definition let x,y be real number;
  func min(x,y) -> real number equals
    x if x <= y otherwise y;
end;
```

Proving theorems involving such functions is rather straightforward and fits to intuition. A prominent exception, though, is the inverse z" of a complex number z, which is usually considered as a partial function, 0" being undefined. In Mizar, however, we find that " is defined as a total function with 0" being equal to 0.

```
definition let z be complex number;
  func z" -> complex number means :: XCMPLX_0:def 7
    z * it = 1 if z <> 0 otherwise it = 0;
end;
```

The point is that defining " as a partial function using **assume** z <> 0 would require to prove this each time the definition is used; so in order to avoid this the author decided to base the development on this "slightly different definition". Note that in Mizar division z/y of complex numbers is defined as z*y". This means that / is a total function too, and in particular that one can prove z/0 = 0 for every complex number z (including z = 0). We will see in section 4 some more implications of this definition.

Of course it is easy to "abuse" these language constructs by introducing unnecessary assumptions, the probably most prominent example is using non-empty sets where this is not necessary. So the question is not only how to provide assumptions that can be reasonably used later, but also how to avoid unneccessary assumptions in a repository.

3   How Many Elements Has the Empty Subset?

To start the discussion we present in this section some issues of the empty set and its elements. Though rather trivial at first sight, this illustrates well the problems arising when moving from "imprecise" descriptions to "complete formal" ones. We will see that though the definition is almost trivial, using it in the environment of a mathematical repository – that is combining the definition with other notations from set theory – needs some care.

The empty set is the set which contains no elements. Thus it is straightforward to define something like

```
definition
  func {} -> set means :: XBOOLE_0:def 1
    not ex x being set st x in it;
end;
```

Though not exactly in the scope of defining functions, we like to mention the following problem here: In mathematical repositories definitions do not stand alone; they have to be considered in the context of other notations, here for example finite and infinite sets. Obviously, the empty set is finite. But in a repository that is not true in advance, it's just obvious in the "mathematical vernacular". Thus in principle it is possible to have objects such as

```
let X be infinite empty set;
```

Generally speaking that's no harm, because such a phrase is meaningless: it includes a contradiction, hence everything stated (and proved) for such objects is of no use. On the other hand the acceptance of a repository in which this is possible is at least questionable. Thus such "contradictable" objects should be ruled out. Therefore Mizar does not allow empty types: before using an object of type `infinite empty set` its existence has to be shown in an existential cluster registration.

Now let's have a look at the elements of the empty set. In Mizar we find the definition of the type `Element of X`, where X is a set. It is "clear" that x is an `Element of X`, if $x \in X$. There are no problems if X is non-empty: There exists an element x in X, so the type `Element of X` is non-empty. If X is empty, however, there is no $x \in X$. Of course one can define the type `Element of X` for non-empty sets X only, ruling out the type `Element of {}`. But then each time the type `Element of` is used, one has to show that its argument is non-empty. Therefore in Mizar the type `Element of {}` is defined to be the empty set:

```
definition let X be set;
  mode Element of X means :: SUBSET_1:def 2
    it in X if X is non empty otherwise it is empty;
end;
```

This, however, does not fit to mathematical vernacular, because the empty set is not an element of the empty set; but has the advantage that the type `Element of X` is well-defined for arbitrary sets X, hence usable without any assumptions. We mention that though the empty set {} is of type `Element of {}`, this does not imply that $\{\} \in \{\}$ is provable in Mizar, that is {} is still the empty set. Furthermore, the Mizar checker itself infers that $x \in X$ holds if x is an `Element of X` and X is non-empty. So we see that even a notion as "obvious" as the empty set calls for basic decisions – especially concerning types and their implications for later proving – when being formalized, that is when moving from mathematical vernacular into a mathematical repository.

4   Special Functions

In this section we consider mainly the definition of trigonometric functions in Mizar. Interestingly, we can find different approaches, one following the intuition and another one using that the inverse of 0 is 0. First, the logarithm of real numbers a and b is defined using exponentiation, in Mizar defined as a functor `to_power` (see section 5.3). Here, the usual "problematic" values for a and b have been ruled out using an assumption:

```
definition let a,b be real number;
  assume that a > 0 & a <> 1 and b > 0;
  func log(a,b) -> real number means :: POWER:def 3
    a to_power it = b;
end;
```

Consequently, theorems have to take these values into account, because the equality a to_power log(a,b) = b is valid only if the assumptions about a and b are fulfilled. We thus find theorems of the following kind.

```
theorem :: POWER:61
  a>0 & a<>1 & b>0 & c>0 implies log(a,b) + log(a,c) = log(a,b*c);
```

This approach follows what Davenport called the "conditional equation approach" in [4]. The advantage is that it is close to textbook mathematics (though assumptions in a book are often not stated thoroughly) and makes the necessary assumptions explicit. On the other hand long lists of assumptions both decrease readability of theorems and require of course re-stating them when using such theorems in other proofs.

What can improve things a bit here is the technique using "default values" as presented in the definition of the inverse function ". Remembering that / is a total function (compare section 2), the tangent function for real numbers can be defined simply as follows.

```
definition let th be real number;
  func tan(th) -> real number :: SIN_COS4:def 1
    sin(th) / cos(th);
end;
```

which actually means that tan(Pi/2) is defined to be 0. Note that given th we can now get the value tan(th) without proving th <> Pi/2. This also implies that a number of theorems can be stated using no assumptions, so for example

```
theorem :: SIN_COS4:2
  tan(-th) = - tan(th);
```

This may seem irritating at first sight for a reader not familiar with the basic definitions of the repository; but has the advantage that this theorem can be used without further prerequisites to be shown. Of course not all theorems can be stated this way, because z * z" = 1 holds only if z <> 0. Here Mizar formalizations fall back to the conditional approach, so for example we find

```
theorem :: SIN_COS4:8
  cos(th) <> 0 implies sin(th) = cos(th) * tan(th);
```

5   Arithmetics and Related Issues

5.1   The Greatest Common Divisor

As the greatest common divisor $GCD(a, b)$ is the largest number dividing both integers a and b, according to our intuitions such number does not exist in case of $a = b = 0$. Indeed, a quick tour through mathematical services available via WWW confirms these convictions: Wolfram's MathWorld's[1] definition of gcd takes only positive integer numbers as arguments; according to Wikipedia[2] both should not be zero simultaneously, similarly is the PlanetMath's[3] opinion, but we can read in Wikipedia that "it is useful to define $\gcd(0,0) = 0$".

In the Mizar library, there are two definitions of the greatest common divisor: hcf for natural numbers and gcd with integer arguments which uses the notion of hcf and the absolute value in its definiens.

```
definition let k, n be Nat;
  func k hcf n -> Nat means :: NAT_1:def 5
    it divides k & it divides n &
      for m st m divides k & m divides n holds m divides it;
end;
```

Based on the above, we can easily prove that

$$a = 0 \wedge b = 0 \iff a \text{ hcf } b = 0$$

for all natural a and b, and similarly for integers. Furthermore, claiming such definition we keep the connectedness with commutative rings, we also obtain a lattice of naturals with gcd and lcm as binary operations to be both distributive and complete.

5.2 The Integer Division

When inspecting the integer division in the Pascal programming language, the FreePascal compiler returns 'division by zero' error both with div and with mod. Since the Mizar system itself is coded in Pascal (and as one can easily see, some Mizar language constructions have been influenced by this programming language), we could expect a similar behaviour for the functions div and mod in the MML.

Since both are defined usually (see Wolfram's MathWorld) as:

$$m \text{ div } n = \lfloor m/n \rfloor, \quad m \text{ mod } n = m - n\lfloor m/n \rfloor, \tag{1}$$

both share the restriction of $n \neq 0$ as usual in the literature.

This is not violation of intuitions, but the MML contains the following definitions, somewhat closer to Euclid's *Elements*:

```
definition let k, l be natural number;
  func k div l -> Nat means :: NAT_1:def 1
    ( ex t being Nat st k = l * it + t & t < l ) or it = 0 & l = 0;

  func k mod l -> Nat means :: NAT_1:def 2
    ( ex t being Nat st k = l * t + it & it < l ) or it = 0 & l = 0;
end;
```

---

[1] http://mathworld.wolfram.com
[2] http://www.wikipedia.org
[3] http://planethmath.org

The above definition is a variant of which we wrote earlier (something like the if-clause), but with a slightly different (but equivalent classically) formulation.[4] There is an agreement in the MML that 0 is an element of $\mathbf{N}$ (to have both functions natural-valued), but there isn't any within mathematics in general: MathWorld writes that "Unfortunately, 0 is sometimes also included to the list of 'natural' numbers" (as Bourbaki and Halmos do), quoting Ribenboim's as the opposition ("...whenever convenient, it may be assumed that $0 \in \mathbf{N}$").

In [8] they state that extending mod to omit the assumption of the division by zero is important, but they do not explain explicitly which one should be taken: the divided number or 0. Inspecting the book we discovered that if we accept the equations (1) as the new definitions of the integer division functions in the MML, we obtain $x = x \bmod 0$, and this is also claimed in [8] in many more places than the alternative $0 = x \bmod 0$.

There are contexts in which division by zero can be considered well-justified. For example, in the extended complex plane $\mathbf{C}^*$ it is defined to be a quantity known as complex infinity. This definition expresses the fact that, for $z \neq 0$, $\lim_{w \to 0} z/w = \infty$ (i.e., complex infinity). However, even though the formal statement $1/0 = \infty$ is permitted in $\mathbf{C}^*$, this does not mean that $1 = 0 \cdot \infty$, so zero does not have a multiplicative inverse. On the other hand, although $\mathbf{R} \subseteq \mathbf{C}$, it is not clear which way to go with the extensions (since to the extended set of real numbers both $+\infty$ and $-\infty$ are added and this is the case of the MML).

As a good example of the other way of definition extending we can quote min* as an opposition to an ordinary min function.

```
definition let A be finite non empty real-membered set;
  redefine func min A means    :: SFMASTR3:def 1
  it in A & for k being real number st k in A holds it <= k;
end;
```

```
definition let A be set;
  func min* A -> Nat means    :: HENMODEL:def 1
    (it in A & for k st k in A holds it <= k) if
      A is non empty Subset of NAT
        otherwise it = 0;
end;
```

These two objects are defined completely independently, but the latter became apparent to be useful when proving the Gödel's Completeness Theorem in Mizar. Theoretically, generalizing min* we can replace an original min to simplify the library a bit. Generalizing can be also interesting from a purely scientific point of view (as e.g., formalizing rough sets with tolerances as described in [7] or [6] instead of equivalence relations). But usually the loci of a definition cannot be just generalized because the information contained in it may be necessary to give the proper meaning of an introduced object.

---

[4] The difference between natural number and Nat (with the latter expanding to Element of NAT) which has origins in various treatment of element of the empty set has to be recalled here. All Nats obtain the attribute natural automatically due to the conditional cluster mechanism.

5.3   The Power Operator

The consequence of introducing in parallel of similar notions (motivated by the need of having their definitions close to the literature) can be observed in the case of the definition of the power function, which is composed with the help of various power operators defined earlier in MML (#R is defined as the limit of sequence of rational powers of a given real number – with the assumption of the positive base, #Z is a integer power, with arbitrary real base).

```
definition let a, b be real number;
  func a to_power b -> real number means :: POWER:def 2
    it = a #R b if a > 0,
    it = 0 if a = 0 & b > 0,
    it = 1 if a = 0 & b = 0,
    ex k st k = b & it = a #Z k if a < 0 & b is Integer;
end;
```

Any efforts to change this definition should be made carefully, because the article with this definition is referenced in 46 other MML items 1407 times. Similar data for the other power operators: 415 references in 39 articles.

Note that this definition is an example of a definition of a partial function (and keyword otherwise is not used there), e.g. according to this definition we still don't know which is the value of $(-1)^{-\frac{1}{2}}$, but it gets the type real number.

5.4   Polynomials

Consider polynomials as a last example. The head term (HT) – and hence the head coefficient (HC) – of a polynomial are usually defined for non zero polynomials only (see for example [2]. From a theorem proving point of view, however, it seems convenient to define a "head term" for the zero polynomial also as follows: The head term of the zero polynomial equals the smallest term with respect to the given order. This is can be seen as an extension of the head term functor found in the literature.

```
definition
  let n be Ordinal, T be connected TermOrder of n,
      L be non empty ZeroStr, p be Polynomial of n,L;
  func HT(p,T) -> Element of Bags n means :: TERMORD:def 6
  (Support p = {} & it = EmptyBag n) or
  (it in Support p &
   for b being bag of n st b in Support p holds b <= it,T);
end;
```

This allows us to formulate theorems about head terms for arbitrary polynomials. As a consequence, when later reusing such theorems the user need not always bother that the actual polynomial is not equal 0 – just like mathematicians. For example, we get

```
theorem :: TERMORD:22
  for n being Ordinal, T being connected TermOrder of n,
    L being non trivial ZeroStr, p being Polynomial of n,L holds
  term(HM(p,T)) = HT(p,T) & coefficient(HM(p,T)) = HC(p,T);
```

if also HC(p,T) is defined appropriately (as head monomial HM(p,T) is), e.g. equals the zero element of the underlying coefficient domain.

6   Modularity and Ambiguity

6.1   Modularity of the Library

Although the fundamentals of set theory in Mizar are established in rather un-flexible way (some of them are built into the verifier, e.g. the Axiom of Choice can be proved – and it is in the Mizar article [1]), the user can also modify his/her (e.g. set-theoretical) framework at very low axiomatic level. At first glance it is not strictly connected with function definitions, but certain preferences can substantially change the need of conditional definitions. As a perfect example in arithmetics of alephs we can cite the Generalized Continuum Hypothesis intro-duced by Josef Urban in [15].

```
definition
  pred GCH means   :: CARD_FIL:def 12
    for N being Aleph holds nextcard N = exp(2,N);
end;
```

```
theorem :: CARD_FIL:31
  GCH implies ( M is inaccessible implies M is strongly_inaccessible );
```

where M is again of the type Aleph.

This trick may be used, e.g. to state the Brouwer Fix Point Theorem for disks on the real euclidean plane as an assumption to prove the famous Jordan Curve Theorem[5]. As the bright side of this approach to the development of the library we can point out the possibility of development of the authors' favorite parts of mathematics in which they are experts, instead of spending most time on bridging the gap between the current and the desired state of the formalization of the theory. This could attract more mathematicians and as we believe it is one of the vital aims of math-assistants and also of the MKM project. Also the research frontier could be so reached faster – which could make the machine codification of recent mathematics more egalitarian.

The modular maintenance of systems could be a solution for someone's wishes to have some meta-assumptions, but the care is advised (e.g., the Axiom of Determinacy contradicts the Axiom of Choice which is proven in the MML, so the earlier should not be accepted as such an assumption). Probably something

---

[5] Actually it is meaningless since Korniłowicz and Shidama proved this version of the Brouwer Fix Point Theorem in February 2005 as the BROUWER article accepted to the MML. The one-dimensional case is pretty old.

like the **requirements** directive with more human-friendly access and giving possibility of defining author's own modules of this type could be an attractive solution.[6]

Clearly, this can also have some impact on the knowledge exchange between different systems, according to the Sacerdoti Coen's advice in [14]: "Make implicit information explicit". Note however that the logical system standing behind the Mizar system is fixed, and Mizar developers rather do not anticipate change of this policy (e.g. from the classical into the constructive logic) in the future. Another drawback is that stating some significant or influential theorems without proofs and using them later as a starting point for further computer-checked reasoning we allow for a gray area of practically machine-unverified mathematics. This is hardly acceptable if we aim at building a knowledge repository as a block, not as a loose collection of solved problems.

Having this idea in mind one might understand the encoding of the solution of the Robbins problem just as proving set of equations given by Huntington can be derived from those of Robbins. Similarly, the problem of Sheffer-stroke-based short single axiomatization of Boolean algebras can be seen as such, involving only "|" operator and showing the equivalence with the 3-axiom system given by Sheffer.

In informal mathematics it is natural to explain that both approaches for Boolean algebras, this using disjunction and negation, and that with the Sheffer stroke are equivalent with the classical one, in terms of two binary operators and a unary complement. Authors can have different ideas for the same concept just as various books on the same topic do. But in the Mizar Mathematical Library, and – as we imagine – in an arbitrary large formal repository of mathematical knowledge it requires some work to provide a proper justification for this equivalence.

6.2   Ambiguities

In a distributed knowledge repository it is hard to establish a high unification level (compare $1/0 = 0$, which is provable in HOL/Mizar, its negation is provable in IMPS, both the formula and its negation are not provable in Coq, or being not a correct formula in PVS), so there is a need to exchange information about the mathlore (as in QED Manifesto they wanted to call "knowledge that is neither taught in classes nor published in monographs") which is accepted (or rather where it was rejected). As a mathlore we understand here not only basic facts which are commonly accepted, but also the formulation of definitions of basic notions.

But what to do with freshmen which are not acquainted well with the mathlore? Anyone remembers from the school that the division by zero is not allowed as a rule, and no one complained, so why think about the motivation to have

---

[6] As of version 7.0.04 of the Mizar system, there are five modules of this type available: BOOLE, SUBSET, HIDDEN, ARITHM, NUMERALS, REAL, where first two introduce automatization of boolean operations on sets, the latter three – calculations on numbers. The detailed exposition of the topic is included in [10].

some value for 0 divided by 0? The answer given in the MML is not obligatory in didactics: the recent policy of computer-aided instruction with Mizar is not to use the whole repository (MML), but to prepare small working environments built from scratch as described in [11], in which decisions do not depend on the Library Committee taking care of the MML. The reason is also that in this way it is independent from the Mizar library which evolves rapidly so the update during a semester could be hardly acceptable.

One of the conservative choices is to keep different definitions (and theorems, consequently) of the same notion in parallel, not to favour any distinct approach. Even then we can measure how often each of them is taken and – based on this quantitative measures – let researchers develop only the one which is used most often (as it can be done e.g. in case of min and min*) via consequently replacing other undesirable occurrences.

7   Improving the Library

As mathematics assistance systems are designed as a tool offering machine help for human researchers, many of the decisions about chosen approach are taken on the user's side depending on the various (subjective as a rule) criteria: elegancy, faithfulness to mathematical standards, feasibility, etc.

However, especially if the cooperation between various systems is taken into account, much improvement of a repository can be done in a highly automatic way. The quality is to be measured by statistical, so quantitative means. As it is clear however, "short" does not mean "readable" and this is a serious drawback when thinking about reusability of proofs and their clarity for people. The de Bruijn factor, which is defined by Wiedijk as the quotient of a size of formal representation by its informal original can be a dead end sometimes.

All Mizar distributions contain the bunch of programs aiming at reviewing a Mizar article and which hence may lead to the enhancing of a human work done by hand. The Library Committee of the Association of Mizar Users uses a collection of editing versions of the mentioned programs.

The software inspects a Mizar text focusing on three main activities:

- shortening and clarifying proofs;
- improving definitions' and theorems' level of generality;
- marking block and items which are just not used anymore.

Since the Mizar Mathematical Library contains knowledge which is not only declared but its correctness is also proved, there is a need for controlling of the necessity of some parts of the proofs written by human. Conditional definitions can be introduced to reflect closely the sense of the original (as the aforementioned division by zero), in many cases additional assumptions may be consequence of too weak formulation of the theorems used in the proof of its correctness, sometimes unnecessary clauses are just left accidentally (and so, polishing proofs can detect them). In all above cases, enhancing proofs can affect also the formulation of definitions via bottom-up stepwise refinement. There are few stages on which such control (hence improvement) can be performed (and this is the case of the Mizar library):

Irrelevant premises
> This is the most unproblematic and the most popular control which can be performed when writing a Mizar article. `relprem` reviews which references are not needed for the justification of a sentence.

Checking unused labels
> Very often removed unused premises are just library references (for definitions and theorems proven in MML already), but sometimes the calling by a local fact is written accidentally. If any other sentence also does not use this labelled item, after the `chklab` pass such label is marked as unnecessary. Still though, the sentence can be needed in a proof via simple linking by the next one (the reserved word 'then' in such a case).

Inaccessible part of proofs
> The program `inacc` points out sentences which are neither labelled nor linked (elements of a proof skeleton are not marked as erroneous).

Finding trivial proofs
> Although Mizar proofs are hierarchical (in the sense considered e.g., by Lamport), sometimes after the aforementioned transformations nested proofs can be simplified by the program `trivdemo` to a simple justification, that is to a list of references preceded by the keyword 'by'.

Irrelevant suppositions
> As unnecessary assumptions (in the sense of elements of proof skeleton, not just as premises) are not marked by any of the programs mentioned before as vital element of proofs, this software operates on the stage of theorem formulation than proof transformation.
>
> This program (`relsup`) is not freely available in the distribution. Explicit formulation of some assumptions in a proof may be forced by the so-called definitional expansions and hence not used directly. They are needed however and their automatic removal could result in an error in the proof skeleton and marking them as erroneous can be highly confusing, especially for an unexperienced author.

The above ordering of these programs reflects their preferred calling sequence.

The only controversial exception of the reviewing software is `relinfer` program (so it was excluded from our enumeration), which points out the unnecessary steps in a proof (and the references should be added to the next step). It can exceptionally shorten proofs but it may result in poor readability of the text:

- some sentences which are important for the proof technically are marked as irrelevant steps, but their removal may force the user to repetition of the same library reference;
- the removal may be accidental in some sense, that is steps which are crucial for human understanding of the idea of a proof, but are still unnecessary for machine (e.g., unwinding definitions – definitional expansions). Here the tendencies to reduce the de Bruijn factor can be misleading.

We also have software which detects unused variable occurrences, irrelevant private predicates and functions, marks unnecessary type changing statements, etc.

Besides the aforementioned proof transformations which are performed very often, some other checkouts are done occasionally. There is a software which checks if there are equal theorems in the library, and what's more interesting, if a theorem is a consequence of another (although due to the large library, both use a lot of resources). The latter one is often not very unlikely: to formulate statements as equivalences is the usual mathematical practice, very often though some assumptions are needed only for one of the implications.

We still do not have any automatic control if the definitions are repeated (authors would have like to introduce independently e.g. closure operators using different structures), so we can speak about the detection of 'equal theorems' rather than 'equivalent' ones. So the role of careful peer-reviewing of a repository is very important, especially if we take into account a large repository of mathematics, written by many authors, so rather not much unified in style. Quantitative parameters of the MML (some 40 thousand of theorems and lemmas, nearly 8 thousand definitions authored by more than 170 authors) justify the necessity of continuously revising of Mizar articles.

8   Conclusions

In this paper we have considered how the mathematical vernacular can be realized in mathematical repositories, thereby focusing on function definitions. The inspection of the Mizar Mathematical Library has shown that its authors used a number of different styles such as the "conditional" style using partial functions or the "extension" style as used in the definition of ". Sometimes even more than one definition is available. It seems to us that these different styles exist due to a clash between (a) working in a formal language close to the mathematical vernacular which (b) is also used to prove the theorems stated. Strictly following the mathematical vernacular sometimes leads to rather tedious formal proofs, so that some authors decide to modify their definitions in order to ease the proving task.

The problem becomes more evident in a repository with a large number of developers and users: here, of course, it is impossible to have an open system without ending up with different realizations of the mathematical vernacular. Hence, should we seek for a kind of standard, that is a "formal mathematical vernacular"? Though we believe that this can be done in general, it seems hardly possible to fix all the details theorem proving introduces into our repositories. Allowing for different realizations, on the other hand, could of course decrease the acceptance of users by confusing them. Also, extending or reusing developments by other authors gets more complicated in case the vernacular of the new author does not fit to the first author's one.

What we can try to do is organize our repositories in such a way that both authors and potential users have the possibility to identify the basic decisions theories and developments rely on. As we have illustrated this also includes the definition of functions. This is not a trivial task, because as already mentioned the large number of Mizar authors has even led to duplication of definitions or theorems. Consequently, we always have to keep track of the development by permanently revising and cleaning up our repositories. The goal must be to automate this as far as possible. A step into this direction are the Mizar tools presented in section 7. They are, however, in most cases still working on the proof transformation level, so that their further development into a "more intelligent direction" is desirable. As we understand the mathematical vernacular not only as a syntactic language, but also as the way how to shape the real mathematics (the stress on the formalized content, not only on the form – see [16]), we find it hard to establish strict guidelines (and so the question stated in the abstract remains open, although there are known direct formalizations of the traditional approach to undefinedness in the literature, e.g. [5]) for a mathematical vernacular to be feasible. We should, however, always keep in mind – especially if we try to develop systems for working mathematicians – that if we break rules accepted widely by mathematicians, this has to be sufficiently justified.

References

1. G. Bancerek, *Zermelo Theorem and Axiom of Choice*, Formalized Mathematics, 1(2), pp. 265–267, 1990.
2. T. Becker and V. Weispfenning, *Gröbner Bases – A Computational Approach to Commutative Algebra*; Springer Verlag, 1993.
3. N.G. de Bruijn, *The Mathematical Vernacular, a language for mathematics with typed sets,* in P. Dybjer et al. (eds.), Proc. of the Workshop on Programming Languages, Marstrand, Sweden, 1987.
4. J.H. Davenport, *MKM from book to computer: a case study,* in: A. Asperti, B. Buchberger, and J. Davenport (eds.), Proc. of MKM 2003, Lecture Notes in Computer Science 2594, Springer, pp. 17–29, 2003.
5. W.M. Farmer, *Formalizing undefinedness arising in calculus,* in: D.A. Basin and M. Rusinowitch (eds.): Proc. of IJCAR 2004, Cork, Ireland, Lecture Notes in Computer Science 3097, Springer, pp. 475–489, 2004.
6. A. Grabowski, *On the computer-assisted reasoning about rough sets,* in: B. Dunin-Kęplicz et al. (eds.), Monitoring, Security, and Rescue Techniques in Multiagent Systems, Advances in Soft Computing, Springer, pp. 215–226, 2005.
7. A. Grabowski and Ch. Schwarzweller, *Rough Concept Analysis – theory development in the Mizar system,* in: A. Asperti, G. Bancerek, and A. Trybulec (eds.), Proc. of MKM 2004, Lecture Notes in Computer Science 3119, Springer, pp. 130–144, 2004.
8. R.E. Graham, D.E. Knuth, and O. Patashnik, *Concrete Mathematics*, Addison-Wesley, 1994.
9. F. Kamareddine and R. Nederpelt, *A refinement of de Bruijn's formal language of mathematics,* Journal of Logic, Language and Information, 13(3), pp. 287–340, 2004.
10. A. Naumowicz and Cz. Byliński, *Improving Mizar texts with properties and requirements,* in: A. Asperti, G. Bancerek, and A. Trybulec (eds.), Proc. of MKM 2004, Lecture Notes in Computer Science 3119, Springer, pp. 190–301, 2004.

11. K. Retel and A. Zalewska, *Mizar as a tool for teaching mathematics,* in Proc. of Mizar 30 workshop, Białowieża, Poland, 2004, available at http://www. macs.hw.ac.uk/~retel/papers/KRetelAZalewska.pdf.

12. P. Rudnicki and A. Trybulec, *Mathematical Knowledge Management in Mizar;* in: B. Buchberger, O. Caprotti (eds.), Proc. of MKM 2001, Linz, Austria, 2001.

13. P. Rudnicki and A. Trybulec, *On the integrity of a repository of formalized mathematics;* in: A. Asperti, B. Buchberger, and J. Davenport (eds.), Proc. of MKM 2003, Lecture Notes in Computer Science 2594, Springer, pp. 162–174, 2003.

14. C. Sacerdoti Coen, *From proof-asistants to distributed knowledge repositories: tips and pitfalls,* in: A. Asperti, B. Buchberger, and J. Davenport (eds.), Proc. of MKM 2003, Lecture Notes in Computer Science 2594, Springer, pp. 30–44, 2003.

15. J. Urban, *Basic facts about inaccessible and measurable cardinals,* Formalized Mathematics, 9(2), pp. 323–329, 2001.

16. F. Wiedijk, *The Mathematical Vernacular,* unpublished note, available at http:// www.cs.ru.nl/~freek/notes/mv.pdf.

Assisted Proof Document Authoring

David Aspinall[1], Christoph Lüth[2], and Burkhart Wolff[3]

[1] LFCS, School of Informatics, The University of Edinburgh, U.K.
[2] Department of Mathematics and Computer Science,
Universität Bremen, Germany
[3] Department of Computer Science, ETH Zürich, Switzerland

Abstract. Recently, significant advances have been made in formalised mathematical texts for large, demanding proofs. But although such large developments are possible, they still take an inordinate amount of effort and time, and there is a significant gap between the resulting formalised machine-checkable proof scripts and the corresponding human-readable mathematical texts. We present an authoring system for formal proof which addresses these concerns. It is based on a *central document* format which, in the tradition of literate programming, allows one to extract either a formal proof script or a human-readable document; the two may have differing structure and detail levels, but are developed together in a synchronised way. Additionally, we introduce ways to assist production of the central document, by allowing tools to contribute *backflow* to update and extend it. Our authoring system builds on the new PG Kit architecture for Proof General, bringing the extra advantage that it works in a uniform interface, generically across various interactive theorem provers.

1 Introduction

While computer-supported proof assistants are increasingly accepted in computer science, in particular in the field of formal methods, their potential for mathematical practice is only beginning to be recognised [20]. Several substantial proofs reaching hundreds or thousands of pages like the Four Colour Problem [6] or the Prime Number Theorem [11] have been formalised with the aid of systems like Coq or Isabelle, and others like the Kepler Conjecture are currently under development [13]. It has been suggested that computer assistants could be generally accepted in mathematical practice if authors with no prior expertise in theorem could formalise mathematical proof texts at an effectivity estimated at one page of mathematical text per day [6].

This formalisation rate is not reached by contemporary systems, and there are two important areas that need work. First, we need to provide systems with a higher degree of automation so that more trivialities can be discharged with less work. Second, we need to increase the user's productivity by making it easier to construct formal proofs, assisting the writing process. The first point has been a focus for theorem prover development in the last few years, but the second

M. Kohlhase (Ed.): MKM 2005, LNAI 3863, pp. 65–80, 2006.
© Springer-Verlag Berlin Heidelberg 2006

point has received less attention. In general, interface technology has been quite neglected; many interfaces still use arcane command line syntax and basic text editors, which do not reach the same levels of productivity as e.g., integrated development environments (IDEs) used in software development. Modern IDEs for programming provide sophisticated mechanisms to assist writing code, for example, constructing templates automatically from graphical models or helping the user to search for library functions and documentation very swiftly. Clearly, much more could be done to assist the user in proof document authoring.

We start by taking a single proof document as the central purpose of the development: so-called *document-centred authoring*. In our sense, an *authoring system* is a set of tools which assist the user in constructing the central document, maintaining the consistency of the development and documentation under change, and generating the *views* which allow fine-grained interactive exploration of the proof detail, animating proof checking in various ways. A machine-checkable proof script and a human-readable document describing its content are just two different views of one document. The authoring assistance should allow powerful graphical user interface techniques such as drag-and-drop and point-and-click [7, 18, 1], going beyond mere text editing. Moreover, an authoring system assists the user by allowing other tools, in particular the prover itself, to edit the document as well, thus increasing productivity. We call the mechanism for this *backflow*.

The context of this work is a software framework for conducting interactive proof called the *Proof General Kit* (PG Kit). The main new contribution, as presented in this paper, is the extension for assisted authoring with backflow.

Outline. Sect. 2 motivates document-centred authoring and backflow. In Sect. 3 we describe the PG Kit architecture, and its extensions for authoring. To demonstrate the viability of our approach, we develop use cases for literate proving and script generation in Sect. 4 and Sect. 5 respectively. Sect. 6 concludes with a survey of related work and an outlook on future work.

2   Document-Centred Authoring and Backflow

A *proof script* is a formal text which can be run through a proof assistant to mechanically check the validity of the proofs therein. A proof script usually does not contain the proofs themselves, just enough information to construct them. Some formal proof languages do contain structuring mechanisms inspired by human proofs, but, unfortunately, the formal syntax and in particular the level of detail required by a proof assistant usually still precludes a proof written in such a language to be accepted as a "textbook proof" by non-expert human readers. In contrast, we call a proof document *human-readable* if it is aimed at a presentation close to textbook proofs or journal papers; typically it may contain a higher level of abstraction, leave out repetitive arguments, and even omit logical steps considered distracting.

Considerable effort has been devoted to bridging the gap between human-readable and machine-checkable proof, either by making practical mathematical

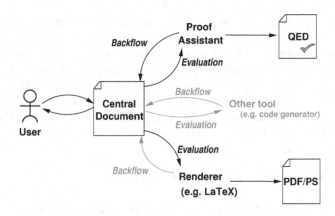

Fig. 1. Control and document flow for document-centred authoring

language more strict and hence machine-checkable, or by making prover input languages less formal, more abstract, and hence more human-readable (see the related work in Sect. 6.1). But the principal dilemma of different requirements from both human readers and proof assistants remains.

An alternative approach is to accept the dichotomy of presentation levels, and adapt techniques similar to those of literate programming [16] to weave structured, human-readable annotations into formal proof scripts. This has already been used, for example, in literate specification environments such as HOL-Z [8] and Isar's integrated LaTeX output mechanism [24], where terms, formulas or proof-states can be generated into the output during the LaTeX-rendering phase.

The underlying idea of these and other approaches, which also underlies our own, is that we have one *central document* from which we can extract both human-readable text and a machine-checkable proof script. This is the *document-centred* approach as depicted in Fig. 1. The user edits the document in a suitable editing environment, and the document can be evaluated by various tools, such as the *proof assistant* which checks that the document contains valid proofs, a *renderer* (e.g., LaTeX) which typesets or renders the document into human-readable documentation readable outwith the system, or other tools, for example a code generator to construct executable versions of specifications.

Our main contribution to this setting is to allow the possibility of *backflow* from each tool into the central document, i.e. each tool can generate text which in turn becomes part of the document. In contrast to the mentioned Isar mechanism, backflow is supported during editing and not during LaTeX-rendering. In the case of the prover, the backflow can generate parts of the central proof document to assist in writing the proof script. We concentrate on this case in Sects. 4 and 5 below, but note that the backflow can equally well originate from tools other than the prover. For example, the LaTeX component may have generated cross-references in previous runs which were offered in a context sensitive way when editing the central document.

Importantly, in contrast to classical literate programming, the backflow assistance has to be *interactive*. The user needs an immediate reaction from the system, such as searching for applicable lemmas or inspecting terms and their types. Interactive development also means that authors can annotate their proofs while developing them, which is easier than to annotate them afterwards.

Thus, the technical challenges we face when implementing a system to support assisted document-centred proof document authoring are *interactivity*, with the document being developed *incrementally*, *synchronisation* of the different views of the document, and the *coordination* of information flow between the different views. Our system architecture is designed to meet these challenges.

3   PGIP and the PG Kit Architecture

The *Proof General Kit* (PG Kit) is a software framework for conducting interactive proof. It evolved from the *Proof General* project, which constructed a generic interface to numerous interactive theorem provers in a piecemeal approach, by individual customisation for communication with each proof assistant. PG Kit is instead based on a uniform mechanism, specifying the syntax of messages exchanged between components and the protocol governing message exchanges. This section introduces just what is needed; full details are elsewhere [2, 3, 4].

Fig. 2 shows the component-based PG Kit architecture, which closely mirrors the document flow of Fig. 1. The *broker* middleware component handles the central document; it is responsible for managing the synchronisation between different views. The *display* components on the left-hand side interact with the user. On the right side, we have *proof assistants* or other tools. Each display may implement a different interaction paradigm: a text editor (e.g., Emacs) based on textual input and cryptic key sequences; a GUI (e.g., PGWin [3]) using graphical techniques such as drag-and-drop and point-and-click to construct proofs, or a generic IDE (e.g., Eclipse [26]) with sophisticated navigation and project management, using both graphical and textual interaction.

3.1   The Message Protocol PGIP

The mechanism for directing proof used by PG Kit is known as the **PGIP** protocol, for *Proof General Interactive Proof*. The order of message exchanges is given by an informal specification [2] and enforced dynamically by the central broker component. The syntax of PGIP messages is defined by an XML schema

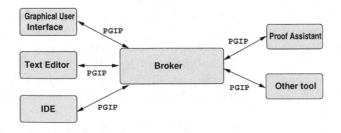

Fig. 2. PG Kit System Architecture

written in RELAX NG [21]. There is a secondary schema called **PGML**, for *Proof General Markup Language*, which is used for annotating concrete syntax within messages (for example, to generate clickable regions) and to represent mathematical symbols.[1] To define the protocol, we distinguish several kinds of messages, including: *display commands* which are sent from the display to the broker, arising from user input; *display messages* sent to the display from the prover or broker, which contain output for the user, and *prover commands* which are sent to the prover and affect the internal (proof-relevant) state of the prover.

Messages are exchanged over channels implemented as Unix pipes or sockets. Compared with simple RPC mechanisms like XML RPC, PGIP message exchange is more permissive, allowing multiple responses. We need this because interactive provers may send a lot of information while a proof proceeds, and a proof may be slow or even diverge (e.g., in proof search). It is essential that this feedback is displayed eagerly so the user can take action as soon as possible.

3.2 The Central Document

The central document is the main artefact of the system. The two principal views on the central document are the machine-checked *proof script* consisting of prover commands, and the human-readable *documentation*. These are extracted from relevant parts of the central document. Note that all document content is in principle free-form and manually generated, but the backflow concept allows tools to assist the user in constructing both proof script and documentation.

PGIP manipulates the central document in an unspecified concrete syntax (subject to a few constraints) by *marking up* the contents with PGIP commands that give the document the structure needed. Fig. 3 shows a proof script in a fictional simple tactical language and its markup in PGIP.[2]

```
goal "length (tl xs) = length xs - 1"
 /* proof by case distinction, then it's trivial */
 case_tac "xs" THEN simp_tac THEN simp_tac
 qed "Simple"
```

```
<opengoal>goal "length (tl xs) = length xs − 1"</opengoal>
<comment> /* proof by case distinction, then it's trivial */</comment>
<proofstep> case_tac "xs" THEN simp_tac THEN simp_tac</proofstep>
<closegoal> qed "Simple"</closegoal>
```

Fig. 3. A proof script and its PGIP markup

The proof script mark-up is more fine-grained than the documentation, because it needs to be evaluated interactively. Typical proof script markup are the elements `<opengoal>`, `<proofstep>` and `<closegoal>`, which start a proof,

---

[1] MathML is another possibility, but PGML should be easier for existing systems.
[2] To save head scratching: this is provable with `tl []=[]` and `(0-1)::nat=0`.

perform a proof step, and end a proof, but also markup for the start and end of a theory etc. Documentation is marked up as a <litcomment> element (not yet shown), and proper comments, for the author's eyes only, are marked up as <comment>. The corresponding (trivial) proof document could read:

Lemma 3.14 (Simple): "length (tl xs) = length xs - 1".
Proof: trivial.

It is not very enlightening for the human reader, neither revealing the main argument, the proof structure, nor the lemmas and definitions used in the proof. We will show refinements of this running example later on.

3.3   Interaction and Authoring

When the broker reads a document or when the user edits the document, it is first parsed, causing the PGIP markup to be inserted. On the marked up document, we can perform the following operations:

– *Interactive evaluation* by the prover. The broker does this on user request, by sending parts of the script to the prover for evaluation (using a simple linear notion of dependency, or a more fine grained dependency analysis if supported by the prover); it corresponds to 'stepping through the proof', and supports the incremental development of proof scripts.
– *Extracting the proof script* by removing all annotation comments and PGIP markup. We obtain a proof script which we can feed directly to the prover, without broker intervention, to *validate* it.
– *Extract the documentation* by extracting all literate comments and (possibly) interleaving formatted prover commands. We obtain a document which we can render to produce a human-readable documentation.

The document may be updated by user editing as usual, or by the *backflow* mechanism. Backflow is characterised by what part of the document it targets: *documentation backflow* contains documentation content, e.g. a display of the current proof state or a named theorem or constant definition, which becomes part of the documentation; whereas *script backflow* contains proof script content, which is provided to assist the user in constructing a proof (typically by the prover itself), and which in turn becomes prover input.

Documentation backflow is treated specially. To help with synchronisation we want to record requests for documentation backflow in the document itself. Then we can regenerate those parts of documentation when the proof is rerun or adjusted: the proof state display which we have inserted previously might have changed, for example. In the same way that the broker tracks the status of prover commands (described in [4]), it can track the status of those parts of the document generated by backflow to see if they are up-to-date or not.

3.4   Proof Commands and Operations

The proof script part of a document contains a sequence of *prover commands* in PGIP, but not all prover commands can appear in a proof script. We distinguish *proper* commands which can appear from *improper* commands which cannot.

The broker does not know about the concrete syntax of the system, so we provide a way to construct them by filling in configurable templates with identifiers and raw text. An <operationsconfig> configuration message provides a prover-specific set of *prover types* and *prover operations*. The prover types (not to be confused with the theorem prover's *logical* notion of type, if it has one!) are used to provide context menus, icons, and drag-and-drop actions (cf [18]). Prover operations may be used to build up commands by textual substitution. They can be bound to input events, and may then be invoked by a menu item or drag-and-drop.

The improper commands are used for controlling and inspecting the prover's state, and cannot appear in the proof script being developed. A standard improper command is <undostep> which undoes the last proof step in a development. In the next section we introduce the idea of allowing configurable improper commands to generate *backflow* for feeding back into the document. This is a much more powerful way of generating prover commands than the <operationsconfig> templates because it can be context sensitive and involve arbitrary external tools.

3.5   Extending PGIP: Interactions for Authoring

The extensions for assisted authoring comprise the <litcomment> element and the backflow. They are not part of the original design, but backwards compatible.

As mentioned above, the documentation is generated from parts of the document which the prover does not see. We call these parts *literate comments* to distinguish them from ordinary comments which are not part of the documentation. A literate comment contains either text or documentation backflow-generating *directives*. A directive contains the PGIP command which generates the documentation (these exist already in PGIP as the proofctxt entity), and the resulting markup in PGML. An example of a proofctxt element is <showproofstate> which embeds the current proofstate in the document. Here is a fragment of the RELAX grammar for the new commands:[3]

```
litcomment = element litcomment { format_attr?, (text | directive)* }
directive  = element directive { (proofctxt, pgml) }
format_attr = attribute format { token }
```

The format attribute can be used to specify the output format if the prover supports more than one output format, e.g. LaTeX, HTML or plain text. We also allow all proper proof commands to have an optional nodisplay attribute, e.g. for <opengoal>:

```
opengoal = element opengoal { display_attr?, thmname_attr?, text }
display_attr = attribute nodisplay { xsd:boolean }
```

The nodisplay attribute allows us to suppress proof commands for document output (e.g., to replace "by simp_tac" with "*Proof is obvious*").

---

[3] See [21] to better understand the format of these rules.

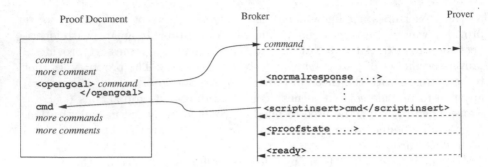

Fig. 4. Backflow

The other new element is <scriptinsert>, for script backflow:

scriptinsert = **element** scriptinsert { metavarid_attr?, text }
metavarid_attr = **attribute** metavarid { token }

To see how <scriptinsert> works, consider the usual PGIP protocol: after a prover command is sent to the prover, the prover may send a number of prover messages such as <normalresponse> or <proofstate>, followed by a final <ready> message to indicate its availability. A <scriptinsert> sent by the prover causes the text of the message to be inserted into the central document at the current point of processing, after being parsed. Fig. 4 illustrates this. The optional metavarid attribute specifies an alternative location in the document.

Having described PGIP and its extensions, we next show use cases which demonstrate the extensions at work, to clarify their use and show their viability.

4   Literate Proving Made Easy

This section demonstrates how documentation backflow can provide literate proving facilities in a generic system architecture.

Consider a prover with a simple tactic language used to write the proof in Fig. 3 shown previously. Neither the structure nor the relevant definitions and lemmas are obvious from this script, and, as typical for LCF-style provers, the intermediate proof states of the underlying reasoning are completely implicit.

We are now going to add literate proving facilities to this prover, based on LaTeX. First, proofs are enclosed in a proof environment, and \com marks literate comments (inside proofs). The content of literate comments is just usual LaTeX code. We also need concrete syntax for the directives, e.g. \proofstate. Finally, there are *pragmas* (comments which have a side-effect while processing the script), such as:

```
%% declare_config cname [= expr]
%% hide [cname]
%% show [cname]
```

Intial user input:

```
\begin{document}
  Here follows a stunning insight from the
  weird and wonderful world of mathematics:
  \begin{proof}
     goal "length (tl xs) = length xs - 1"
     \com{The proof proceeds by case distinction:}
     %% hide
     case_tac "xs"
     \com{If the list is empty, we have to show: \proofstate{}}
       which follows by simplification from \thm[List.tl_def.1]{}
       and \thm[Nat.diff_0_eq_0]{}.}
     THEN simp_tac

     \com{Otherwise, we have: \proofstate{}}
     \com{which is a consequence of \thm[List.tl_def.2]{}
         and arithmetic calculations.}
       THEN simp_tac
     %% show
     qed "Simple"
  \end{proof}
\end{document}
```

Parsing

Central document with PGIP markup:

```
<litcomment>
\begin{document} ...
\begin{proof}
</litcomment>
<opengoal> goal "length (tl xs) = length xs - 1"<opengoal>
<litcomment>\com{The proof proceeds by case distinction:}</litcomment>
<proofstep display="false">case_tac "xs"</proofstep>
<litcomment>\com{If the list is empty, we have to show:
<directive><showproofstate/><proofstate>...</proofstate></directive>
which follows by simplification from <directive><showid name="List.tl_def.1"
<thm>tl []= []</thm></directive> and
<directive><showid name="Nat.diff_0_eq_0"/><thm>0- 1= 0</thm></directive>.}
</litcomment>
<proofstep display="false">THEN simp_tac</proofstep>
<litcomment> . . . </litcomment>
<proofstep display="false">THEN simp_tac</proofstep>
<closegoal>qed "simple"</closegoal>
<litcomment>  \end{proof}  \end{document}   </litcomment>
```

Proof script

```
goal "length (tl xs) = length xs - 1"
  case_tac "xs"
  THEN simp_tac
  THEN simp_tac
  qed "simple"
```

Prover

Documentation

```
\begin{document}
Here follows a stunning insight from the
weird and wonderful world of mathematics:
\begin{proof}
goal "length (tl xs) = length xs - 1"
\com{The proof proceeds by case distinction:}
\com{If the list is empty, we have to show:
  \proofstate{length (tl ([])) = length ([])- 1}
  which follows by simplification
  from \thm[List.tl_def.1]{tl [] = []}
  and \thm[Nat.diff_0_eq_0]{0-n=0}.}
\com{Otherwise, we have:
  \proofstate{length(tl(x::xs'))=length(xs')-1}}
\com{which is a consequence of
  \thm[List.tl_def.2]{tl (x::xs)=xs}.}
qed "simple"
\end{proof}
\end{document}
```

LaTeX

Fig. 5. Literate proving

These can be used to set or reset the `nodisplay` attribute. The `hide` and `show` pragmas may optionally have a *configuration name* like `short` or `detailed` that allows for the generation of different versions of a proof document during rendering. Configuration names may be expressed in terms of other previously declared configuration names, e.g.

```
%% declare_config both = short or detailed
```

The resulting LATEX document and the concrete document flow is shown in Fig. 5. As we can see, the parsing adds the necessary markup to the central document. The proof is run in the broker by just stepping through it, skipping over comments, and filling in the proofstate or the references to lemmas in the literate comment. After that, the second literate comment in the proof reads (we have elided the actual proofstate and displayed theorems):

```
<litcomment>\com{If the list is empty, we have to show:
<directive><showproofstate/><proofstate>...</proofstate></directive>
which follows by simplification from
<directive><showid name="List.tl_def.1"/><term>...</term></directive> and
<directive><showid name="Nat.diff_0_eq_0"/><term>...</term></directive>.
```

Notice that the proofstate (the elided part) is encapsulated by the `<proofstate>` element, such that if we rerun the script, it will be replaced by the then current proofstate. The same holds for lemmas or theorems like `<thm>`. From this document, we can extract both a proof script and LATEX documentation easily.

Our approach is quite generic: we just need to integrate the parser for some concrete syntax; in principle, it should be possible to generate the XML-formats used by OpenOffice, for example. Of course, this kind of literate programming is enhanced if the prover can generated typeset output to embed in the document.

5   Script Backflow

Here is our sample proof again, this time in Isabelle/Isar, which makes the structure of the case distinction quite explicit:

```
lemma "length (tl xs) = length xs − 1":
proof (cases xs)
  case Nil thus ?thesis by simp
  txt{* If the list is empty, we have to show: @{proofstate}
       which follows from simplification with @{thm Nat.diff_0_eq_0}
       and @{thm{List.tl_def.1}.*}
next
  case (Cons y ys) thus ?thesis by simp txt{* ... *}
qed
```

However, the text is much longer than the tactical proof above, and even though this verbosity is exactly what makes Isar proofs easier to read and more stable to maintain, proofs become laborious to type at the outset. The only input inherently *required* of the user (after starting the proof) is the decision to perform a case distinction (`cases`), and the name of the variable "xs". Given this, the prover can choose the right case distinction rule according to the type of the variable; this results in the patterns, their local variables, etc., which are transferred to the PG broker as a template via backflow:

proof (cases xs)
      case Nil **thus** ?thesis <**proof**>
  next
      case (Cons y ys) **thus** ?thesis <**proof**>
done

Afterwards, the user can continue to fill in the two actual proofs (for the placeholder <proof>) in the case branches. For the subproof shown, the prover can also yield the list of lemmas used in the simplifications included in the template:

txt{* which follows **from** simplification **with** @{**thm** Nat.diff_0_eq_0}
    and @{**thm**{List.tl_def.1}. *}

(the prover could even be more clever and try to fill in obvious proofs by checking whether simplification or other automatic proof patterns would succeed, and documenting appropriately). In a graphical interface, the proof template above would be generated by a mouse-click for the selection of the proof method **cases** and two keystrokes to type **xs**. These three user actions replace the tedious typing of the complete text. Other proof methods such as proof by induction could be generated using the same backflow mechanisms.

The protocol for script backflow. In detail, the above interaction between prover, broker and display proceeds as follows. Suppose a user event such as a menu select, mouse click or drag-and-drop has occurred. The prover operation (as described in Sect. 3.4) triggered by this event causes the display to send a prover command, which is marked as proper or improper, to the broker. If it is a proper prover command, the broker inserts it into the proof script; if it is an improper prover command, it is transferred to the prover directly.

We need the prover to configure the displays to bind events to prover operations; this is done once, in the initial startup phase (operation configurations in PGIP are intended for displays). Then whenever the event occurs, the display evaluates the operation. This may require more input from the user, e.g. the variable name in the case distinction example above; to this end, operations can have *inputforms* configured which describe this additional input, e.g. here a one-line string input with prompt "Name of variable". Finally, the prover evaluates the command, which results in a backflow to the broker.

Fig. 6 shows the resulting flow of messages, slightly abridged. In the configuration phase, the prover sets up an operation **casedist_op**, which requires a string (the name of the variable) as user input. The operation is bound to a menu entry *Case distinction* in the menu *Prove by* (there will be other menus and sub-menus). When users activate that menu, the operation is executed and they will be asked to input the variable name, e.g. by typing in an input form or by pointing to it in the proofstate display. The placeholder %**casevar** in the operation command is replaced with the input, and the command is sent to the broker. According to the configuration (the attribute **improper**), it is marked

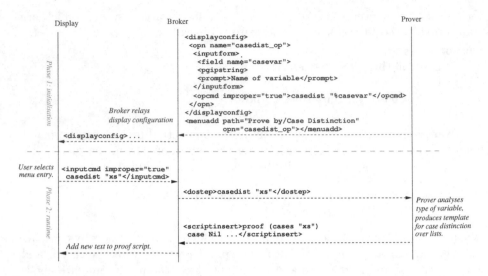

Fig. 6. Message exchange for backflow in the case distinction situation

as an improper command, so the broker relays the command to the prover. The prover analyses the type of the variable, decides case distinction on lists is appropriate, with one case for the empty and non-empty lists each, and generates the corresponding backflow. The generated proof template text is inserted into the proof script.

PGIP has a generic display model, where simpler displays (e.g. text-based ones) are free to ignore configurations which only make sense for graphical displays. The only adjustment needed here over the description in [3] is to allow improper commands as well as proper ones. Note how only the prover needs knowledge about the logical structure of the proof, the types involved and so on; from the broker's and displays' point of view, the protocol is completely generic.

5.1   Calculational Proof

Calculational proof is probably the most well-known proof presentation paradigm as it is taught in school mathematics. Here is an example in Isar (which uses Isabelle's axiomatic type class mechanism to restrict instances of the type variable 'a to those satisfying the group axioms):

```
theorem group_right_one: "x * one = (x::'a::group)"
proof −
  have "x * one = x * (inverse x * x)"
    by (simp only: group_left_inverse)
  also have "... = x * inverse x * x"
    by (simp only: group_assoc)
  also have "... = one * x"
    by (simp only: group_right_inverse)
```

also have "... = x"
    by (simp only: group_left_one)
    finally show ?thesis .
qed

This is already quite readable, and the generation of proof presentations that abstract the proof technical details fully or up to the names of the used lemmas in each step are straightforward. As an Isar proof text, this proof pattern requires the user to type all the intermediate proof stages; they may be abbreviated by meta-variables ?X1,... , ?Xn, but it is still cumbersome. GUI-supported backflow helps here substantially: the user states only the overall goal, selects *calculational proof*, and sets a *focus* on a subterm (e.g., x * inverse x) serving as the redex of a theorem, and a theorem (e.g., group_left_inverse).

In this scenario, the construction of backflow is quite complex and requires the development of specialised tactic support. The main problem is to generate proof scripts that are as *general* and *reusable* as possible, ideally avoiding positional referencing by using general methods such as

    by (simp only: group_assoc)

instead of a left-to-right one-step application such as

    by (rule_tac P=% x. x * (inverse x * x) = x in subst[OF group_assoc.assoc]).

The technique of *proof abstraction* is based on generate-and-test heuristics for successful proof attempts with the fall-back of the least general proof method.

5.2 Window Inferencing

Logically, calculational proof depends on the transitivity of equality which allows us to string together a sequence of lemmas the form $t_i = t_{i+1}$ for $i = 1, \ldots, n$ to one theorem $t_1 = t_{n+1}$. Window inferencing [12] is a generalisation of calculational proof where instead of an equation we have a non-disjoint *family* of binary relations. Window inferencing also allows us to apply rules to subterms of the current proof state; this is referred to as *opening a window* on that subterm, and it may produce additional assumptions (e.g. opening a window on the positive branch of a conditional adds the condition as an assumption). When closing a window, implicit monotonicity reasoning is executed to validate replacing a focus with the result of the sub-derivation in a window at the next higher level.

Previous work has shown how window inferencing can be implemented as a tactic in Isabelle, using a dedicated GUI for window inferencing [19]. Here, we can achieve the same thing using *annotated terms* in PGML and backflow in PGIP. The special input field %selected can be used in operations to denote the selected subterm (on displays that do not support subterm selection, these operations will be ignored). The operation to open a window then sends the command open_win %selected, which causes the command open_win p to be sent to prover (via the broker, as in Fig. 6), and the prover constructs the relevant subterm and context from p. The path p is in the prover's internal abstract syntax representation of the term, it only makes sense to the prover and needs to be post-processed to render a PGML string. Again, with only modest support from the prover, we can add a very useful high-level feature for assisting document authoring.

6   Conclusions

We have presented a new component-based system architecture for authoring mathematical documents together with formal proofs. It extends the generic PG Kit infrastructure for interactive proof. The novel concept in extending proof script editing to authoring is the support of component backflow on the protocol level PGIP as well as in its implementation in the broker.

The implementation of our design is ongoing. The broker architecture, with an Emacs-based and an Eclipse-based display, has been developed and is available as a prototype [3, 4]. The authoring extensions have been added to this prototype and support from provers (in particular, Isabelle) is anticipated in future development versions.

6.1   Related Work

The basic idea of the document-centred approach can be traced back to Knuth's work on literate programming [16]. In the context of formal proof and formalisation of mathematics, the field can be divided into two fundamentally different approaches: one tries to make formal proofs more human-readable, or one tries to make textbook-proofs more formal or at least intuit their underlying formal structure. In the former line of research stands Automath [9], Mizar [23], and its descendants like Isabelle/Isar [25] or Coq's integrated documentation facility coqdoc that can extract a document offline in various formats. Théry's approach [22] bridges the gap by defining an XML format for manually annotating statements in mathematical papers to link them to formal counterparts, wherein proofs must be supplied; consistency is checked in a prover. Similar approaches include Weak Type Theory [15], MathLang [14]), or the DIALOG project [5]. In a sense in the opposite direction, Kohlhase [17] works on the existing mathematical corpus of LaTeX papers and tries to capture their semantic content automatically with additional markup.

Although we take formal proof as the starting point, our document-centred approach eases the task of reconstructing a human-readable format during formal proof development, using the information available via backflow from the presentation of terms and proofstates, or the information from certain automated proof strategies or advanced techniques like proof planning [10]. Of course, the resulting annotations are merely organised text, kept consistent by using references to theorems, etc., which are resolved late in the presentation process. Integrating with the complementary approach of [15] with respect to these annotations is worth investigating.

6.2   Outlook and Future Work

This paper describes authoring facilities on the document level. An important future direction is to study large and richly connected developments, spanning multiple proof script files and proof modules, and supporting reordering in producing the human-readable documentation. The framework partly addresses this at the moment because there are PGIP elements describing file-level commands and dependencies between prover commands (relying on information from the theorem prover), so to extend the example in Sect. 4 we can add commands like \openscriptfile{example.thy} and \closescriptfile to indicate destination script files; several files may then be produced on processing.

Another interesting use case for our architecture would be to have the prover insert proof objects into the document via backflow. Here, a proof object would just be formal object which can be reconstructed by the prover on demand to show the validity of the proof. This would allow a proof to be more or less completely informal except for the embedded proof objects, which could be used to validate the formal content.

Finally, we want to conduct usability studies to substantiate the claim that assisted authoring increases productivity compared to unassisted editing. A good evaluation methodology would be to investigate usability for mid-sized proofs using well-known HCI techniques (e.g., keystroke-measures), as well as to collect subjective experience reports from larger proof authoring projects.

Acknowledgements. We are grateful to our collaborators on the Proof General Kit project and the developers of Isabelle for their support of our experiments in interface development. For this paper, several useful suggestions for improvement were made by the MKM'05 referees.

References

1. J.-R. Abrial and D. Cansell. Click'n'prove: Interactive proofs within set theory. In D. Basin and B. Wolff, editors, *TPHOLs 2003*, LNCS 2758, p. 1–24. Springer Verlag, 2003.
2. D. Aspinall and C. Lüth. Commentary on PGIP. Available from http://proofgeneral.inf.ed.ac.uk/kit/, September 2003.
3. D. Aspinall and C. Lüth. Proof General meets IsaWin. In D. Aspinall and C. Lüth, editors, *User Interfaces for Theorem Provers UITP'03. ENTCS* 103:C, 2003.
4. D. Aspinall and C. Lüth. Parsing, editing, proving: The PGIP display protocol. In *User Interfaces for Theorem Provers UITP'05*, Apr. 2005.
5. S. Autexier, C. Benzmüller, A. Fiedler, H. Horacek, and Q. Bao Vo. Assertion level proof representation with underspecification. In F. Kamareddine, editor, *Proc. MKM Symposium MKM'2003*, Edinburgh, Nov. 2003.
6. J. Avigad. Notes on a formalization of the prime number theorem. Technical report, Carnegie Mellon, 2004.
7. Y. Bertot, G. Kahn, and L. Théry. Proof by pointing. In M. Hagiya and J. C. Mitchell, editors, *Proce. of the International Symposium on Theoretical Aspects of Computer Software*, LNCS 789, p. 141–160. Springer Verlag, 1994.

8. A. D. Brucker, F. Rittinger, and B. Wolff. HOL-Z 2.0: A proof environment for Z-specifications. *Journal of Universal Computer Science*, 9(2):152–172, 2003.
9. N. G. de Bruijn. A survey of project AUTOMATH. In J. P. Seldin and J. R. Hindley, editors, *To H. B. Curry: Essays in Combinatory Logic, Lambda Calculus and Formalism*, p. 589– 606. Academic Press, 1980.
10. L. Dixon and J. Fleuriot. A proof-centric approach to mathematical assistants. *Journal of Applied Logic: Special Issue on Mathematics Assistance Systems*, 2005. To appear.
11. G. Gonthier. A computer-checked proof of the four colour theorem. Technical report, Microsoft Research Cambridge, 2004. `http://research.microsoft.microsoft.com/ gonthier/4colproof.pdf`.
12. J. Grundy. Transformational hierarchical reasoning. *Computer Journal*, 39:291– 302, 1996.
13. T. C. Hales. The Flyspeck project page. `http://www.math.pitt.edu/ thales/flyspeck/index.html`.
14. F. Kamareddine, M. Maarek, and J. B. Wells. Flexible encoding of mathematics on the computer. In A. t. Asperti, editor, *Mathematical Knowledge Management MKM 2004*, LNCS 3119, p. 160– 174. Springer Verlag, 2004.
15. F. Kamareddine and R. Nederpelt. A refinement of deBruijn's formal language of mathematics. *Journal of Logic, Language and Information*, 13(3):287– 340, 2004.
16. D. E. Knuth. Literate programming. *The Computer Journal*, 27(2):97–111, 1984.
17. M. Kohlhase. Semantic markup for TeX/LaTeX. In *Informal Proc. Mathematical User Interfaces, Math UI '04*, 2004.
18. C. Lüth and B. Wolff. Functional design and implementation of graphical user interfaces for theorem provers. *Journal of Functional Programming*, 9(2):167– 189, 1999.
19. C. Lüth and B. Wolff. TAS — a generic window inference system. In J. Harrison and M. Aagaard, editors, *Theorem Proving in Higher Order Logics: 13th International Conference, TPHOLs 2000*, LNCS 1869, p. 405–422. Springer Verlag, 2000.
20. D. Mackenzie. What in the name of Euclid is going on here? *Science*, 307:1402–1403, 2005.
21. RELAX NG XML schema language, 2003. Home page at `http://www.relaxng.org/`.
22. L. Théry. Formal proof authoring: An experiment. In *Informal Proc. User Interfaces for Theorem Provers, UITP '03*, 2003.
23. A. Trybulec et al. The Mizar project, 1973. See web page hosted at `http://mizar.org`, University of Bialystok, Poland.
24. M. Wenzel. Isar — a generic interpretative approach to readable formal proof documents. In Y. Bertot, G. Dowek, A. Hirschowitz, C. Paulin, and L. Thery, editors, *Theorem Proving in Higher Order Logics, 12th International Conference, TPHOLs'99*, LNCS 1690. Springer Verlag, 1999.
25. M. Wenzel. *Isabelle/Isar — a versatile environment for human-readable formal proof documents*. PhD thesis, Technische Universität München, 2001.
26. D. Winterstein, D. Aspinall, and C. Lüth. Proof General/Eclipse. In *User Interfaces for Theorem Provers UITP'05*, 2005.

A Tough Nut for Mathematical Knowledge Management

Manfred Kerber[1] and Martin Pollet[2]

[1] School of Computer Science, University of Birmingham,
Birmingham B15 2TT, England
http://www.cs.bham.ac.uk/~mmk
[2] Fachbereich Informatik, Universität des Saarlandes,
66041 Saarbrücken, Germany
http://www.ags.uni-sb.de/~pollet

Abstract. In this contribution we address two related questions. Firstly, we want to shed light on the question how to use a representation formalism to represent a given problem. Secondly, we want to find out how different formalizations are related and in particular how it is possible to check that one formalization entails another. The latter question is a tough nut for mathematical knowledge management systems, since it amounts to the question, how a system can recognize that a solution to a problem is already available, although possibly in disguise. As our starting point we take McCarthy's 1964 mutilated checkerboard challenge problem for proof procedures and compare some of its different formalizations.

1 Introduction

Mathematical colloquial language as well as languages of formalized mathematics offer a large variety of ways to formalize a problem. If a problem is given in an informal way, the first question is, how to formalize it, that is, how to write it down. While this is already the first question in a mathematical vernacular as well, it is more acute in a formal system, be it a proof development environment, or a mathematical knowledge representation system. Even within one such formal system, it is typically possible to represent the same problem in a large variety of ways. The adequacy of a representation depends of course on its intended purpose (e.g., information retrieval, tutoring system, automated problem solving). In this contribution we focus mainly on the representation of a problem in order to find a proof. The obvious question is then, which representations are appropriate and which ones not. The choice of a good formalization depends on the formal language itself as well as the available tools and system support available for certain representations. In consequence, users of different proof development systems will use formalizations which are particularly adequate for their system. This leads to the question how we can know that two different formulations are equivalent, or in a weaker version how we can know that one formulation entails another.

M. Kohlhase (Ed.): MKM 2005, LNAI 3863, pp. 81–95, 2006.

We will use the so-called "mutilated checkerboard problem" to study the relationship between different problem formalizations. This problem was introduced by McCarthy as a challenge problem for proof procedures [1]. While the challenge was mastered by different proof procedures, it is still a challenge for mathematical knowledge management.

In McCarthy's original paper [1] we can find two different formalizations. McCarthy [2] himself presented at the second QED workshop a different formalization which makes – compared to the original paper – use of set-theoretical notions and basic integer arithmetic. We will also look at the formalizations in Isabelle [3] and Coq [4].

There are various proofs of the problem available. Some are close to one of McCarthy's formalizations (e.g., [5]), others are using formalizations which are significantly different and need creative thought to understand that they are related to the original problem. Even the close formalizations of [2] and [5] need some adaptation in order to see that they can be mapped into each other. While McCarthy numbers the checkerboard from 0 to 7, Bancerek uses 1 to 8. A shift of an index in an array by 1 is a trivial re-representation to anybody with mathematical training, but it has to be recognized as an index shift. Neither do the strings directly match, nor will simple unification do, since 0 is not 1 and 7 not 8.

Why would not everybody just take McCarthy's original formalization? Obviously, different formal systems have different strengths and limitations. The first formalization by McCarthy is a first order formalization which does not make use of equality and function symbols. While this is a very restricted formalism, some systems can deal with just this formalism only. McCarthy's second formalism contains function symbols and equality. A reasoner which can deal with function symbols and equality will have special procedures how to do that and would probably not live up to its strengths if given the first formalization. Likewise Isabelle and Coq, which have very powerful representation languages, have strengths which these systems could not use if they had to stick to one of the original formalizations. For this reason it should be considered as legitimate that each system user chooses a representation which suits their system best. However, when we take a closer look at the formalizations, it is neither trivial to see that they solve the original problem nor to see that they can be identified or subsumed by one another.

While problems such as a mutilated checkerboard problem exhibit a big gap between their informal description in natural language and their various formal representations, the problem is universal in mathematics. Typically even in a single very strict formal system it is possible to say the same thing in a wide variety of ways and not for all of these possibilities it is obvious that they are equivalent on some level. The reason why one formalization entails another may involve a simple syntactic modification, or may rest on a deep semantic connection.

2   The Mutilated Checkerboard Problem(s)

Let us first introduce the problem and proof from McCarthy's original memo [1, p.1]:

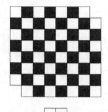

"It is impossible to cover the muti-lated checkerboard shown in the figure with dominoes like the one in the figure. Namely, a domino covers a square of each color, but there are 30 black squares and 32 white squares to be covered."

While the original challenge of the problem was the size of the search space in various formalizations – and still is for a naive usage – the much bigger challenge for proof procedures is that a good proof contains a creative invention, namely the colouring of the arrays of the board and the domino so that the squares with different colours can be counted. For the solution we can differentiate between the following phases:

1. Formalization of the problem.
2. Formalization of a concept representing the creative invention.
3. Realization of the proof on the basis of the creative invention.

The formulation of step 2 depends on the problem description given in step 1, and that the proof idea consists of both step 2 and step 3.

At each step there are several possibilities, colouring the arrays is only one of several creative inventions. Most publications which have taken up the challenge start with the motivation that their problem formalization is adequate with respect to the informal description and concentrate on steps 2 and 3.

In this contribution, we want to investigate the different possibilities and choices which can be made in step 1. The mathematical problem formalization assumes certain background knowledge that is available to the reader. Only with this background knowledge it is possible to understand why the problem formalization actually covers the original problem. The main objects, notions, and properties used (explicitly or implicitly) in the proofs are the following.

- Board: a rectangular structure containing squares. The squares are oriented in vertical/horizontal lines.
- Domino: a domino can be associated with two adjacent squares, either vertically or horizontally.
- Adjacent: for a square the square to the left, to the right, below, or on top.
- Sets: for collecting dominoes.
- Covering: all squares of the board are associated with exactly one domino.
- Numbers: numbers can be used to identify the different squares of the board, and to express the structure in which the squares are related to each other.
- Pairs and Cartesian product: pairs which contain numbers as representation of squares, Cartesian product as board.

Later, the creative invention of colouring corresponds to mapping squares, represented by pairs of numbers, to the set {*black, white*}. Sets can be counted, mapped to numbers, so we have cardinality and relations for the comparison of cardinality.

2.1   Challenge Problems by McCarthy

McCarthy gives already two problem formalizations in his initial paper [1] and a set-theoretic description thirty years later [2]. Whereas the early formalizations describe the problem completely, the latter builds on existing concepts of set theory, including cardinality, and integer arithmetic.

M64a

The language is a predicate logic without function symbols and without equality. The signature consists of constants $1, \ldots, 8$, and the binary relations given here with their intended meaning.

$S(x, y)$ $y = x + 1$

$L(x, y)$ $x < y$

$E(x, y)$ $x = y$

$G^1(x, y), \ldots, G^4(x, y)$ square (x, y) and the top/right/bottom/left neighbour square are covered by a domino

$G^5(x, y)$ square (x, y) is uncovered

The problem is stated in form of unsatisfiable axioms.

1. $S(1, 2) \wedge S(2, 3) \wedge S(3, 4) \wedge S(4, 5)$
   $\wedge\, S(5, 6) \wedge S(6, 7) \wedge S(7, 8)$ ⎫
2. $S(x, y) \Rightarrow L(x, y)$
3. $L(x, y) \wedge L(y, z) \Rightarrow L(x, z) \wedge \neg S(x, z)$ ⎬ properties of numbers
4. $L(x, y) \Rightarrow \neg E(x, y)$
5. $E(x, x)$ ⎭
6. $G^1(x, y) \vee G^2(x, y) \vee G^3(x, y) \vee G^4(x, y) \vee G^5(x, y)$ ⎫
7. $G^1(x, y) \Rightarrow \neg(G^2(x, y) \vee G^3(x, y) \vee G^4(x, y) \vee G^5(x, y))$
8. $G^2(x, y) \Rightarrow \neg(G^3(x, y) \vee G^4(x, y) \vee G^5(x, y))$ ⎬ placement of dominoes
9. $G^3(x, y) \Rightarrow \neg(G^4(x, y) \vee G^5(x, y))$
10. $G^4(x, y) \Rightarrow \neg G^5(x, y)$ ⎭
11. $G^5(1, 1) \wedge G^5(8, 8)$ ⎱ uncovered squares
12. $G^5(x, y) \Rightarrow (E(1, x) \wedge E(1, y)) \vee (E(8, x) \wedge E(8, y))$ ⎰
13. $S(x_1, x_2) \Rightarrow (G^1(x_1, y) \Leftrightarrow G^3(x_2, y))$ ⎱ adjacency of dominoes
14. $S(y_1, y_2) \Rightarrow (G^2(x, y_1) \Leftrightarrow G^4(x, y_2))$ ⎰
15. $\neg G^3(1, y) \wedge \neg G^1(8, y) \wedge \neg G^2(x, 8) \wedge \neg G^4(x, 1)$ } border of the board

M64b

The second problem is formalized in a predicate logic with function symbols and equality. The five predicates $G^1, \ldots, G^5$ are represented by the values of the function g. The problem is again given in form of axioms.

1′. $s(s(s(s(s(s(s(s(8)))))))) = 8$ ⎱ eight distinct numbers
2′. $\neg s(s(s(s(s(x))))) = x$ ⎰
3′. $g(x, y) = 5 \Leftrightarrow x = 8 \wedge y = 8 \vee x = 1 \wedge y = 1$ } uncovered squares
4′. $g(x, y) = 1 \Leftrightarrow g(s(x), y) = 3$ ⎱ adjacency of dominoes
5′. $g(x, y) = 2 \Leftrightarrow g(x, s(y)) = 4$ ⎰

$6'$. $g(1,y) \neq 3 \wedge g(8,y) \neq 1 \wedge g(x,1) \neq 4 \wedge g(x,8) \neq 2$ } border of the board
$7'$. $1 = s(8) \wedge 2 = s(1) \wedge 3 = s(2) \wedge 4 = s(3) \wedge 5 = s(4)$ } names for numbers
$8'$. $g(x,y) = 1 \vee g(x,y) = 2 \vee g(x,y) = 3 \vee g(x,y) = 4 \vee g(x,y) = 5$ } covering

M95

The language is predicate calculus and expects a formalization of the set theoretical concepts, like operations on sets, Cartesian product, ordered pairs, and cardinality. Furthermore the formalization uses operations on integers including the absolute value function. The additional concepts are introduced by definitions, the problem is stated as a formula to be proved.

Definitions
$$Board = \mathbb{Z}_8 \times \mathbb{Z}_8$$
$$mutilated\text{-}board = Board \setminus \{(0,0),(7,7)\}$$
$$domino\text{-}on\text{-}board(x) \Leftrightarrow (x \subset Board) \wedge card(x) = 2$$
$$\wedge (\forall x_1\ x_2)(x_1 \neq x_2 \wedge x_1 \in x \wedge x_2 \in x$$
$$\Rightarrow adjacent(x_1, x_2))$$
$$\Leftrightarrow (x \subset Board) \wedge card(x) = 2$$
$$\wedge (\forall x_1\ x_2)(x = \{x_1, x_2\} \Rightarrow adjacent(x_1, x_2))$$
$$adjacent(x_1, x_2) \Leftrightarrow |c(x_1,1) - c(x_2,1)| = 1 \wedge c(x_1,2) = c(x_2,2)$$
$$\vee |c(x_1,2) - c(x_2,2)| = 1 \wedge c(x_1,1) = c(x_2,1)$$
$$\Leftrightarrow |c(x_1,1) - c(x_2,1)| + |c(x_1,2) - c(x_2,2)| = 1$$
$$c((x,y),1) = x$$
$$c((x,y),2) = y$$
$$partial\text{-}covering(z) \Leftrightarrow (\forall x)(x \in z \Rightarrow domino\text{-}on\text{-}board(x))$$
$$\wedge (\forall x\ y)(x \in z \wedge y \in z \Rightarrow x = y \vee x \cap y = \{\})$$

Theorem
$$\neg(\exists z)(partial\text{-}covering(z) \wedge \bigcup z - mutilated\text{-}board)$$

Note that McCarthy defines *domino-on-board* and *adjacent* in two equivalent ways. Which definition to prefer depends on the context of its usage.

2.2 Formalization with Inductive Definitions

Paulson presented his formalization and proof of the checkerboard problem in the Isabelle system [3].

P96

The language is Isabelle/HOL which allows inductive definitions and supports reasoning with them. The formalization uses notions from integer arithmetic and set theory.

Definitions
$$lessThan(m) := \{i \in \mathbb{N} | i < m\}$$
$$board := lessThan(2 \cdot s(m)) \times lessThan(2 \cdot s(n))$$

$$tiling(A) : set(set(\alpha)) := \{\} \in tiling(A) \wedge (a \in A \wedge t \in tiling(a) \wedge a \cap t = \{\})$$
$$\Rightarrow a \cup t \in tiling(a)$$
$$domino : set(set(\mathbb{N} \times \mathbb{N})) := \{(i,j),(i,s(j))\} \in domino \wedge$$
$$\{(i,j),(s(i),j)\} \in domino$$

Theorem
$$(board \setminus \{(0,0)\}) \setminus \{(s(2 \cdot m), s(2 \cdot n))\} \notin tiling(domino)$$

2.3 Formalization in Second Order Logic

Huet [4] formulates the problem on a level which is more abstract than all the other formalizations. The formalization is based on properties such as injectivity and surjectivity of functions, and finiteness (characterized on the basis of injectivity and surjectivity). The geometry of the board is not considered at all, numbers or cardinals are not necessary for the argument.

H96

The theorem is formalized in the Coq system, but needs only the expressiveness of second order logic. Given a signature of sets B and W, two functions $Board$: $B \to W$ and $Domino : W \to B$ representing the board and the dominoes, the existence of a tiling for the full checkerboard problem is stated as:

$$injective(Board) \wedge injective(Domino) \wedge finite(B) \Rightarrow surjective(Domino).$$

We can generate from the full checkerboard a mutilated one by taking B' as a proper subset of B. The theorem is then that for an injective function $Board'$ with $Board' : B \to W$ and $finite(B')$ there is no function $Domino : W \to B'$ which is injective (i.e., a – possibly partial – covering) and surjective (i.e., total).

2.4 A Very Abstract Formalization

Let us add here another formalization which is even more abstract than Huet's and just reasons about the cardinality of sets (finite or infinite).

KP05

Two sets B and W cannot at the same time have the same cardinality and a different cardinality, that is, not $|B| = |W|$ and $|B| \neq |W|$.

More concretely we can say, if $|B| < |W|$ (that is, we have strictly fewer black than white elements) then we cannot have $|W| = |B|$ (that is, equal numbers which would follow from a covering with 2×1-dominoes).

3 Relationships Between Formalizations

Different systems make use of different problem representations. Fig. 1 gives an overview (which is not complete with respect to problem representations as well as solutions). The initial formalization $M64a$ was verified by the model generator Mace, only the symbols L and E together with the corresponding axioms were replaced by built-in concepts [6]. The proof in the TPS system uses exactly

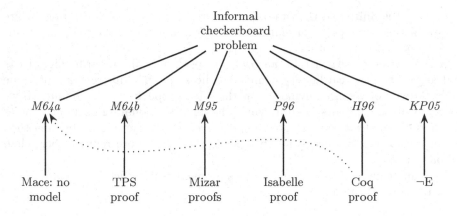

Fig. 1. Overview of formalizations of the checkerboard problem and proofs

the formalization given as *M64b* [7]. There are two proofs in the Mizar system for *M95* [5,8]. Paulson and Huet constructed proofs for their formalizations in Isabelle [3] and Coq [4], respectively. *KP05* can be proved in almost any formal system by calling the corresponding calculus level rule.

Every given solution is a formal proof for the corresponding problem representation and thus for the informal representation of the checkerboard problem. While it is difficult to be more rigorous with respect to the informal statement – and probably impossible to be fully rigorous – we can ask whether the Coq proof is a solution for the formalization given as *M64a* (represented as a dotted line in Fig. 1) and ask for a formal justification. In extreme cases (relationship between *M64a* and *KP05*), such a formal justification may be as difficult as a proof of the one problem (here *M64a*) from first principles. We will come back to the implications of this observation for mathematical knowledge management, but first we look at concrete problem transformations.

3.1 Problem Formalizations and Their Generalizability

Each formalization can be looked at from different perspectives. *M64a* is very concrete and its proof captures exactly the original problem and is difficult to generalize. *M64b* is slightly more general (also capturing tori). *M95* although concrete in the formulation, allows for an easy generalization of the argument.

P96 and *H96* are much more general and go beyond the concrete 8 × 8-mutilated checkerboard. In Paulson's formulation the problem is shown for any $2m \times 2n$ rectangular structure in which two opposite corners are missing, while Huet's formulation entails that arbitrary finite sets of black and white squares with an injective mapping from black to white sets, which is not surjective, cannot allow for an injective mapping form the white to the black squares. This proof is very abstract and many different aspects of the problem are irrelevant: the relative relationship of black and white squares, how many fewer black squares than white squares we have, and the shape of the dominoes. Important is only that the set is finite and that a domino covers exactly one black and one white square (actually the problem representation does not speak of coverings, so this is yet another interpretation).

KP05 is even more general than *H96*. We speak only about cardinality of sets and argue that the cardinality cannot be the same and different at the same time. We do not require anymore that the sets we speak about are finite. If we cover, for instance, pairs (a, b) in which the as are rational numbers and the bs real numbers so that each rational number is covered exactly once by an a then the bs cannot cover the real numbers since there are more real numbers than rational numbers.

Even more general than *KP05* is the statement true, which is logically equivalent to all true statements.

3.2 Proofs and Formalizations

If we revisit the different representations then the last one, true, does not require any proof. It is *trivially* satisfied, but it does not help us anything in understanding the problem.[1] Let us imagine the situation of a teacher and a student and the teacher wants to convince the student that the mutilated checkerboard cannot be covered by 2×1-dominoes. She may say "This is trivial." to which the student may answer either "Yes, it is trivial, I learnt to know this problem a while back already." or "No, I don't believe it."

In the latter case she would need to give him a better argument (based on *KP05*) such as "Look at a partition of the checkerboard into two disjoint sets so that there are arguments that the two sets must have the same number of elements and at the same time must not have the same number of elements." Again the student may now believe the statement and accept this as a proof, or not.

If the previous argument is not accepted *H96* can further refine it, and so on. Each formulation will get closer to the representation of a concrete checkerboard on the one hand. On the other hand, the proof of why each statement holds becomes increasingly more difficult.

3.3 Translations Between Different Formalizations

A more formal way to look at the relationship between the different formalizations is to see how one can be translated into another. We will discuss now some of these translations to a varying degree of depth.

M64a⟶*M64b*

Let us assume now that the predicative formulation *M64a* is given and we want to generate from this a functional formulation of the type *M64b*. To this end we would need to map different objects of the first signature to the second.

---

[1] Richard Feynman states only half jokingly that "mathematicians can prove only trivial theorems, because every theorem that's proved is trivial." [9, p.84].

$M64a$	$\mapsto$	$M64b$	
$S(x,y)$	$\mapsto$	$s(x) = y$	(1)
$1, 2, \ldots, 8$	$\mapsto$	$1, s(1), \ldots, s(s(s(s(s(s(s(1))))))) $	(2)
$G^i(x,y)$	$\mapsto$	$g(x,y) = i$ for $i \in \{1, \ldots, 5\}$	(3)
$E(x,y)$	$\mapsto$	$x = y$	(4)

Why is this translation justified? (1) means that we can translate the binary predicate symbol S into a unary function symbol s. In order to justify such a translation we need to show that S not only represents a general relation, but a left total and right unique one. Let us first observe that it suffices to look at the numbers 1 through 8. S is right unique for these numbers since from the axioms 1.–5. follows that the usual relationships between these numbers hold, for instance, $1 < 3$ from $S(1,2)$ and $S(2,3)$ with the second and third axiom. With the third axiom follows also $\neg S(1,3)$. Likewise $\neg S(1,4)$ and so on, that is, S is right unique.

The left totality is a more subtle issue. While it is intentionally clear that we have to consider only the eight different numbers 1, 2, ..., 8, the formalization $M64a$ does not warrant that only eight different numbers exist. More seriously, we get from the axioms 1.–5. as a consequence that for all $x \in \{1, 2, \ldots, 8\}$ holds $\neg S(8, x)$.[2] The argument why S can be translated to a function requires either to extend the range to all natural numbers or to give up the idea of a strict order and to consider a cyclic structure in which $s(8) = 1$.

Both approaches are formally possible and amount to the same covering problem, assumed we never make use of $s(8)$. Axiom 6' states that dominoes do not stick out, that is, that only the dominoes in a range from 1 through 8 are to be considered. The first approach means we view the mutilated checkerboard as a subset of the infinite plane. When putting dominoes down, we never cross the boundaries. The second approach means that we consider the checkerboard as a torus, axiom 6' puts up boundaries between the 8th row and the 1st row (and the columns likewise) which we may not cross in covering the checkerboard. Strictly speaking, axiom 6' is not necessary to prove the theorem, since the mutilated torus, that is, the full torus without the images of the two deleted corners of the checkerboard, cannot be covered by dominoes either, even if it is possible to cross the invisible boundaries, since the mutilated torus contains two white squares more than black ones.

In $M64b$ McCarthy chooses the second view (of a torus). The transition from $M64a$ to $M64b$ cannot be syntactically established at a low level (but only almost), since the original binary predicate symbol S is not left-total. This does not matter, however, since $s(8)$ has never to be used. Formally the second approach has the advantage that we have only finitely many (eight) different elements in the Herbrand universe, and hence has a finite search space, while the other formalization has a potentially infinite search space. Depending on which procedure one uses, it is advantageous to be more restrictive and to include 6' (for instance, when using Mace to show that no model exists), or to allow coverings of the torus which cross the magic boundary and have no correspondence of cov-

---

[2] If we had $S(8,x)$ for some x we would get with the second axiom $L(8,x)$. Together with $L(x,8)$ (which follows from axioms 1.–3. for $x \neq 8$, and from $L(8,x)$ for $x = 8$) we get $L(8,8)$, hence $\neg E(8,8)$, which contradicts $E(x,x)$.

erings in a real checkerboard with 2×1-dominoes and exclude 6' (for instance, when proving the theorem with TPS).

As replacing the predicate symbol S by a function symbol s, it is possible to replace the predicate symbols G^i (with $i = 1, \ldots, 5$) by a function symbol g so that $G^i(x, y)$ goes to $g(x, y) = i$. Note that initially the G^i are different names for binary predicate symbols (and could have been called P, Q, R, S, T instead). Since the names do not matter we can use a single ternary predicate symbol G instead which takes the index i as an argument, that is, $G(x, y, i)$ instead[3] of $G^i(x, y)$. $G(x, y, i)$ corresponds for a left-total and right-unique G to $g(x, y) = i$. The left-totality and right-uniqueness of the $G(x, y, i)$ follows from axioms 6.–10.

Axioms 11 and 12 mean that precisely the squares (1,1) and (8,8) are missing. They are translated as $g(1, 1) = 5 \wedge g(8, 8) = 5$ and $g(x, y) = 5 \rightarrow (E(1, x) \wedge E(1, y)) \vee (E(8, x) \wedge E(8, y))$ which is with the translation of $E(x, y)$ to $x = y$ logically equivalent to axiom 3'.

Axioms 13 and 14 can be translated directly to 4' and 5' (with the simplification to rename x_1 to x, and get rid of x_2 by replacing it by $s(x)$). Axiom 6 goes to 8', and 15 to 6'.

Note that the translation from S to s and G to g leads to proof obligations, namely that the functions s and g, used to represent the relations S and G are well-defined. While this should be a straightforward syntactical proof and actually is for G, matters are more subtle with S, and intuitively need an argument why a mutilated checkerboard cannot be covered when a mutilated toroid checkerboard cannot. The translation of E to equality strictly also leads to proof obligations that the usual properties of reflexivity, symmetry, transitivity, and substitutivity are satisfied. Only the first of the four is actually formally given (in axiom 5), since only this one is needed.

$M64b \longrightarrow M64a$

The reverse transformation is much simpler. Functional expressions such as $f(x_1, \ldots, x_n) = y$ are translated to predicative ones such as $F(x_1, \ldots, x_n, y)$. The additional axioms for the left-totality and right-uniqueness of F may be added. Nested functional expressions such as $g(s(x), y) = 3$ require the introduction of new variables as in $S(x_1, x_2) \rightarrow G(x_2, y, 3)$.

$M95 \longrightarrow H96$

Let us see now how the solution of Huet formalized as $H96$ can be applied to the problem formalization given by McCarthy as $M95$. We start again to relate the objects in the different formalizations.

$M95$	$\mapsto$	$H96$	
$\{(x, y) \mid a \in Board \wedge 0 = (x + y) \mod 2\}$	$\mapsto$	B	(1)
$\{(x, y) \mid a \in Board \wedge 1 = (x + y) \mod 2\}$	$\mapsto$	W	(2)
$Board$	$\mapsto$	$Board_H : B \to W$	(3)
set D with	$\mapsto$	$Domino : W \to B$	(4)
$partial\text{-}covering(D) \wedge W \subset \bigcup D$			

---

[3] Note also that the order of the arguments does not matter, provided any re-ordering is done consistently throughout the whole problem description.

None of the objects in the formalization in *H96* exists in *M95*. The sets B and W have to be constructed. The mapping representing the board *Board*$_H$ has to be defined using the set *Board* and the sets corresponding to B and W. The same holds for the mapping *Domino*, which additionally needs the set of dominoes covering all white squares.

In terms of the three steps describing the process of solving the problem given in Sec. 2, the formalization *M95* is the problem formalization of step 1, the proof given for *H96* corresponds to the realization of the proof in step 3, and the transformation which identifies objects in *M95* with objects in *H96* is the creative invention of step 2.

For the mutilated board of *M95* the set $B' = B \setminus \{(0,0),(7,7)\}$ corresponds to the restriction of *Board*$_H : B \rightarrow W$ to *Board*$'_H : B' \rightarrow W$, and the theorem of *M95* expressed with respect to the notions in *H96* translates then to $\neg(\exists Domino : W \rightarrow B'\ injective(Domino))$.

M95⟷*P96*

The correspondence between the objects in *M95* and *P95* is more direct. In both formalizations the board is represented by the Cartesian product of the integer interval $[0, 2n - 1]$, the squares of the board are elements of the Cartesian product, and dominoes are sets containing exactly two squares which are adjacent.[4]

We see here that even when there is an agreement how to represent the objects, there can be a difference how properties of the objects are represented. For example, the predicate *partial-covering* in *M95* is defined for the given board, and includes the definition of adjacent dominoes. The predicate *tiling* in *P96* defines the covering independently from dominoes. Here we see one dimension in the choice for a formalization: the generality of the introduced concepts.

In *M95* the general concept of tiling is expressed by the formula $tiling'(z) \Leftrightarrow (\forall x\ y)(x \in z \wedge y \in z \Rightarrow x = y \vee x \cap y = \{\})$ which is equivalent to *tiling* in *P96*. Even for equivalent concepts there is a choice in the formalization. This choice may depend on the assistance provided by the different proof systems. McCarthy supposes a 'heavy-duty' set theory prover, whereas the choice of Paulson is motivated by the support for inductive definitions available in Isabelle.

M64b⟶*M95*

When we compare the problem formalizations *M64b* and *M95*, then we find that the latter contains explicit objects to model the board, the squares of the board as pairs of coordinates, and the dominoes as sets with exactly two pairs. In *M64b* we find a representation which formalizes the different situations for each square (x, y) of the board as values of the function $g(x, y)$. Abstract concepts, like dominoes, size of the board, and covering are expressed as restrictions on the values of g and thus are given only indirectly.

---

[4] Although *M95* speaks about a concrete 8×8 checkerboard, the argument is as general as for the more general formalization of *P96*.

KP05⟶*H96*

The most basic argument in *KP05* is very simple. If we have two sets, then we cannot have $|B| = |W|$ and $|B| \neq |W|$. The slightly more expanded argument is: If we have two sets with $|B| < |W|$ then we cannot have $|W| \leq |B|$ (which is a slightly stronger version than the original formulation, but maps much better to Huet's argument).

How can we apply this statement to the case of *H96*? We need to know that the conditions are given, that is, we assume an injective but not surjective mapping *Board* : $B \to W$, and the finiteness of B. The injectivity entails $|B| \leq |W|$. The non-surjectivity entails together with the finiteness of B, $|B| \neq |W|$. The two facts together give us $|B| < |W|$. Hence *KP05* is applicable and we can conclude $\neg |W| \leq |B|$. Hence there is no injective mapping *Domino* : $W \to B$.

3.4   Soundness Considerations

The relationship between different formulations can in some cases be easily seen on the basis of a syntactic transformation. In other cases it may be necessary to make use of more complicated reasoning. Consider, for instance the relationship between *M64a* and *M64b*. The first describes the mutilated checkerboard as a subset of the plane, while the second describes it as a mutilated toroid. The reasoning why the one relationship entails the other is subtle.

In general we can compare the process of formalization of a problem to a projection from the *problem per se* to different formalizations as in Fig. 2. Since the *problem per se* is not given in a formal way, it is difficult to reason about the correctness of a formalization. However, it should always be possible the establish the relationship between different formalizations such as **F1** and **F2** in Fig. 2.

The relationship between different formalizations should follow the commutative diagram in Fig. 3. Note that the relationship may be non-trivial since the different formalizations may use different formal systems. A proof for the original problem $\Gamma \models \varphi$ will be

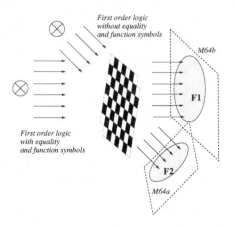

Fig. 2. Representation as projection

obtained by transforming the assumptions Γ to Γ' from which a tranformed theorem φ' is derived in a potentially different calculus. Then φ' is translated back to the original theorem φ. Assumed that the translations and the calculus π are sound, then the overall construction is a sound argument why φ follows from Γ.

$$\Gamma \quad \vdash_{\pi_\varphi \circ \pi \circ \pi_\Gamma}^{!} \quad \varphi$$

$$\top_{\pi_\Gamma} \qquad \qquad \bot_{\pi_\varphi}$$

$$\Gamma' \quad \vdash_\pi \quad \varphi'$$

Fig. 3. Commutative diagram for proving by reformulation

Note that typically one formalization represents not only a single problem, but as discussed in Sec. 3.1 it represents a whole class which can be considered as the inverse image of different projections as displayed in Fig. 4.[5]

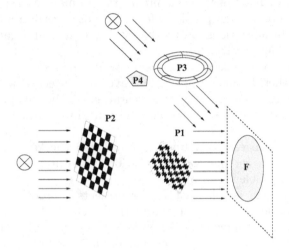

Fig. 4. Generalizability as inverse projection

4   Conclusion

It is a standard task when applying mathematical or (more stringently) formal methods to problems that we have to formalize them within a system. It is indeed an important aspect in the application of formal systems in safety critical areas, since any proof can start only once the formalization is given. If different people formalize the same problem they will typically come up with different formulations. Assumed the two formulations cannot mapped to each other in a sound way, then at least one does not capture the original problem adequately.

---

[5] One of the reviewers pointed out the relationship to Benjamin Whorf's hypothesis that languages constrain thought. We do not want to go into the deep philosophical discussion surrounding this hypothesis in general. However, it seems suggestive that not only languages, but more so formulations within languages may facilitate creative reasoning and generalizations, or not. While the same problem can be formalized in various ways, it matters significantly for problem solving which formalization is used.

In order to study the process of relating different formulations to each other and to understand how to represent one example we have looked at an old problem, McCarthy's mutilated checkerboard problem. This problem is interesting since it has been represented and proved in a large variety of systems. Of course, any representation has to abstract from unimportant details, such as the size of a square on the checkerboard and its material. Other details may or may not be represented, for instance, whether we speak of a checkerboard of size 8×8, or arbitrary size $2n \times 2n$, or arbitrary finite size, or arbitrary size. Perhaps astonishingly, this easy problem exemplifies a wide range of problems involved.

Different formalizations are used all the time, and certain standard translations, such as different versions of relativization in sorted logic, representations of equality in a logic without equality, reification of higher order expressions in first order logic, or functional versus predicative formalizations, are used quite regularly when representing problems formally. Often certain relationships between different formalizations are proved by a meta-logical argument (e.g., the sort theorem for the relativization).

There is also a very general mechanism how different representations can be linked to each other. Farmer et al. [10,11] have introduced a theory mapping in their little theories approach. One representation can be mapped faithfully to another if the axioms of the one become theorems in the other in a systematic translation.

Two major challenges follow from the mutilated checkerboard problem for mathematical knowledge management. Obviously the problem is not restricted to this particular problem, but a general one.

Firstly, if a system is told that *H96* represents a proof for the mutilated checkerboard problem, how can it automatically check that this is so indeed. Secondly – even more difficult – if a system is given a problem such as the mutilated checkerboard problem, how can it find relevant information which may be given in forms so diverse as *M64a*, *M64b*, *M95*, *P96*, *H96*, or *KP05*.

The second problem may be eased by proper annotations. The first in its full generality can be arbitrarily hard. While retrieving and relating information given in different shapes seems a core activity not only in mathematical knowledge processing, the ways in which mathematical information is transformed is very rich. A good representation is one that is simple and at the appropriate level so that it captures the main ideas of the proof.

What is appropriate is, however, not independent of the actual user. An argument about cardinalities, for instance, requires at least some basic knowledge of the concepts by the user. Huet's elegant proof remains obscure to somebody who does not know anything about the relationships between injective and surjective mappings on finite sets of the same cardinality, and so on.

As Ayer [12, p.85f] puts it "A being whose intellect was infinitely powerful would take no interest in logic and mathematics. For he would be able to see at a glance everything that his definitions implied, and, accordingly could never learn anything from logical inference which he was not fully conscious of already." Everything is trivial for such a being. Experts may do with some easy arguments on a high level, while beginners need low-level lengthy arguments at a low level. The relationships between the different levels are often syntactic, but can be complicated in detail.

A truly helpful system would not only find relevant information but also present it to a user on an appropriate level. A good understanding of typical transformations is an important step toward such a system.

References

1. McCarthy, J.: A tough nut for proof procedures (1964) Stanford Artificial Intelligence Project Memo No. 16, 1964. Available from `http://www-formal.stanford.edu/jmc/`.
2. McCarthy, J.: The mutilated checkerboard in set theory. [13] 25–26. Available from `http://www.mcs.anl.gov/qed/index.html`.
3. Paulson, L.C.: A simple formalization and proof for the mutilated chess board. Logic Journal of the IGPL **9** (2001) 475–485. Also published as Technical Report Computer Laboratory, University of Cambridge, 394, May 1996.
4. Huet, G.: The mutilated checkerboard (Coq library) (1996) `http://coq.inria.fr/contribs/checker.html`
5. Bancerek, G.: The mutilated chessboard problem – checked by Mizar. [13] 37–38 Available from `http://www.mcs.anl.gov/qed/index.html`.
6. McCune, W.: Another crack in a tough nut. Association for Automated Reasoning Newsletter **31** (1995) 1–3
7. Andrews, P.B., Bishop, M.: On sets, types, fixed points, and checkerboards. In Miglioli, P., Moscato, U., Mundici, D., Ornaghi, M., eds.: Theorem Proving with Analytic Tableaux and Related Methods. 5th International Workshop. (TABLEAUX '96), Terrasini, Italy, Springer Verlag, LNAI 1071 (1996) 1–15
8. Rudnicki, P.: The mutilated checkerboard problem in the lightweight set theory of Mizar. Technical Report TR96-09, Department of Computing Science, University of Alberta (1996)
9. Feynman, R.P.: Surely you're joking Mr. Feynman. Vintage, London, UK (1985)
10. Farmer, W.M., Guttman, J.D., Thayer, F.J.: Little theories. In Kapur, D., ed.: Proceedings of the 11th CADE, Saratoga Springs, New York, USA, Springer Verlag, LNAI 607 (1992) 567–581
11. Farmer, W.M.: An infrastructure for intertheory reasoning. In McAllester, D., ed.: Proceedings of the 17th CADE, Pittsburgh, Pennsylvania, USA, Springer Verlag, Berlin, LNAI 1831 (2000) 115–131
12. Ayer, A.J.: Language, Truth and Logic. 2nd edition, 1951 edn. Victor Gollancz Ltd, London, UK (1936)
13. Matuszewski, R., ed.: The QED Workshop II. (1995) Available from `http://www.mcs.anl.gov/qed/index.html`

Textbook Proofs Meet Formal Logic – The Problem of Underspecification and Granularity

Serge Autexier and Armin Fiedler

Saarland University & German Research Centre for Artificial Intelligence
(DFKI GmbH), Stuhlsatzenhausweg 3, 66123 Saarbrücken, Germany
{autexier, afiedler}@ags.uni-sb.de

Abstract. Unlike computer algebra systems, automated theorem provers have not yet achieved considerable recognition and relevance in mathematical practice. A significant shortcoming of mathematical proof assistance systems is that they require the fully formal representation of mathematical content, whereas in mathematical practice an informal, natural-language-like representation where obvious parts are omitted is common. We aim to support mathematical paper writing by integrating a scientific text editor and mathematical assistance systems such that mathematical derivations authored by human beings in a mathematical document can be automatically checked. To this end, we first define a calculus-independent representation language for formal mathematics that allows for underspecified parts. Then we provide two systems of rules that check if a proof is correct and at an acceptable level of granularity. These checks are done by decomposing the proof into basic steps that are then passed on to proof assistance systems for formal verification. We illustrate our approach using an example textbook proof.

1 Introduction

Unlike computer algebra systems (CASs), mathematical proof assistance systems have not yet achieved considerable recognition and relevance in mathematical practice. Clearly, the functionalities and strengths of these systems are generally not sufficiently developed to attract mathematicians on the edge of research. For applications in e-learning and engineering contexts their capabilities are often sufficient, though. However, current systems suffer from several major drawbacks. First, instead of supporting the language the mathematician is used to, most systems impose their own formal language on the user and require a machine-oriented formalization of the mathematical content to allow for powerful automatic inference capabilities. As a result, the line of reasoning is often unnatural and obscured. Next, the proofs are at a level of excruciating detail spelling out many logically necessary steps, which a human would nevertheless consider trivial or obvious. Thus, the proofs are often illegible and incomprehensible. Finally, the acceptance of mathematical assistant systems would be increased by integrating them with scientific WYSIWYG text editors. Indeed, current word processors regularly employ spell checkers to check the correct

M. Kohlhase (Ed.): MKM 2005, LNAI 3863, pp. 96–110, 2006.

spelling of the words and sometimes grammar checkers to check the correct application of the grammar rules in the sentences. Our aim is to support the practice of mathematical paper writing in the scientific text editor by employing proof assistance systems that provide definitions, lemmas and theorems from mathematical databases and automatically check the derivations spelled out in a mathematical document. The vision is to achieve the possibility to verify mathematical documents fully automatically. We envision a scientific text editor that allows the author to write semantically annotated mathematical content. The semantic annotations can then be exploited to generate a formal representation of the mathematical content (cf. Autexier *et al.* [2]), which allows for further automatic processing. The first step towards this end is to provide the formal language that can represent human-authored mathematical content.

Since the 1960ies, the AUTOMATH project [9] has been addressing the problem of developing a formal language with a natural-language-like syntax that allows both for the exact formalization of mathematical content and for the easy reading and authoring of the documents by mathematicians [8]. Whereas the original AUTOMATH language is very mechanical and thus tedious to author, its derivatives Mathematical Vernacular [4], Weak Type Theory [10] and MathLang [7] are close to a natural language. Since the 1970ies, the Mizar[1] project aims at supporting mathematical publications by means of a formal language that allows for automatic consistency checks of documents and for references to other articles published in the same formal language. A similar, more recent approach is taken in Isabelle/Isar [11], where proofs can be entered in a formal language of mathematics, which are readable for both human and machine (in fact, the Isar language is very similar to Mizar's language [14]). In another approach in the same tradition, Abel and colleagues [1] present a formal language for first-order intuitionistic logic used in a tutorial system for intuitionistic logic. The student writes proofs directly in this language, which are then automatically checked using a system of proof checking rules for intuitionistic logic.

A more sophisticated approach with respect to human readability is taken in the grammatical framework [12], a formalism based on a typed λ-calculus that allows for the definition of context-free grammars for fragments of natural language. However, only simple linguistic structures can be captured in this approach.

The major drawback of these approaches is that they do not sufficiently succeed in combining two diverging requirements, namely automatic processibility and readability. Automatic processing requires exact formalization, which in turn requires many details that humans consider obvious or trivial. Whereas detailed steps can be abstracted from in formal proofs by using lemmas, all steps must be included in the formal proof, even if they are easily inferable by the human user. Conversely, because of the omission of easily inferable steps, human-authored derivations often turn out to have gaps when scrutinized formally. The mentioned systems sacrifice the readability in favor of the processibility.

Therefore, we suggest a formal representation language for human-authored proofs where the gaps are filled in by underlying proof assistance systems

---

[1] http://www.mizar.org

(without committing ourselves to a specific prover or proof procedure). This formal language mediates between the semantically-annotated natural language representation of the mathematical document in the scientific text editor, where the user enters his input, and the logic representation required by proof assistance systems that check the mathematical content of the documents and fill in gaps. The core idea is to define a formal language that allows for underspecified parts and two systems of rules that check if a proof is correct and at an acceptable level of granularity. These checks are done by decomposing the proof into basic steps that are passed on to proof assistance systems for formal verification.

Clearly, which level of granularity is acceptable depends on many factors, the most prominent ones being the knowledge and skills of the intended audience, the mathematical theory the proven theorem belongs to and the personal style of the author. We do not cope with these factors in this paper, but define one specific level of granularity based on Hilbert's ideas [5] to demonstrate how a specific notion of granularity can be captured by constraining the proof checking rules.

This paper is organized as follows: We start with an overview of our approach. Then, we introduce our formal language for human-oriented proofs, and define a proof checking system for the proofs in that language. Finally, we experiment with means to formally capture notions of granularity. More specifically, we investigate how granularity can be defined as a restriction of the proof checking system.

2   Our Approach

Since many details of a proof, although logically required for a correct derivation, are considered obvious or trivial by human beings, they should be omitted from the proof. For our purposes, we capture the *level of granularity* using the following two distinct aspects:

First, the *level of concept*, at which a proof is done, can be described in terms of the definitions and theorems that can be used in the proof. More precisely, we can identify a mathematical *theory*, to which the theorem that is being proved belongs, as a logical collection of axioms, assumptions, definitions, lemmas and theories (collectively called *assertions*) as well as proofs. We consider a whole hierarchy of theories maintained in a mathematical database, where one theory draws on underlying theories. Now, if the theorem we want to prove belongs to some theory T, then we define the level of concept of the proof as the collection of assertions in theories underlying T plus the assertions of T that logically come before the theorem to be proved.

Second, the *amount of detail* in a proof refers to all facts and inference rules that are explicitly mentioned in the proof. Human-authored proofs are often imprecise in several respects, namely the used inference rule is not mentioned, some of the premises needed for a step in the derivation are not mentioned, and some steps of the derivation are completely omitted. That natural language texts and utterances are inherently imprecise (i.e., several readings of a sentence are possible) is a well-known phenomenon in linguistics, called *underspecification*.

Thus, automated processing of the content of human-authored texts requires the resolution of underspecification by singling out one possible reading.

Our work has been inspired by the work of Abel and colleagues [1], who worked on a tutorial system for intuitionistic logic. In their approach, they defined a linear syntax to represent first-order natural deduction proofs at the assertion level in intuitionistic logic and combined it with a deductive system of proof checking rules for that logic. Thus, the student can write proofs directly in this language and the proofs are automatically checked for correctness. The representation language, however, allows for one possible reading only.

In our approach, we adopt this idea of separating the representation language from the set of checking rules. However, we extend the approach in two dimensions:

First, we suggest a formal representation language for mathematical content detached from any particular logic or calculus. This allows us to represent arbitrary content regardless of the underlying logic. Moreover, the language allows us to represent both different levels of concept and underspecification and is thus particularly well-suited to represent proofs that are authored in a natural way by humans.

Second, we add two deductive systems, namely one for checking the correctness of proofs and one for checking the level of concept. The former decomposes the proof into basic steps, which either can be verified directly by one of the rules of the system or is passed on to an external proof assistance system that checks its correctness, and, if it is successful, provides a correctness proof. As a side effect, underspecified parts of such a basic proof step are resolved. The second deductive system similarly decomposes the proof into basic steps, but now checks if the steps are justified using acceptable inference rules.

We envision that our approach can serve as a first step towards an integration of a scientific text editor with mathematical proof assistance systems. In particular, the deductive systems show how mathematical proof assistance systems can be employed. To achieve the overall goal, however, many additional problems must be tackled, most notably a natural language analysis component that transforms the human-authored proofs into proofs in our representation language. Thus, we require for the time being that the author enter semantically annotated text by using LaTeX-style macros. These macros can then be expanded into a formal representation (cf. [2]), such as our representation language.

3   A Formal Representation Language

In this section we present the formal representation language for proofs (cf. Fig. 1). The language accommodates the mostly linear structure of textual proofs by representing complete proofs as a ";"-separated sequence of proof steps. In order to account for the internal structure of the proofs, the language allows for complex structures such as the introduction of subgoals or hypotheses, case analysis and induction. In each proof step, which either introduces subgoals or derives a fact, we distinguish in the syntax between the used concepts to justify

$$
\begin{aligned}
\text{S} &::= \text{A; S} \mid \textit{Trivial} \mid \epsilon \\
\text{A} &::= \textit{Fact } \text{ N : F } \textit{by } \text{ R}^* \textit{ from } \text{ R}^* \\
&\quad \mid \textit{Subgoals } (\text{N : F})^+ \textit{ in } \text{S}^+ \textit{ to obtain } \text{ N : F } \textit{by } \text{ R}^* \textit{ from } \text{R}^* \\
&\quad \mid \textit{Assume } \text{H}^* \textit{ in } \text{S} \textit{ to obtain } \text{ N : F } \textit{by } \text{ R}^* \textit{ from } \text{R}^* \\
&\quad \mid \textit{Assign } (\text{VAR} := \text{TERM} \mid \text{CONST} := \text{TERM}) \\
&\quad \mid \textit{Or}(\text{S}\| \ldots \|\text{S}) \\
&\quad \mid \textit{Cases } \text{F}^+ : (\textit{Case } \text{N : F : S } \textit{End})^+ \textit{ to obtain } \text{ N : F}
\end{aligned}
$$

H ::= N:F	CONST ::= *const* N
| CONST: TYPE?	VAR ::= *var* N
| VAR: TYPE?	
R ::= (N, F, P)	N ::= STRING | .
F ::= FORMULA | .	P ::= POSITION | .

Fig. 1. The grammar of the formal proof language

that proof step (denoted by the keyword *by* in the language) and to which premises or goals the concepts have been applied (denoted by *from*). Finally, in order to support the linguistic analysis of mathematical documents, which is not always able to uniquely categorize a given text fragment, we introduce a nondeterministic branching over possible proofs (Or) to represent the different alternative interpretations.

For the definition of the language we assume languages for formulas, terms, and types referred to by the nonterminal grammar symbols FORMULA, TERM,

---

If A and B are sets such that $x \in A$ implies that $x \in B$ (that is, every element of A is also an element of B), then we shall say that A is **contained** in B, or that B contains A, or that A is a **subset** of B, and we shall write $A \subseteq B$ or $B \subseteq A$. [...]
1.1.1 Definition   Two sets A and B are **equal** if they contain the same elements. If the sets A and B are equal, we write $A = B$.
[...]
1.1.4 Theorem   *Let A, B, C, be any sets, then*
[...]
(d) $A \cap (B \cup C) = (A \cap B) \cup (A \cap C)$, [...]
[...]
In order to give a sample proof, we shall prove the first equation in (d). Let x be an element of $A \cap (B \cup C)$, then $x \in A$ and $x \in B \cup C$. This means that $x \in A$, and either $x \in B$ or $x \in C$. Hence we either have (i) $x \in A$ and $x \in B$, or we have (ii) $x \in A$ and $x \in C$. Therefore, either $x \in A \cap B$ or $x \in A \cap C$, so $x \in (A \cap B) \cup (A \cap C)$. This shows that $A \cap (B \cup C)$ is a subset of $(A \cap B) \cup (A \cap C)$.

Conversely, let y be an element of $(A \cap B) \cup (A \cap C)$. Then, either (iii) $y \in A \cap B$, or (iv) $y \in A \cap C$. It follows that $y \in A$, and either $y \in B$ or $y \in C$. Therefore, $y \in A$ and $y \in B \cup C$ so that $y \in A \cap (B \cup C)$. Hence $(A \cap B) \cup (A \cap C)$ is a subset of $A \cap (B \cup C)$.

In view of Definition 1.1.1, we conclude that the sets $A \cap (B \cup C)$ and $(A \cap B) \cup (A \cap C)$ are equal.

Fig. 2. A textbook example

and TYPE, respectively. In the proof language, ϵ denotes an empty (sub)proof while *Trivial* indicates that the (sub)proof should be completed now, for instance, if there is a formula that occurs both as a goal and a hypothesis. The step *"Fact N : F by R* from R*"* indicates that a fact F has been derived from the objects referenced in the *from* slot using the objects referenced in the *by* slot and has been assigned the name N. A reference consists of three parts: the *name* of a formula, a *formula* and a *position* denoting a sub-object of that formula or the one referenced by the name; each component of the reference can be left open, which is made explicit by a period ("."). Thus, the sub-language for references explicitly allows for underspecification. A proof step *"Subgoals (N : F)$^+$ in S$^+$ to obtain N : F by R$_1^*$ from R$_2^*$"* represents the fact that we introduced a list of subgoals $(N : F)^+$ for some previous goals R_2^* and the proofs in S^+ are the subproofs for these subgoals. Note that the facts used to perform that goal reduction may be given in R_1^*. A proof step *"Assume H* in S to obtain N : F by R$_1^*$ from R$_2^*$"* is used to decompose goals R_2^* into the new hypotheses H* and the new goal F of name N. The hypotheses can be either named formulas N : F, or new constants and variables, possibly with some type. A proof step *"Assign var $x := t$"* allows us to assign a value t to some variable x and *"Assign const $c := t$"* encodes the introduction of an abbreviation c for some expression t. The expression *"$Or(S_1 \parallel \ldots \parallel S_n)$"* describes a situation, where the linguistic analysis identifies several possible interpretations resulting in different possible proofs. Finally, case distinctions can be introduced by the *Cases* construct, where for each formula φ in F$^+$ there is exactly one case $n : \varphi$. However, we consider case analysis as a derived construct that can be encoded by *Subgoals* and *Assume* proof steps. Analogously, induction proof steps can be defined.

To examine an example, let us consider an excerpt from Chapter 1 of the undergraduate analysis textbook *Introduction to Real Analysis* [3], which is shown in Fig. 2. For the purposes of this paper, we neglect the representation of the notation, the definition and the theorem, and focus only on the given proof,

---

1. *Assume* $. : x \in A \cap (B \cup C)$ *in*
   1.1 *Fact* $. : x \in A \wedge x \in B \cup C$ *by* $.$ *from* $.$;
   1.2 *Fact* $. : x \in A \wedge (x \in B \vee x \in C)$ *by* $.$ *from* $.$;
   1.3 *Fact* $. : (x \in A \wedge x \in B) \vee (x \in A \wedge x \in C)$ *by* $.$ *from* $.$;
   1.4 *Fact* $. : (x \in A \cap B) \vee (x \in A \cap C)$ *by* $.$ *from* $.$;
   1.5 *Fact* $. : x \in (A \cap B) \cup (A \cap C)$ *by* $.$ *from* $.$;
   to obtain $. : A \cap (B \cup C) \subseteq (A \cap B) \cup (A \cap C)$ *by* $.$ *from* $.$;
2. *Assume* $. : y \in (A \cap B) \cup (A \cap C)$ *in*
   2.1 *Fact* $. : y \in A \cap B \vee y \in A \cap C$ *by* $.$ *from* $.$;
   2.2 *Fact* $. : y \in A \wedge (y \in B \vee y \in C)$ *by* $.$ *from* $.$;
   2.3 *Fact* $. : y \in A \wedge (y \in B \cup C)$ *by* $.$ *from* $.$;
   2.4 *Fact* $. : y \in A \cap (B \cup C)$ *by* $.$ *from* $.$;
   to obtain $. : (A \cap B) \cup (A \cap C) \subseteq A \cap (B \cup C)$ *by* $.$ *from* $.$;
3. *Fact* $. : A \cap (B \cup C) = (A \cap B) \cup (A \cap C)$ *by* Def1.1.1 *from* $.$;
4. *Trivial*

Fig. 3. An example representation

which starts with "Let x be an element..." The proof can then be represented in our language as depicted in Fig. 3. Note that the labels of the proof steps are only added for convenience and are not part of the representation language.

4   Proof Checking

Now, having a means of representing proofs, we also want to check the correctness of the represented proofs. To this end, we propose a deductive system consisting of eight rules that allow us to check the encoded proofs by recursively checking each individual proof step starting from the first. For each individual proof step we need to know all declared types and constants, collected in the *signature*, all declared variables, collected in the *context*, and all visible hypotheses and previous goals, which both are lists of named formulas. The result of a successfully checked proof step S is a set of facts derived by the subproof with S as its root.

The deductive system does not directly encode any specific calculus, but collects proof obligations, called *lemmas*, for proof steps. These lemmas need to be verified in order to establish the validity of the corresponding steps. For example, a *Trivial* proof step gives rise to a lemma $\Gamma \Longrightarrow_{\text{Triv}} \Delta$, which states that from the hypotheses Γ some goal in Δ follows *"trivially"*. Thus, we also have to provide a specific proof strategy that decides if a proof step is trivial or not. An example would be a simple check if some goal in Δ also occurs in Γ. In general, we allow for specific *strategies* strat to establish the validity of a lemma $\Gamma \Longrightarrow_{\text{strat}} \Delta$. For the purposes of this paper, however, we will not go into the details of the strategies, but consider them as given (e.g., by a call of an automated theorem prover such as the proof planner ΩMEGA [13]).

Formally, a signature Sig consists of a list of type declarations $const\ \tau : \textbf{type}$ and constant declarations $const\ c : \tau$:

$$Sig ::= \epsilon \mid const\ \tau : \textbf{type}, Sig \mid const\ c : \tau, Sig$$

A context Ctx consists of a list of variable declarations:

$$Ctx ::= \epsilon \mid var\ x : \tau, Ctx$$

Now let Sig be a signature, Ctx a context, and S a proof. Furthermore, Γ and Δ denote sequences consisting of named formulas $N : F$, abbreviations $c \equiv t$ and substitutions $x \leftarrow t$. The judgments are:

– $Sig;\ Ctx;\ \Gamma \langle S \rangle\ \Delta \hookrightarrow \Gamma'$
  Given the signature Sig, the context Ctx, the hypotheses in Γ and the open goals in Δ, the (partial) proof S derives the facts Γ'.
– $P(\Gamma'; \Delta') : \Gamma \Longrightarrow_{\text{strat}} \Delta$
  The proof strategy strat proves the lemma $\Gamma \Longrightarrow_{\text{strat}} \Delta$ and returns the proof object $P(\Gamma'; \Delta')$. This notation for the proof object indicates that the proof requires the subsequences Γ' and Δ' of Γ and Δ, respectively.

The deductive system for proof checking is given in Fig. 4, where an expression $\bar{e}$ means a sequence of expressions $e_1, \ldots, e_n$ and $\uplus$ stands for the disjunctive

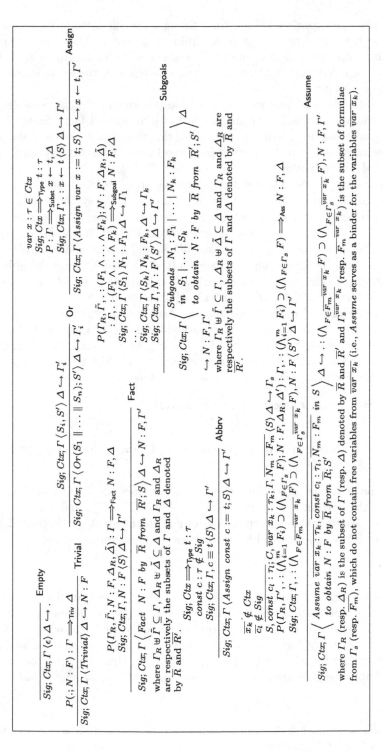

Fig. 4. Proof Checking Rules

union. Note that we explicitly refrain from fixing a specific logic in these rules, as we envision the use of our language in different domains. Therefore, we parameterize the proof checking system over the calculus for the specific logic. The connection to the calculus is established via the strategies for the lemmas arising during proof checking.

The individual kinds of lemmas are Triv to establish proof steps considered as trivial by the author, Fact to ensure the validity of derived facts, Subgoal to prove valid goal reductions, Ass to show that the stated fact can be inferred from some assumptions, Type to verify type correctness, and Subst to ensure the admissibility of a substitution.

To illustrate some of the lemmas arising during proof checking, let Sig_G denote the general signature, which contains, among others, the symbols $const\ A :$ set, $const\ B :$ set, $const\ C :$ set, $const\ \cup :$ set $\times$ set $\rightarrow$ set, $const\ \cap :$ set $\times$ set $\rightarrow$ set, and $const\ \subseteq:$ set $\times$ set $\rightarrow o$. The initial context Ctx is empty, Γ_G denotes the logical context, in which the proof takes place, and contains, among others, the Definition 1.1.1, the definitions of $\cup$ and $\cap$ (Def($\cup$), Def($\cap$)), and the abbreviating notation $\subseteq$, which we denote by Abbrv($\subseteq$)[2]. Furthermore, let Δ_G be the conclusion $A \cap (B \cup C) = (A \cap B) \cup (A \cap C)$ of Theorem 1.1.4.

In order to check Step 1 of the proof, the rule Assume requires to first check the subproof 1.1–1.5 where the context is augmented by the local variable x of type elem and the local set of assertions contains the hypothesis $. : x \in A \cap (B \cup C)$. The proof checking of Step 1.1 is then as follows: Assume S' are the proof steps 1.2-1.5, then the Fact proof checking rule is invoked as

$$Sig_G; const\ x : \text{elem}; \Gamma_G, . : x \in A \cap (B \cup C)$$
$$\langle Fact\ \ . : x \in A \land x \in (B \cup C)\ by\ .\ from\ .; S' \rangle \Delta_G$$

Checking that proof step requires to check the following judgments (cf. the rule Fact from Fig. 4):

1. We first have to check whether the given fact formula is actually derivable from the current assertions ($\Gamma_G, . : x \in A \cap (B \cup C)$) and the current goal ($\Delta_G$). We pass the corresponding lemma to the strategy (indicated by $\Longrightarrow$Fact) which we use to establish that lemma. The lemma is:

$$\Gamma_G, . : x \in A \cap (B \cup C) \Longrightarrow_{\text{Fact}} . : x \in A \land x \in (B \cup C), \Delta_G$$

From the strategy (for instance a prover or a tactic) we require not only to prove that lemma, but also to return a proof object P. From that proof object we require that it must "rely" on the named formulas provided by the references in the "by" and "from" (denoted as Γ_R and Δ_R) slots as well as on the given fact ($. : x \in A \land x \in (B \cup C)$). The fact that they must occur in the proof object is indicated by making them part of the arguments of P, i.e.

$$P(\Gamma_R, \tilde{\Gamma}; , \Delta_R, \tilde{\Delta})$$

---

[2] Assuming, for instance, a higher-order logic language for formulas and terms, the abbreviation $\subseteq$ could be written as $\subseteq \equiv \lambda A, B :$ set $. \forall x :$ elem $. x \in A \supset x \in B$.

In the present case there are no references, and hence Γ_R and Δ_R are empty. The list $\tilde{\Gamma}$ (resp. $\tilde{\Delta}$) denotes all further assertions from Γ_G (resp. goals from Δ_G) on which the proof object P relies. Those lists provide us with the missing references. In our case the proof object is $P(\mathsf{Def}(\cap), . : x \in A \cap (B \cup C); . : x \in A \wedge x \in B \cup C)$ which provides the non-specified reference to the definition of $\cap$.

2. After checking the lemma, the proof checking recurs over the proof steps S' by adding the new fact to the list of usable assertions, which is expressed by $Sig_G; Ctx_G, const\ x : \mathsf{elem}; \Gamma_G, . : x \in A \wedge x \in (B \cup C) \langle S' \rangle \Delta_G$. This returns a list of facts Γ' derived in that subproof S', which is denoted by $\hookrightarrow \Gamma'$. In our case Γ' is $. : x \in (A \cap B) \cup (A \cap C), . : (x \in A \cap B) \vee (x \in A \cap C), . : (x \in A \wedge x \in B) \vee (x \in A \wedge x \in C), . : x \in A \wedge (x \in B \vee x \in C)$ (in that order).

From that last list the result of the proof checking of the fact Step 1.1 are all the named formulas derived in the subproof S' plus $. : x \in A \wedge x \in B \cup C$. This is expressed by $\hookrightarrow N : F, \Gamma'$, which in the present case is $. : x \in A \wedge x \in B \cup C, . : x \in (A \cap B) \cup (A \cap C), . : (x \in A \cap B) \vee (x \in A \cap C), . : (x \in A \wedge x \in B) \vee (x \in A \wedge x \in C), . : x \in A \wedge (x \in B \vee x \in C)$. This list is the result of the subproof inside the *Assume* part and we denote this list by Γ_s. Validating the *Assume* step then requires to prove the lemma

$$\Gamma_G, . : x \in A \cap (B \cup C) \supset \begin{pmatrix} . : x \in A \wedge x \in B \cup C \\ \wedge . : x \in (A \cap B) \cup (A \cap C) \\ \wedge (. : (x \in A \cap B) \vee (x \in A \cap C)) \\ \wedge (. : (x \in A \wedge x \in B) \vee (x \in A \wedge x \in C)) \\ \wedge . : x \in A \wedge (x \in B \vee x \in C) \end{pmatrix}$$
$$\Longrightarrow_{\mathsf{Ass}} . : A \cap (B \cup C) \subseteq (A \cap B) \cup (A \cap C), \Delta_G.$$

The proof object for that lemma is $P(. : x \in A \cap (B \cup C) \supset (\bigwedge_{F \in \Gamma_s} F), \mathsf{Abbrv}(\subseteq); . : A \cap (B \cup C) \subseteq (A \cap B) \cup (A \cap C))$, and provides the missing references to the abbreviation of $\subseteq$ and the used premise $. : x \in A \cap (B \cup C) \supset (\bigwedge_{F \in \Gamma_s} F)$. The result of proof checking this *Assume*-proof step consists of (1) the result Γ' obtained from checking the remaining proof, (2) the obtained (named) formula $. : A \cap (B \cup C) \subseteq (A \cap B) \cup (A \cap C)$, and (3) all formulas derived in the subproof of *Assume*, which are not dependent on any local variables. The latter is expressed by the (schematic) formula $(\bigwedge_{F \in \overline{\Gamma_m}^{var\ x_k}} F) \supset (\bigwedge_{F \in \Gamma_s^{var\ x_k}} F)$, where $\Gamma_s^{var\ x_k}$ expresses the filtering.[3]

Proof checking of the next *Assume*-proof step 2—inclusive its subproof—also succeeds using the rules in Fig. 4. The set of derived facts up to before Step 3 is

- From *Assume*-proof step 1 we obtain $\Gamma_1 := . : A \cap (B \cup C) \subseteq (A \cap B) \cup (A \cap C), \Gamma_G$.
- From *Assume*-proof step 2 we obtain $\Gamma_2 := . : (A \cap B) \cup (A \cap C) \subseteq A \cap (B \cup C), \Gamma_{1.1}$.

---

[3] Note that this way any substitution—expressed by an equation $x = t$—for a local variable x inside the subproof of *Assume* is also removed.

Checking the *Fact*-proof step 3 requires to establish the lemma

$$\Gamma_2 \Longrightarrow_{\mathsf{Fact}} . : A \cap (B \cup C) = (A \cap B) \cup (A \cap C), \Delta_G$$

Furthermore, we require from the proof of this lemma which is returned by the strategy, that (i) it uses the goal formula $. : A \cap (B \cup C) = (A \cap B) \cup (A \cap C)$ and (ii) it uses the indicated Definition 1.1.1. This is expressed by

$$P(\mathrm{Def}1.1.1, \tilde{\Gamma}; . : A \cap (B \cup C) = (A \cap B) \cup (A \cap C), \Delta_R, \tilde{\Delta})$$

In this case, the used facts from Γ_2 are those obtained in the *Assume*-proof steps, i.e. $\tilde{\Gamma}$ is $. : A \cap (B \cup C) \subseteq (A \cap B) \cup (A \cap C), . : (A \cap B) \cup (A \cap C) \subseteq A \cap (B \cup C)$ (note that $\tilde{\Delta}$ is empty). Checking the final *Trivial*-proof step reduces to establish the lemma

$$\Gamma_2, . : A \cap (B \cup C) = (A \cap B) \cup (A \cap C) \Longrightarrow_{\mathsf{Triv}} \Delta_G$$

The lemma is trivially provable, since $\Delta_G := \mathrm{Thm}(1.1.4) : A \cap (B \cup C) = (A \cap B) \cup (A \cap C)$, which completes the proof checking of our example proof.

5   Granularity

In this section we investigate the problem how we can check that a proof is at a specific level of granularity. Our focus here is how at all notions of granularity could be captured formally, in order to design proof procedures that not only check the correctness of a (partial) proof, but also if it is at a given level of granularity. Knowing how granularity can be defined is a necessary prerequisite before we can move on to analyze which level of granularity is appropriate in which context and how it could be made user-adaptive.

In general, proof sketches can be at some specific, appropriate level of granularity, without being correct proofs. Conversely, a proof sketch can be correct, but not at a desired level of granularity. In other words, the notion of granularity and correctness do only overlap, but there is not necessarily a subsumption relation in whichever direction. Comparing both notions with respect to the existing means for their formalization, there is a long tradition and a large class of formalisms to represent correct proofs — one of them has been presented in the previous section. On the contrary, there are to our knowledge neither formalizations to check whether a proof is at some specific level of granularity (aside from, for instance, being a calculus-level proof), nor any other means to reduce the inspection of the granularity to computation.

As we pointed out earlier, there is an overlap between correctness and granularity. In this section we consider this overlap, since there we can hope to exploit the available formalisms for correctness and adapt them to accommodate some notion of granularity. However, it is not obvious a priori, whether any notion of granularity can be captured simply by refining the notion of correctness. With the work presented in the following we explore some aspects of that problem by defining a notion of granularity through restricting the notion of correctness.

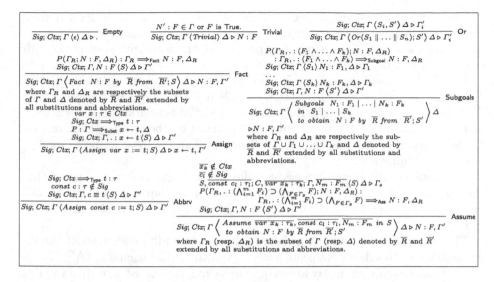

Fig. 5. *What-You-Need-Is-What-You-Stated* Granularity Checking Rules

The set of proofs which are accepted by the proof checking rules strongly depends on the strength of the strategies used to discharge the arising lemmas *and* on the knowledge, that is, facts and subgoals, the strategies can use during the proof attempt. However, the strength of the strategies can not be controlled at the level of our proof representation language. Moreover, finding and implementing the *right* strategies is difficult as it strongly depends on the individual authors. The only information we can control is which knowledge is actually passed to the strategies, which in turn can be influenced via two criteria: (1) The selection of *locally* available knowledge for the strategies and (2) the *global* flow of knowledge between different parts of the proof. So far in the proof checking rules (1) all locally visible assumptions and subgoals are used, and (2) all possible derived facts from earlier proof steps are passed to subsequent proof steps.

We now show how a specific, intuitive notion of granularity can be formalized by imposing restrictions for (1) and (2). The notion of granularity we define here is inspired by the assertion level proofs by Huang [6] and by a description of what a proof is by Hilbert [5]. This level of granularity can intuitively be described by *What-You-Need-Is-What-You-Stated Granularity*, that is, all necessary facts, assertions and rules are stated explicitly in the proof and the proof is performed at the assertion level. For instance, when we use some fact, we should have stated it before explicitly and not assume it is inferable from the context.

To formally define granularity, we introduce the judgment

$$Sig; \, Ctx; \, \Gamma \, \langle S \rangle \, \Delta \triangleright \Gamma'$$

which intuitively means that given the signature Sig, the context Ctx, the hypotheses in Γ and the open goals in Δ, the proof S derives the facts Γ' at the given level of granularity.

The deductive system to check the granularity (cf. Fig. 5) is derived from the proof checking system, by imposing restrictions with respect to criterion (1) for Trivial, Fact, and Subgoals proof steps, and restrictions with respect to both criteria (1) and (2) for Assume proof steps.

For the Fact and Subgoals proof steps, we restrict the rules by selecting from Γ and Δ only those formulas that are explicitly referenced in the proof step description, as well as all substitutions $x \leftarrow t$ and abbreviations $c \equiv t$. Furthermore, we require that the proof object P returned by the strategy strat *rely* on all the referenced formulas, that is, the deletion of any of these formulas renders the lemma unprovable.

For instance, the Fact and Subgoals rules are strengthened by requiring that the references $\overline{R}$ and $\overline{R'}$ denote *exactly* the set of assumptions and conclusions necessary to derive the new fact. Note that this requires all references to be defined.

The Assume rule is strengthened similarly, but in addition we restrict the information flow to subsequent proof parts by omitting the formula . : $(\bigwedge_{i=1}^{m} F_i) \supset \Gamma_s^{var \ x_k}$, as it is only implicitly known and not explicitly stated, and thus violates the intuitive *What-You-State-Is-What-You-Need* condition.

The Trivial rule is restricted by removing the call to the Triv strategy. Instead we require that either there is a trivially valid formula $N : F$ in the accumulated goals or one of the goal formulas occurs as an assumption. This makes this rule analogous to an Axiom rule in a sequent-style calculus.

Let us consider again our example proof to illustrate the checking of the granularity. If we check the granularity of the proof object given in Fig. 3, the check fails. For instance, in Step 1.1, a reference to the definition of $\cap$ is missing, and in Step 3 the references to the two derived subgoals are missing. Note that in Step 1 and 2 no information about $\subseteq$ is needed, since Bartle and Sherbert [3] introduced $\subseteq$ as an abbreviating *notation* and not as a defined

---

In order to give a sample proof, we shall prove the first equation in (d). Let x be an element of $A \cap (B \cup C)$, then $x \in A$ and $x \in B \cup C$ *by the definition of* $\cap$. This means that $x \in A$, and either $x \in B$ or $x \in C$ *by the definition of* $\cup$. Hence we either have (i) $x \in A$ and $x \in B$, or we have (ii) $x \in A$ and $x \in C$ *by the distributivity of "and" over "or"*. Therefore, either $x \in A \cap B$ or $x \in A \cap C$ *by the definition of* $\cap$, so $x \in (A \cap B) \cup (A \cap C)$ *by the definition of* $\cup$. This shows that $A \cap (B \cup C)$ is a subset of $(A \cap B) \cup (A \cap C)$.   (1)

Conversely, let y be an element of $(A \cap B) \cup (A \cap C)$. Then, either (iii) $y \in A \cap B$, or (iv) $y \in A \cap C$ *by the definition of* $\cup$. It follows that $y \in A$, and either $y \in B$ or $y \in C$ *by the definition of* $\cap$ *and the distributivity of "and" over "or"*. Therefore, $y \in A$ and $y \in B \cup C$ *by the definition of* $\cup$ so that $y \in A \cap (B \cup C)$ *by the definition of* $\cap$. Hence $(A \cap B) \cup (A \cap C)$ is a subset of $A \cap (B \cup C)$. *(2)*

In view of Definition 1.1.1, we conclude *from (1) and (2)* that the sets $A \cap (B \cup C)$ and $(A \cap B) \cup (A \cap C)$ are equal.

Fig. 6. The patched textbook proof example

concept, and abbreviations are globally visible to prove the lemmas arising during granularity check.

Patching the proof object from Fig. 3 by including the missing references as suggested by the granularity checker consists of: (1) the inclusion of the references to the definitions of $\cup$ and $\cap$ for all but the last *Fact* proof steps, and (2) for the last *Fact* proof step the inclusion of the references to the used premises. Exploiting the relationship of the proof steps to individual sentences in the text, we can propagate the additional information back into the textual representation. The resulting proof is shown in Fig. 6, where the added text fragments are set in ***boldface italics*** font.

6   Conclusion

In this paper, we presented a calculus-independent formal representation language for human-authored proofs and two deductive systems that allow for checking the correctness and the level of granularity of the proofs. This formal language mediates between the semantically-annotated natural language representation of the mathematical document in a scientific text editor, where the user enters his input, and the logic representation required by proof assistance systems that check the mathematical content of the documents and fill in gaps. Using an example textbook proof, we showed how this proof is represented in our language and checked its correctness and granularity. We based a first notion of granularity on Hilbert's approach that demands that everything that is needed in the proof must be stated explicitly. This, however, resulted in the failure of the granularity check of the example textbook proof, such that a patch was required that added missing references to used definitions and previously derived facts.

Putting the patched proof under scrutiny, we see that it is now easier to follow the line of reasoning in the proof. However, one could argue that the proof now contains too many details. Therefore, we plan to enhance our notion of granularity in order to allow for implicit references as well. To do so, we need a flexible model of granularity that captures when a reference must be explicit and when it can be implicit, which can only be obtained via empirical studies.

In addition to granularity, a notion of conciseness would be desirable, which could be used to check if a proof is more involved than necessary. It is unclear, though, how such a notion could be captured formally.

Yet, we envision the approach presented in this paper to be a first step towards an integration of a scientific text editor with mathematical proof assistance systems. In particular, the deductive systems show how mathematical proof assistance systems can be employed via the strategies to prove raised lemmas. To achieve the overall goal, however, many additional problems must be tackled, among them the connection to a scientific text editor, and, most notably, a natural language analysis component that transforms the human-authored proofs into proofs in our representation language.

References

1. Andreas Abel, Bor-Yuh Evan Chang, and Frank Pfenning. Human-readable machine-verifiable proofs for teaching constructive logic. In Uwe Egly, Armin Fiedler, Helmut Horacek, and Stephan Schmitt, editors, *Proc. of the Workshop on Proof Transformation, Proof Presentations and Complexity of Proofs (PTP-01)*, pages 37–50. Universitá degli studi di Siena, 2001.
2. Serge Autexier, Christoph Benzmüller, Armin Fiedler, and Henri Lesourd. Integrating proof assistants as reasoning and verification tools into a scientific WYSI-WYG editor. In David Aspinall and Christoph Lüth, editors, *User Interfaces for Theorem Provers (UITP 2005)*, pages 16–39, 2005.
3. Robert G. Bartle and Donald Sherbert. *Introduction to Real Analysis*. Wiley, 2 edition, 1982.
4. N. G. de Bruijn. The mathematical vernacular, a language for mathematics with typed sets. In Nederpelt et al. [9], pages 865 – 935.
5. David Hilbert. Die Grundlegung der elementaren Zahlenlehre. *Mathematische Annalen*, 104:485–494, December 1930.
6. Xiaorong Huang. *Human Oriented Proof Presentation: A Reconstructive Approach*. Number 112 in DISKI. Infix, Sankt Augustin, Germany, 1996.
7. Fairouz Kamareddine, Manuel Maarek, and Joe Wells. MathLang: An experience driven language of mathematics. *Electronic Notes in Theoretical Computer Science*, (93C):138–160, 2004.
8. Fairouz Kamareddine and Rob Nederpelt. A refinement of de Bruijn's formal language of mathematics. *Logic, Language and Information*, 13(3):287–340, 2004.
9. R. P Nederpelt, J. H. Geuvers, and R. C. de Vrijer, editors. *Selected Papers on Automath*, volume 133 of *Studies in Logic and the Foundations of Mathematics*. Elsevier, 1994.
10. Rob Nederpelt. Weak Type Theory: A formal language for mathematics. Computing Science Report 02-05, Eindhoven University of Technology, Department of Math. and Comp. Sc., May 2002.
11. Tobias Nipkow. Structured Proofs in Isar/HOL. In Herman Geuvers and Freek Wiedijk, editors, *Types for Proofs and Programs (TYPES 2002)*, volume 2646 of *LNCS*, pages 259–278. Springer, 2003.
12. Aarne Ranta. Grammatical framework — a type-theoretical grammar formalism. *Journal of Functional Programming*, 14(2):145–189, 2004.
13. Jörg Siekmann, Christoph Benzmüller, Vladimir Brezhnev, Lassaad Cheikhrouhou, Armin Fiedler, Andreas Franke, Helmut Horacek, Michael Kohlhase, Andreas Meier, Erica Melis, Markus Moschner, Immanuel Normann, Martin Pollet, Volker Sorge, Carsten Ullrich, Claus-Peter Wirth, and Jürgen Zimmer. Proof development with ΩMEGA. In Andrei Voronkov, editor, *Proceedings of the 18th International Conference on Automated Deduction (CADE-18)*, number 2392 in LNAI, pages 144–149, Copenhagen, Denmark, 2002. Springer.
14. M. Wenzel and F. Wiedijk. A comparison of the mathematical proof languages Mizar and Isar. *Journal of Automated Reasoning*, (29):389–411, 2002.

Processing Textbook-Style Matrices

Alan Sexton and Volker Sorge

School of Computer Science, University of Birmingham, UK
{A.P.Sexton, V.Sorge}@cs.bham.ac.uk
http://www.cs.bham.ac.uk/~aps|vxs

Abstract. In mathematical textbooks matrices are often represented as objects of indefinite size containing abbreviations. To make the knowledge implicitly given in these representations available in electronic form they have to be interpreted correctly. We present an algorithm that provides the interface between the textbook style representation of matrix expressions and their concrete interpretation as formal mathematical objects. Given an underspecified matrix containing ellipses and fill symbols, our algorithm extracts the semantic information contained. Matrices are interpreted as a collection of regions that can be interpolated with a particular term structure. The effectiveness of our procedure is demonstrated with an implementation in the computer algebra system Maple.

1 Introduction

Mathematical texts often employ intuitive and diagrammatic representations to adequately describe mathematical objects. Correctly interpreting these mathematical representations and making them available to electronic mathematical knowledge management is a crucial task for bridging the gap between informal and formal mathematics. For instance in every day mathematical practice, matrices are often not given as fully specified objects, but rather are of indefinite dimension and are depicted with abbreviations and underspecified parts such as ellipses. In this paper, we give an algorithm that enables the semantic interpretation of a large class of underspecified matrices, and facilitates their translation into a formal logical representation as well as their use in symbolic computation.

Our work is inspired by [7], which gives a taxonomy of different types of matrices occurring in the literature with a treatment of some of them in an automated reasoning context. While [7] focuses mainly on the proper representation of diverse matrix representations as formal lambda-calculus expressions it also contains a preliminary study on how a limited number of elliptical constructs can be correctly interpreted as lambda expressions. In this paper we now show how a much more general class of underspecified matrices can be represented as a mutually constrained set of linearly interpolated regions. We present a novel parsing algorithm that can handle elliptical expressions, which reflect those that can be found in mathematical textbooks. It returns a data structure called an *Abstract Matrix* that captures the constraints implied by the textbook representation. The resulting data structure can subsequently be used to generate a corresponding formal term in lambda-calculus. We have implemented

M. Kohlhase (Ed.): MKM 2005, LNAI 3863, pp. 111–125, 2006.

our approach as a package in the computer algebra system Maple [1, 4]. As an example of the kind of matrices our method can handle, consider the two matrices below. While matrix (1) is relatively straightforward to deal with, matrix (2) poses slightly more complicated challenges for its correct interpretation. In particular, our algorithm will deduce that the matrix is, indeed, square and that the diagonally opposite triangles are of equal size.

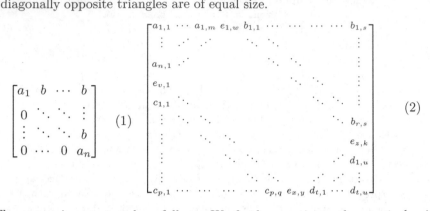

$$\begin{bmatrix} a_1 & b & \cdots & b \\ 0 & \ddots & \ddots & \vdots \\ \vdots & \ddots & \ddots & b \\ 0 & \cdots & 0 & a_n \end{bmatrix} \quad (1)$$

$$\begin{bmatrix} a_{1,1} & \cdots & a_{1,m} & e_{1,w} & b_{1,1} & \cdots & \cdots & \cdots & \cdots & b_{1,s} \\ \vdots & & & & & & & & \vdots \\ a_{n,1} & & & & & & & & \vdots \\ e_{v,1} & & & & & & & & \vdots \\ c_{1,1} & & & & & & & b_{r,s} \\ \vdots & & & & & & & e_{z,k} \\ \vdots & & & & & & & d_{1,u} \\ \vdots & & & & & & & \vdots \\ c_{p,1} & \cdots & \cdots & \cdots & \cdots & c_{p,q} & e_{x,y} & d_{t,1} & \cdots & d_{t,u} \end{bmatrix} \quad (2)$$

The paper is structured as follows: We further motivate the particular issues arising when interpreting abstract matrices in Sec. 2. We then give a brief overview of the elliptical constructs and the input syntax we support (Sec. 3). The algorithm described in Sec. 4 performs the initial parsing and uses a weighted directed graph to construct a set of constraints on the lengths of the ellipses. A second algorithm identifies all closed polyline regions bounded by ellipses and concrete terms (Sec. 5). These in turn can be expressed by a set of constraints and often be made more precise by taking into account information that can be gained from terms defining endpoints of ellipses. The information is extracted using a unification-like algorithm to generate *generalised terms* for the terms on the boundaries of each region (Sec. 6). Finally, in Sec. 7 we present how the results of our algorithm can be employed for constructing formal lambda expressions.

2   Analysing Abstract Matrices

While computing with only partially specified matrices is mathematically routine, there is very limited automated support for this. Computer algebra systems very efficiently deal with numerical or symbolic computations on concrete matrices of nearly arbitrary size, but they offer little functionality in the area of specifying matrices in terms of their general patterns without supplying all the concrete values immediately. Maple [1, 4], for instance, provides functionality to easily specify matrices with some predefined shapes, such as anti-symmetrical or hermitian. However, their dimension has to be concretely given and the arbitrary specification of shapes or combinations of shapes is not possible.

We introduce a new data type, which we call an *Abstract Matrix* that extends the usual matrix data types to the style of matrix objects and expressions that frequently appear in mathematical texts: namely the dimensions of the matrices may be unknown and the matrix may include ellipses and fill terms. An abstract matrix, therefore, can be seen as a template for a class of concrete matrices that could be instantiated from the abstract matrix. For example matrix (1) can be considered as a template of the class of all square matrices of the above shape. The concrete instantiations for dimension n where $1 \leq n \leq 4$ are then

$$
(a_1), \quad
\begin{pmatrix} a_1 & b \\ 0 & a_2 \end{pmatrix}, \quad
\begin{pmatrix} a_1 & b & b \\ 0 & a_2 & b \\ 0 & 0 & a_3 \end{pmatrix}, \quad
\begin{pmatrix} a_1 & b & b & b \\ 0 & a_2 & b & b \\ 0 & 0 & a_3 & b \\ 0 & 0 & 0 & a_4 \end{pmatrix}.
$$

It is not sufficient to attempt to capture the nature of the expandability of an abstract matrix merely by considering the ellipses in isolation. The regions bounded by the ellipses must also be dealt with. We therefore choose a representation for abstract matrices that consists of a set of regions. These regions correspond to the expandable triangles or rectangles that are commonly seen in matrix terms in mathematical text books. However, for full generality, we allow regions to correspond to arbitrary closed polylines in the input matrix expression and support the degenerate cases of single ellipsis (i.e., one-dimensional) and single term (i.e., no ellipsis) regions.

The defining characteristic of a region is that there is a single template, which we call a *generalised term*, for the terms that can fill the region as it is expanded. If the generalised term is a simple constant, it can be entered in all the matrix cells created as the region is expanded. However, the situation is more complex if the contents varies over the region. Here, common mathematical practice is to specify a sequence of terms, following a simple pattern, by specifying the terms at either end of an ellipsis, e.g., $1, \ldots, n$, or $a_n^1, \ldots, a_1^n$. These patterns are almost always in the form of a general term with sub-terms which vary linearly as the ellipsis is expanded.

To cope with such patterns, we allow generalised terms to contain variables called *unification variables*. These generalised terms must be unifiable with the concrete terms at the ends of the ellipsis. Thus, as an ellipsis is expanded, the resulting cell contents can be filled with a generalised term whose unification variables are instantiated by interpolating values for them from the relative position of the cell in the expanded ellipsis and the corresponding values that the unification variables must have at the start and end of the ellipsis in order to match the pattern specified.

This is satisfactory for individual ellipses, but there remains a problem for regions enclosed by a closed polyline of ellipses. Here the same generalised term should be used to fill in the expanding region as fills in the cells along each boundary ellipsis. This is only possible if the generalised term is unifiable with all concrete terms on the boundaries of the region and, furthermore, if all the ellipses along the boundary are mutually compatible with the generalised term. Since we restrict these patterns to linear sequences, this corresponds to requiring, for each unification variable in a region, all 3-dimensional points defined by the row number and column number of the expanded matrix cell and the interpolated value of the unification variable, to fit on a common plane. Determining if the boundary ellipses of a region are mutually compatible consists of checking, for each unification variable in the generalised term for the region, that the 3-dimensional points obtained as described above for the terms on the ends of the boundary ellipses do, in fact, sit on the same plane.

In an abstract matrix, not only do we not know the length of the ellipses, but we often do not concretely know the location of either of the end cells of an ellipsis because those locations will typically depend on the length of other ellipses in the matrix. For this reason, our cell locations have to be defined as integer expressions on ellipsis length variables. These locations we call *generalised positions*.

Thus our regions have three parts: (a) a closed polyline of generalised positions, (b) a generalised term with a number (possibly 0) of unification variables, and (c) for each concrete term on the boundary of the region, a structure containing the generalised position of the term and the list of instantiation values that corresponds to each unification variable of the generalised term.

We have implemented a parsing algorithm that constructs the abstract matrix data structure from the input syntax defined in the next section. The algorithm has to (a) identify the ellipses in the input matrix, (b) construct the generalised positions of each concrete term in the input matrix as functions of the lengths of the ellipses, (c) identify regions by finding minimal closed polylines in the graph of ellipses and concrete terms, (d) apply a unification-like algorithm to all the concrete terms on the region boundary to construct disagreement sets from which the generalised terms and corresponding instantiation values can be derived, and (e) package the results up into an abstract matrix object. The lambda calculus expression generation algorithm then takes an abstract matrix, finds appropriate interpolation functions for its regions and generates a lambda calculus expression suitable for use in other mathematical knowledge management tools such as theorem provers, mathematical knowledge bases etc.

3   Syntax for Abstract Matrices

In this section we define our concept of an abstract matrix by specifying what types of ellipses are valid in our context. In the literature, there are many inconsistent ways of describing classes of matrices by patterns involving ellipses and concrete terms. For our algorithm, and in particular for its Maple implementation, we have chosen one that is general enough to allow us to express most matrix classes we have seen in a variety of texts on linear algebra and matrix analysis (see [5] for example). However, it is restricted enough to allow for as little ambiguity in the interpretation of the input as possible.

More formally we define an *Abstract Matrix* input term to be a rectangular arrangement of symbols in rows and columns, where each symbol is either

1. an ordinary, or *concrete* term
2. a vertical, horizontal, diagonal, or anti-diagonal ellipsis
3. a single dot, or
4. a fill term.

We generally refer to symbols of the type 2–4 as elliptic constructs. Ellipses of type 2 are the ordinary suspension points "...". Every ellipsis must be terminated at both ends by a concrete term within the confines of the input matrix. Furthermore, the line of ellipsis symbols separating the two terminating concrete terms of an ellipsis must all be the same and compatible with the direction of the said line (e.g., a set of horizontal ellipsis suspension points could not be followed directly by those of a vertical ellipsis). Any diagonal or anti-diagonal ellipsis must be instantiated to a sequence of cells which have the same vertical and horizontal extent.

A single dot can occur in exactly two different situations: (a) Inside a closed polyline of ellipses and concrete terms it signifies that the region is filled homogeneously with terms determined by the boundary of that region. (b) If it occurs next to a fill term or is only separated from a fill term by other single dots, it denotes the expansion of that fill term into the region occupied by single dots.

A fill term, finally, denotes that a certain region in a matrix contains only this particular term as elements. In the literature it is usually denoted as an enlarged term. Fill terms can fill entire regions without explicit boundaries or can fill a region that is bounded by terms dissimilar to the fill term. It is illegal to have two different fill terms within the same region.

As examples for the different forms of elliptical constructs allowed in our syntax we consider the following three matrices, where matrix (3) is the same as matrix (1), only written with fill terms.

$$
\begin{bmatrix} a_1 & \mathbf{b} \\ & \ddots & \\ \mathbf{0} & & a_n \end{bmatrix} \quad (3)
\qquad
\begin{bmatrix} 1 & \cdots & \cdots & 1 \\ \vdots & & \cdot^{\cdot} & 0 \\ \vdots & \cdot^{\cdot} & \cdot^{\cdot} & \vdots \\ 1 & 0 & \cdots & 0 \end{bmatrix} \quad (4)
\qquad
\begin{bmatrix} 1 & \cdots & \cdots & 1 \\ \vdots & \mathbf{0} & \cdot^{\cdot} & 0 \\ \vdots & \cdot^{\cdot} & \cdot^{\cdot} & \vdots \\ 1 & 0 & \cdots & 0 \end{bmatrix} \quad (5)
$$

In a concrete implementation we have to fix the syntax of our elliptical constructs as well as to impose certain requirements on the overall form of the matrix. For our Maple implementation we still require the input matrix to be rectangular but it may contain concrete terms as well as distinguished ellipsis symbols. The ellipsis symbols are **hdots**, **vdots**, **ddots**, and **adots**, which correspond, respectively, to horizontal, vertical, diagonal and anti-diagonal ellipses. The single dot can be given by **dot** and a fill term such as **0** is specified by the distinguished *fill* function: **fill(0)**. Thus the above matrices could be respectively entered in Maple as:

```
[[a(1)   ,dot  ,fill(b)],    [[1      ,hdots,hdots,  1  ],   [[1      , hdots ,hdots,  1  ],
 [dot    ,ddots,dot    ],     [vdots, dot ,adots,  0  ],      [vdots,fill(0),adots,  0  ],
 [fill(0),dot  ,a(n)   ]]     [vdots,adots,adots,vdots],      [vdots, adots ,adots,vdots],
                              [1    ,  0  ,hdots, 0  ]]       [1      ,  0   ,hdots, 0  ]]
```

4 Ellipsis Lengths

Initial parsing will have checked that the input abstract matrix is well formed and will have identified all ellipses. However, we need to determine the lengths of all ellipses as functions of each other and of the dimensions of the intended concrete instantiations of the abstract matrix. For example, the 3 ellipses in the upper left triangular region of abstract matrix (4) are of length which is the same as the width of the matrix, which is the same as height of the matrix. The other ellipses are of length one less than the width of the matrix.

We call these equations *structural constraints*, as they involve equations on ellipsis length and matrix dimension variables only and not on sub-terms of terms in the cells of the matrix.

To discover this set of structural constraints, we consider all paths through the matrix. Each path can traverse an ellipsis, paying a cost represented by the length of the ellipsis, or can traverse a concrete term, paying a cost of 1. There may be more than one sub-path found between any two positions in the matrix and the equation extracted by identifying the costs on these different paths contribute to a set of equations which, when simplified, provide the constraints we seek.

The process works in two phases. The first phase is to construct a weighted directed graph which represents all possible paths, as described above, through the abstract matrix. In the second phase we analyse the graph to find pairs of different paths between the same vertices.

4.1   Phase 1: Graph Construction

Consider the topmost row in abstract matrix (2). One might be tempted to write the horizontal path across that row as $a_{1,1} \rightarrow a_{1,m} \rightarrow e_{1,w} \rightarrow b_{1,1} \rightarrow b_{1,s}$ and define the horizontal cost of the path as $e_1 + 1 + 1 + e_2$ where e_1 and e_2 are constraint variables representing the costs of traversing the two ellipses. However, then the cost, when the constraint variables are eventually bound to integers, would not correspond to the number of cells in the row because, for concrete terms, we have counted the number of transitions from one cell to the next instead of the number of cells (i.e. 1 for a non-expandable cell). Hence we need to consider the costs as those of traversing between edges or corners of cells instead of between centres of cells.

We construct a graph whose vertices represent the corners of the cells in the input matrix. Thus cell (i, j) in the matrix has its upper left corner associated with vertex $\langle i, j \rangle$ in the graph and its lower right corner associated with vertex $\langle i + 1, j + 1 \rangle$. In our implementation we create the vertices lazily and do not create any that have no edges incident on them. Each edge in the graph will carry two weights: the horizontal cost of traversing the edge and the vertical cost of traversing it. We then add edges as follows:

1. For each cell (i, j) in the input matrix containing a concrete term, we add four edges: from top left to top right and from bottom left to bottom right, both with horizontal weights 1 and vertical weights 0, and from top left to bottom left and from top right to bottom right, both with horizontal weights 0 and vertical weights 1.
2. For each horizontal or vertical ellipsis, we add two edges: For a horizontal ellipsis $(i, j), \ldots, (i, k)$ where $k > j$, we add the edges: $\langle i, j \rangle \rightarrow \langle i, k + 1 \rangle$ and

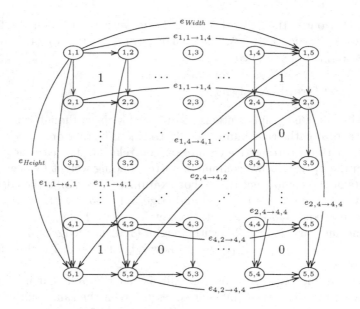

Fig. 1. Graph constructed for ellipsis length analysis for (4) (Edge weights not shown)

$\langle i+1,j \rangle \to \langle i+1,k+1 \rangle$, both with vertical weight 0 and horizontal weight encoded as a fresh variable $e_{i,j \to i,k}$.

For a vertical ellipsis $(i,j), \ldots, (k,j)$ we construct two edges $\langle i,j \rangle \to \langle k+1,j \rangle$ and $\langle i,j+1 \rangle \to \langle k+1,j+1 \rangle$, both with vertical weight 0 and horizontal weight encoded as a fresh variable $e_{i,j \to k,j}$.

3. For each diagonal or anti-diagonal ellipsis, we add a single edge: For a diagonal ellipsis $(i,j), \ldots, (m,n)$ where $m > i$ and $n > j$, we add a single edge connecting the extreme vertices: $\langle i,j \rangle \to \langle m+1,n+1 \rangle$ with both vertical and horizontal weights encoded as the same fresh variable $e_{i,j \to m,n}$.

The anti-diagonal is constructed similarly as an edge from the extreme top right corner to the bottom left corner except that the horizontal weight is the negative of the vertical weight.

4. We add two extra edges to capture the width and height of the matrix. If the input matrix has n rows and m columns, we add the edge $\langle 1,1 \rangle \to \langle 1, m+1 \rangle$ with horizontal weight e_{Width} and vertical weight 0 and another edge $\langle 1,1 \rangle \to \langle n+1,1 \rangle$ with horizontal weight 0 and vertical weight e_{Height}.

In Figure 1, we show the graph constructed for the abstract matrix input (4). The vertices are labelled with their names inside ovals. To show the relationship with the original input matrix, the matrix cell terms are shown in their appropriate position relative to their cell corners (which correspond to the graph vertices). To improve readability, we have not shown the weights on the edges but simply labelled the edges with their variable names if they have variable names. However, edge weights are easily recoverable by applying the steps of the graph construction. Thus if we describe the weights as a pair (v, h) where v is the vertical weight and h the horizontal, then edge $1,1 \rightarrow 1,2$ has weights $(0,1)$, edge $1,2 \rightarrow 2,2$ has weights $(1,0)$, edge $1,2 \rightarrow 5,2$ has weights $(e_{1,2\rightarrow5,2}, 0)$, edge $1,5 \rightarrow 5,1$ has weights $(e_{1,5\rightarrow5,1}, -e_{1,5\rightarrow5,1})$ etc.

4.2 Phase 2: Graph Analysis

The resulting directed graph contains no directed loops; although it is not a tree as it will contain paths that bifurcate only to rejoin further on.

We next perform a transitive closure on the relation. For each new edge we produce in the operation, we compute a corresponding weight by accumulating the weights of the component edges. For example, we combine the edge $1,5 \rightarrow 5,1$ with weight $(e_{1,5\rightarrow5,1}, -e_{1,5\rightarrow5,1})$ and edge $5,1 \rightarrow 5,2$ with weight $(0,1)$ to produce a new edge $1,5 \rightarrow 5,2$ with weight $(e_{1,5\rightarrow5,1}, 1 - e_{1,5\rightarrow5,1})$. Similarly, we combine the edge $1,5 \rightarrow 2,5$ with weight $(1,0)$ and edge $2,5 \rightarrow 5,2$ with weight $(e_{2,5\rightarrow5,2}, -e_{2,5\rightarrow5,2})$ to produce a new edge $1,5 \rightarrow 5,2$ with weight $(1 + e_{2,5\rightarrow5,2}, -e_{2,5\rightarrow5,2})$.

We then produce equations on the ellipsis lengths and matrix dimensions by identifying the weights on different edges between the same pair of vertices. From the previous example, we have two edges $1,5 \rightarrow 5,2$ so we obtain the two equations $e_{1,5\rightarrow5,1} = 1 + e_{2,5\rightarrow5,2}$ and $1 - e_{1,5\rightarrow5,1} = -e_{2,5\rightarrow5,2}$ by equating the respective vertical and horizontal weights.

Finally we simplify the resulting set of equations to get our required constraints on the relative lengths of all the ellipses and the dimensions of the matrix.

5 Ellipsis Regions

As described in Sec. 2, an abstract matrix is a collection of regions where a region is a closed polyline of generalised positions, with a generalised term, possibly containing unification variables, and, for each concrete term on the boundary of the region, the generalised position of the term and the instantiation values, corresponding to each unification variable. In this section we describe the algorithm to construct regions given the graph analysis results of the previous section.

The polylines of a region correspond to the minimal ellipsis cycles that appear in the input matrices. For example, in abstract matrix (4), there are two regions: the upper left 1 triangle and the lower right 0 triangle. We also interpret single cells that are not terminal cells of any ellipsis as regions as well as any single ellipsis that is not part of a closed polyline such as, for example, the major diagonal ellipsis of abstract matrix (1). The size and position of a concretised instantiation of a region can (and usually does) vary with the size and other parameters specified when an abstract matrix is concretised. A region corresponding to a cell that is not a terminal cell to an ellipsis can not vary in size, but can vary in position. A single ellipsis region can vary in size only along its ellipsis direction whereas other regions can vary in size in two dimensions.

The algorithm starts by identifying, for each cell containing a concrete term, a *generalised position* in the abstract matrix input. The generalised position is simply a pair of expressions over integers, ellipsis length and dimension variables (essentially one half of a structural constraint) that define the row and column position index of the cell. The algorithm is essentially the same for finding the row and for finding the column components of the generalised position, with the obvious modifications, so we describe only that for the row. There are three cases we must consider for the algorithm:

First — the column component of the generalised position is normally obtained by finding, in the graph relation, any edge from a vertex on the left boundary of the matrix to a vertex on the left boundary of this cell. The horizontal weight on the edge is the required column component.

Second — under some circumstances, there may be no such edge: consider, for example, a matrix consisting of two rectangular blocks side by side, where the left block is defined by a fill term. In this situation, there may be no path in the graph from the left edge of the matrix to a concrete term of the right hand block in the matrix. In such a case, we can use any edge from a vertex on the left boundary of the cell to a vertex on the right boundary of the matrix and subtract the associated horizontal weight from the width of the matrix (which is always available as the variable e_{Width}).

Finally — there is one case where there might be no suitable edge at all, e.g., 3 side by side rectangular blocks where the two outside blocks are fill terms: in this case the inner block "floats" horizontally in the matrix. Thus we need to introduce a new unknown to capture the horizontal location of the cell. We do this in the same way we introduced the e_{Width} and e_{Height} variables which capture the width and height of the matrix: add a new edge to the graph terminating at the top left vertex of the cell and with horizontal width set to a fresh variable. This edge is slightly unusual, however, because it can not be anchored at any specific vertical position. Otherwise it imposes constraints on the relative vertical position of this cell and other cells in the matrix for whom no such constraint should be imposed. Hence we provide a vertical weight of *null* on the edge, which is interpreted by the graph transitive closure algorithm as a non-existent vertical edge. Now we can safely anchor the source of the edge on the vertex $\langle 1, 1 \rangle$. Finally, we have to re-execute the transitive closure operation and consequent constraint extraction and simplification as described in Sec. 4 to integrate this new variable into our system of equations and associate any other cells in the same floating section of cells with the new variable.

Once we have obtained the required generalised positions, we need to identify all regions in the input matrix. This is done by, for each ellipsis in the input matrix, traversing the path obtained by taking the anti-clockwise-most sequence of ellipses until we return to the starting point, then doing the same for the clockwise-most path. The resulting collection of paths must have duplicate paths removed and then form the set of region boundaries. Some ellipses may not participate in closed paths and, instead, form one-dimensional regions.

One final complication is that of regions defined by fill terms. Here we simply start from a cell containing a fill term and proceed along a single direction until we find a region boundary or the edge of the input matrix. We then use the above path traversal algorithm to find the region boundaries as before but with two modifications. The first is that, in the absence of an ellipsis along an edge of the matrix which bounds the region, the edge of the matrix itself is taken as a valid region boundary: this is not the case for non-fill regions where the absence of a suitable bounding ellipsis is considered to be an error.

The second modification is that a fill region extends up to but not including its boundary ellipses, unlike normal regions which share ellipses along their boundaries. In the latter case a shared ellipsis belongs to both regions. This, of course, does not cause a conflict because both regions would interpolate the same terms for the cells instantiated from such an ellipsis. Finally it is an input error if any region contains more than one fill term.

6   Generalised Terms

Regions contain a generalised term which may contain unification variables and which must be unifiable with all concrete terms on the boundaries of the region.

Generalised terms are constructed using a unification-like algorithm, which is an extended version of the one already mentioned in [7]. We impose certain restrictions on generalised terms that are in line with those already mentioned in Sec. 3. In particular, generalised terms are created purely with respect to syntactical similarity in order to avoid the ambiguity of more sophisticated conditions.

The construction of generalised terms is based on a general unification algorithm for first order logic [3,9]. But, unlike regular unification, our algorithm does not try to compute a most general unifier, but rather produces a minimal disagreement set for the terms under consideration. Given a list of input terms $[t_1, \ldots, t_n]$ (i.e., all the boundary terms of a given region) the algorithm recursively constructs a disagreement set D and a generalised term t.

We first initialise the disagreement set $D = \emptyset$. We then recursively and simultaneously traverse the terms $[t_1, \ldots, t_n]$ as follows:

1. If $t_i = x$ for all $i = 1, \ldots, n$ and some term x, return x and D.
2. If $t_1 = x_1, \ldots, t_n = x_n$ where all the x_i, $i = 1, \ldots, n$ are either integers or variables (i.e., not functional terms), return a new variable symbol α and $D := D \cup \{\alpha \mapsto [x_1, \ldots, x_n]\}$.
3. If $t_1 = f(x_{1_1}, \ldots, x_{1_m}), \ldots t_n = f(x_{n_1}, \ldots, x_{n_m})$ for some function symbol f of arity m and terms x_{i_j}, $i = 1, \ldots, n$, $j = 1, \ldots, m$, then compute generalised terms $y_1, \ldots, y_m$ together with disagreement sets $D_1, \ldots, D_n$ for the term lists $[x_{1_1}, \ldots, x_{n_1}], \ldots, [x_{1_m}, \ldots, x_{n_m}]$, respectively, and return $f(y_1, \ldots, y_m)$ and $D := D \cup D_1 \cup \ldots \cup D_m$.
4. In all other cases fail.

If the algorithm succeeds, the returned generalised term can contain new variable symbols of the form α, which we call *unification variables*. Each of these unification variables has a corresponding mapping in the disagreement set, relating it to the different instantiations for the original terms. We call this mapping the *range* of the unification variable. Note that we implicitly assume our unification variables to be of integer type.

As an example, suppose we want to compute the generalised terms for the following triangular region:

$$a_{1,1} \cdots a_{1,m}$$
$$\vdots \quad \cdot \cdot$$
$$a_{n,1}$$

The boundary terms $a_{1,1}, a_{1,m}, a_{n,1}$ would be given in our term representation as a(1,1), a(1,m), and a(n,1). When applying our algorithm to these terms it first considers case 3, since all terms are of functional type and have the same head symbol a. Thus in the next recursion the algorithm tries to compute generalised terms for the two sets [1,1,n] and [1,m,1]. For the former set we can apply case 2, which introduces a new unification variable α. and the disagreement set $D_1 = \{\alpha \mapsto [1,1,n]\}$. Similarly, for the latter set we get a unification variable β and a disagreement set $D_2 = \{\beta \mapsto [1,m,1]\}$. Thus the algorithm yields the generalised term a(α,β) and $D = D_1 \cup D_2$ as result.

The algorithm automatically also provides us with information on how to instantiate the unification variables in order to regain the original boundary terms. We call these the *instantiation values* for a particular term. For instance, the instantiation values for a(1,m) would be [1,m].

After the generalised term for a region has successfully been computed, the resulting disagreement set is checked for trivial inconsistencies, that is ranges of unification variables in the disagreement set that are purely numerical and would lead to non-integer indices for interpolating terms. For example, an ellipsis of the type a(1,2) $\cdots$ a(3,1) would be considered not valid, since the α of an interpolated a(2,α) could not be integral. If any inconsistency is found the matrix is rejected as invalid. If there are no inconsistencies, we take the instantiation values for each boundary term and associate them with the generalised position for the respective term.

Case 2 of our algorithm determines what kind of terms are allowed for the range of an ellipsis. In its current form it only allows integers and single variables but no functional terms, even those containing only arithmetic operations. For instance, we do not allow ellipses of the form $x_1 \ldots x_{n+m}$. In the future, it could be desirable to extend case 2 in order to allow a larger spectrum of terms. However, the extensions would have to be done carefully in order to avoid ambiguities that could make it impossible to compute the actual size of ellipses and regions.

7   Formal Interpretation of Abstract Matrices

At this point, the structure we have is a complete, fully analysed abstract matrix. It contains a set of structural constraints and a set of regions, where each region has been refined to include the instantiation values of the generalised term.

If we now want to transform the matrix into a formal expression we have to interpret the semantics of an abstract matrix by specifying the actual content of each region. We do this by interpolating the terms of each regions in a manner consistent with the terms along the boundary of the region and the generalised positions in the matrix. This is trivial in the case where a region consists of only a single (constant) term. However, in the case when the region has term constraints attached (i.e., the generalised term has actual unification variables that have to be instantiated correctly), interpolation is a non-trivial affair. The result of the interpretation process is an expression in lambda calculus involving *if-then-else*, which are of the general form presented in [7]. While the expressions produced by our algorithm are simply typed, for clarity we explain them here in untyped form. For instance, matrix (1) can be translated into the lambda term:

$$\lambda i \lambda j. \quad \begin{array}{lll} \text{if} & i = j & \text{then } a(i) \\ \text{if} & i < j & \text{then } b \\ \text{else} & & 0 \end{array}$$

Thus the matrix is a function in the index variables i and j, the main diagonal consists of a expressions, where a is a function in one index variable, and above and below the diagonal we have the terms b and 0, respectively.

While the above representation of matrix (1) is relatively simple, in the general case both the restrictions of the index function (the conditions in the if-then-else expression) and the functions in the index variables (representing the terms of the region) are far more complicated. Both can be treated independently. In the following we describe the general idea of these algorithms using the upper left triangle of matrix (2) as an example. The parsing algorithm constructs a graph with 48 vertices and 82 edges and subsequently discovers that the diagonally opposite triangles in the matrix are of the same size and that the matrix is square. The relevant part of the shape structure for the triangle in question consists of the following set of structural and sub-term constraints:

$$\left\{ \begin{array}{l} e_{1,1\to1,3} = e_{1,1\to3,1}, \\ e_{1,3\to3,1} = e_{1,1\to1,3} \end{array} \right\} \quad (6) \qquad \left[\begin{array}{l} [\,[1,1],\ [\alpha,\beta] \mapsto [1,1]\,], \\ [\,[1,r],\ [\alpha,\beta] \mapsto [1,m]\,], \\ [\,[q,1],\ [\alpha,\beta] \mapsto [n,1]\,] \end{array} \right] \qquad (7)$$

Here q, r represent the end points of the horizontal and vertical ellipses in the respective generalised positions. In fact, from the structural constraints it immediately follows that $q = r$ and we can therefore restrict ourselves to only using q in both expressions from now on. $e_{1,1\to1,3}$, etc. stand for the length of the ellipses involved as explained in Sec. 4.

Restricting the index function: For each ellipsis we can give the index function for the regions on either side of it with respect to the generalised positions for the endpoints of that ellipsis. The following diagram depicts the corresponding general inequalities for our four different ellipses. The generalised positions are denoted by the variables q_1, q_2, r_1, r_2:

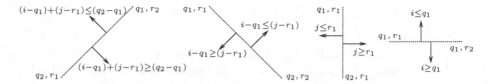

The index function can then be restricted to the region of a convex closed polygon by simply conjoining the single inequalities on the interior of all ellipses on the boundary (concave closed polygons pose other problems discussed below). For our example region we can combine the inequalities for the three enclosing polygons to obtain the condition $j \geq 1 \wedge i \geq 1 \wedge i + j - 2 \leq q - 1$.

Interpolating the region: There still remains the question of how to interpolate over a two dimensional region. In our example, we have a triangle, but we might have an arbitrarily shaped region with an arbitrary number of terms specified on its boundary. We use a plane fitting algorithm to interpolate the values independently for each unification variable of the generalised term. More precisely, we treat each point of the region shape as a point in 3-d space, where the row and column expressions correspond to the x and y dimensions and the difference sub-term corresponds to the z dimension. We then solve for any three non-colinear points of our shape the following standard plane equation:

$$\begin{vmatrix} x - x_1 & y - y_1 & z - z_1 \\ x_2 - x_1 & y_2 - y_1 & z_2 - z_1 \\ x_3 - x_1 & y_3 - y_1 & z_3 - z_1 \end{vmatrix} = \begin{vmatrix} i - x_1 & j - y_1 & p - z_1 \\ x_2 - x_1 & y_2 - y_1 & z_2 - z_1 \\ x_3 - x_1 & y_3 - y_1 & z_3 - z_1 \end{vmatrix} = 0$$

Solving this equation for p yields an interpolation function in the two index variables i and j. In the case when we have more than one unification variable, we have to find a different interpolation function for each variable by solving the corresponding equations. If we have a shape that is determined by more than three points, we compute interpolation functions using 3 non-colinear points and see if they are consistent with the remaining points. If not, the given abstract matrix is not valid.

In our example region we have three points given and have to compute two interpolation functions, one for each of the unification variables α and β. Below, the table on the left relates the x and y coordinates to the values of the unification variables. Solving the equation of the plane on the right then yields the first interpolation function $p_1(i, j)$.

x	y	α	β
1	1	1	1
1	q	1	m
q	1	n	1

for α: $\begin{vmatrix} i - 1 & j - 1 & p_1 - 1 \\ 1 - 1 & q - 1 & 1 - 1 \\ q - 1 & 1 - 1 & n - 1 \end{vmatrix} = 0 \Rightarrow p_1 = \dfrac{in - i - n + q}{q - 1}$

Similarly we can compute an interpolation function $p_2(i, j) = \frac{jm - j - m + q}{q - 1}$ for the unification variable β. This then gives us the part of the lambda expression for the matrix that describes the upper left triangle in our matrix (2)

$\lambda i \lambda j.$ if $(j \geq 1 \wedge i \geq 1 \wedge i + j - 2 \leq q - 1)$ then $a(\frac{in-i-n+q}{q-1}, \frac{jm-j-m+q}{q-1})$
if ...

Here q and n are free variables. Note that $n = q$ does not necessarily hold since we could, for instance, set $q = 3$ and $n = -1$ and would then get indices $1, 0, -1$. Subsequently we can compute the entire lambda term but omit it here to preserve space. The computed lambda term is the most general function characterising the given abstract matrix and contains all possible interpretations of the elliptical constructs with respect to the variables given in the sub-term constraints. It is therefore entirely possible that not all instantiations of q and n in the expression will be meaningful in an intuitive sense (e.g., some instantiations might lead to fractions as indices).

The algorithm can return a lambda expression for most matrices accepted by our parsing algorithm, specifically, for all those where the non-degenerate regions are convex. The interpolation function for a concave region is not a problem as it depends only on being able to find 3 non-colinear points. However, the expression for describing the interior of such a region does pose difficulties because of the implied disjunctions which may depend on under-constrained generalised positions. Our algorithm currently handles a number of cases for concave regions. For example, if there is only one concave region, it can be left to an **else** clause of the expression. We also handle some other special cases but a general solution awaits further work.

Even if we cannot generate lambda expressions for every possible abstract matrix, we can exploit our ability to interpolate over arbitrary regions by using abstract matrices as templates for concrete matrices in Maple. This process, called *concretisation*, consists of stepwise constraining the free variables and interpolation function until a set of fully solved constraints is obtained. The set is consistent if all regions can be interpolated such that all unification variables are only instantiated by integer values of increment or decrement of one. For more detail on the concretisation see [8].

8   Conclusion and Future Work

We have presented an approach to interpreting textbook style matrix expressions to make them available for electronic mathematical knowledge management. Our algorithm extracts the semantic information of abstract matrix expressions by analysing the patterns formed by ellipses in the matrix and the terms that they connect. This results in a set of constraints on the relative positions of regions in a matrix, which can be exploited to construct lambda expressions for the abstract matrix for further formal treatment and to compute concrete instances. The implementation of our ideas in Maple demonstrates the feasibility of our approach. It can handle matrices in a syntax that already very closely resembles ordinary mathematical notation. Nevertheless, we do not allow for all possible notational variations that can be found in the literature. One strand of future work will be to extend the input syntax as to allow for more freedom combining ellipsis. More significantly, our system can handle simple block matrices, but matrices where ellipses range over blocks rather than just terms, such as the Jordan normal form, will require further work.

We believe that our work can then serve as a bridge between the direct input of mathematical text via OCR methods and its computational treatment in mathematical software systems such as computer algebra systems or theorem provers. In this respect related to our work is the network based parsing algorithm presented by Kanahori and Suzuki in [6] for the analysis of matrix structures in the context of optical character recognition of mathematical texts. It can analyse structural elements of a matrix, detect separate regions and compute their size relative to each other using a system of simultaneous equations. The algorithm does not, however, use any semantical information given in the concrete elements of the matrix in order to further constrain the size of regions and therefore does not lend itself easily to the type of concretisation of abstract matrix objects as presented in this paper.

We have implemented the abstract matrix data type as well as the algorithm to create concrete instances in a Maple package. The package is intended as a first step towards more comprehensive computational treatment of abstract matrices. In particular we intend to develop the algorithms for elementary operations on and with abstract matrices, such as matrix addition, multiplication, etc. Some preliminary work in this direction has been done by Fateman in Macsyma [2], in which indefinite matrices can be subjected to some basic algebraic manipulations. While his matrices are indefinite in size, their elements are fixed to one particular functional expression and cannot be of arbitrary composition. Although Fateman presents some ideas how to exploit elliptic notation to enhance the display of indefinite matrices by using ellipses, the work does not deal with having unspecified elements and ellipses as input in the first place.

References

1. B. W. Char, K. O. Geddes, G. H. Gonnet, B. Leong, M. B. Monagan, and S. M. Watt. *Maple V: Language Reference Manual*. Springer, 1991.
2. R. Fateman. Manipulation of matrices symbolically. Available from `http://http.cs.berkeley.edu/ fateman/papers/symmat2.pdf`, January 9 2003.
3. M. Fitting. *First-Order Logic and Automated Theorem Proving*. Springer, 1990.
4. A. Heck. *Maple Manuals*. Springer, 3rd ed., 2003.
5. R. Horn and C. Johnson. *Matrix Analysis*. Cambridge University Press, 1990.
6. T. Kanahori and M. Suzuki. A recognition method of matrices by using variable block pattern elements generating rectangular areas. In *Graphics Recognition, LNCS* 2390. Springer, 2002.
7. M. Pollet, V. Sorge, and M. Kerber. Intuitive and formal representations: The case of matrices. In *Proceedings of MKM-2004, LNCS* 3119. Springer, 2004.
8. A. Sexton and V. Sorge. Semantic analysis of matrix structures. To appear in *International Conference in Document Analysis and Recognition*. IAPR, 2005.
9. W. Snyder. *A Proof Theory for General Unification*, volume 11 of *Progress in Computer Science and Applied Logic*. Birkhäuser, 1991.

A Generic Modular Data Structure for Proof Attempts Alternating on Ideas and Granularity

Serge Autexier[1,2], Christoph Benzmüller[1], Dominik Dietrich[1], Andreas Meier[2], and Claus-Peter Wirth[1]

[1] FR Informatik, Saarland University, Saarbrücken, Germany
{autexier, chris, dodi, cp}@ags.uni-sb.de
[2] DFKI GmbH, Stuhlsatzenhausweg, Saarbrücken, Germany
ameier@dfki.de

Abstract. A practically useful mathematical assistant system requires the sophisticated combination of interaction and automation. Central in such a system is the proof data structure, which has to maintain the current proof state and which has to allow the flexible interplay of various components including the human user. We describe a parameterized proof data structure for the management of proofs, which includes our experience with the development of two proof assistants. It supports and bridges the gap between abstract level proof explanation and low-level proof verification. The proof data structure enables, in particular, the flexible handling of lemmas, the maintenance of different proof alternatives, and the representation of different granularities of proof attempts.

1 Introduction

A careful and objective inspector of the history of automated theorem proving in the last fifty years would come to the following hypothesis:

> Stand-alone automated theorem provers will never develop into practically useful mathematical assistant systems.

To achieve the original design goal of a practically useful mathematical assistant system, we aim at *interactive* systems with a high degree of automated support. To combine interaction and automation into a synergetic interplay and to bridge between abstract level proof explanation and low-level proof verification is an enormous task. It requires sophisticated achievements from logic, tactics programming, proof planning, agent-based architectures, graphical user interfaces, and integration of other reasoning tools on the one hand, and a deeper experience in informal and formal human proof construction on the other hand.

The main task of the proof data structure in the center of such a system is to maintain the current states of the proof attempts with their open goals and available lemmas. To further the communication between a theorem proving system on the one hand and a human user, another system, or a proof archive on the other hand, an appropriate representation and transformation of proof

M. Kohlhase (Ed.): MKM 2005, LNAI 3863, pp. 126–142, 2006.
© Springer-Verlag Berlin Heidelberg 2006

attempts is necessary. In this paper we describe a new proof data structure (*PDS*) for this purpose. It generalizes both the existing PDS (with its features for granularity) of the ΩMEGA system [8, 16, 5] and the proof forests (with their alternative proof attempts and lemmatization) of the QUODLIBET system [3] and incorporates the experience gained in the last dozen years.

The basic ideas are:

- Each conjectured *lemma* gets its own *proof tree* (actually a directed acyclic graph (dag)).
- In this *proof forest*, each lemma can be applied in each proof tree; either as a lemma in the narrower sense, or as an induction hypothesis in a possibly mutual induction process, see [18].
- Inside its own tree, the lemma is a goal to be proved reductively. A *reduction* step reduces a *goal* to a conjunction of *sub-goals* w.r.t. a *justification*.
- Several reduction steps applied to the same goal result in alternative proof attempts, which either represent different proof *ideas* or the same proof idea with different *granularity* (or detailedness).

Although the application of a lemma of one tree (*generative* step) results in a *reductive* step inside another tree, we do not overemphasize reduction by this:

- For purely generative abstract theory expansion we may assume some trivial reductions, which can later be refined to the reductions that will be necessarily involved in this generation on the concrete level of a logical calculus.
- All steps in a traditional sequent or tableau calculus as well as *backward* and *forward* steps in Natural Deduction can be realized as reduction steps.

A parallel representation of *different granularities* of proof attempts is necessary for increasing granularity from proof sketches to the actual elaboration of the concrete proofs, and for decreasing the granularity from huge automatically generated proofs to tactical descriptions of a size that can be stored and archived. Moreover, the inspection of proof attempts by human users requires different granularities and the possibility to switch between them for size management and modular focusing according to their varying intentions and different expertise. An important new feature compared to the existing PDS of the ΩMEGA system is that granularity does not have to be linearly ordered: there may be two incomparable subtrees that both represent a more fine-grained version of a reduction step. Note that we do not have well-defined *levels* of granularity because we have no means (yet?) to define such levels from our experience, neither as mathematicians nor as theorem-proving engineers.

Our novel data structure is *generic* insofar as it is parameterized in both the *justifications* of the reductions (ranging from tentative hopes based on insecure knowledge to inference steps in a formal logic calculus) and in the data type of the *goals*, which may reach from sentences in natural languages to the *proof task* data structure of the CORE system [10]. Note, however, that we cannot distinguish yet between *levels* of abstraction realized by different data types for goals. For example, we do not distinguish between different levels of abstraction in the language of our goals and have no means for signature morphisms at the level of our PDS.

Although the above-mentioned tasks of the CORE system are not the subject of this paper, they may help (in form of a concrete instance) to describe the form of our novel PDS: Roughly speaking, a *task* is simply a disjunctive list of formulas (i.e. the simplest form of a sequent in classical logic) with some augmentations for different purposes, such as—among others—a distinction on one formula as the *focus*, rendering the conjugates of the other formulas as *context* formulas to be assumed when reasoning on the focus, such as a weight term for generating the ordering constraints in applications of induction hypotheses, and such as colorings for heuristic guidance.

The paper is structured as follows. We start in Section 2 with a brief summary of the old data structures of the ΩMEGA and the QUODLIBET systems and motivate their unification and generalization in the new data structure. Section 3 provides a formal description of the new generic proof data structure. Its usage is illustrated in Section 4 by a sample proof development. In Section 5 we give our answer to the question on fundamental design alternatives and Section 6 concludes the paper.

2 ΩMEGA's and QUODLIBET's Old Proof Data Structures

ΩMEGA's Proof Data Structure. ΩMEGA (see [16] for an overview and a list of further literature) is a mathematical assistant tool that supports proof development in mathematical domains at a user-friendly level of abstraction. It is a modular system in which supplementary subsystems are placed around a central proof data structure (PDS) such that the subsystems can work together to construct a proof whose status is stored in the PDS. The facilities provided by the subsystems include support for interactive and mixed-initiative theorem proving incorporating the user, proof planning, access to external systems such as automated theorem provers and computer algebra systems, and proof expansion to and proof checking at the basic level of an underlying logic calculus (which, however, is of no interest to the human user of ΩMEGA). These facilities require, in particular, the representation of proof steps at different granularities ranging from abstract human-oriented justifications to logic-level justifications.

Technically speaking, the old PDS [5] is a dag consisting of nodes, justifications and hierarchical edges. Each node represents a sequent and can be open or closed. An open node corresponds to a sequent that is to be proved and a closed node to a sequent which is already proved or reduced to other sequents using an inference rule $R := \frac{A_1 \dots A_k}{B}$. Such a rule says that from $A_1, \dots, A_k$ we can conclude B or reading it the other way round that B can be reduced to $A_1, \dots A_k$. Such an inference step is represented by a justification which connects sequents $A_1, \dots A_k$ stored in nodes $n_1, \dots n_k$ with a node n_b containing B. If a node has more than one outgoing justification, each of them represents a proof attempt of the sequent stored in the source node, but at different granularity. These have to be ordered with respect to their granularity using hierarchical edges.

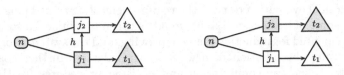

Fig. 1. Possible views of proofs at different granularities inside a PDS

A hierarchical edge connects two justifications j_1 and j_2 with the meaning that justification j_1 represents a more detailed proof attempt than justification j_2.

If proofs of different granularity are linked together by hierarchical edges, the user normally just wants to see one proof at a specific granularity. By selecting the granularity for each node he gets a view onto the graph, called PDS-view.

An example is given in Fig. 1: It shows a node n which has two outgoing justifications j_1 and j_2, which are connected by an hierarchical edge from j_1 to j_2. The user can decide whether to see the more detailed version of the proof given by j_1 (and its subtree t_1) or the more abstract version given by j_2 (and its subtree t_2). The different possible views are indicated by shading the respective nodes and justifications.

QUODLIBET's **Proof Data Structure.** Although ΩMEGA's old PDS can represent proofs at different granularity within one data structure, it still has some weaknesses compared to QUODLIBET's PDS:

- Alternative proof steps cannot be represented. That is, it is not possible for the user to tackle different proof ideas in parallel within the same proof data structure. This holds for both the reduction of a goal to some sub-goals as well as for the expansion of a complex proof step to a lower granularity. For both cases different alternatives may exist whose parallel inspection should be supported.
- An explicit handling of lemmas is not supported by ΩMEGA's old PDS. That is, it is one monolithic dag and lemmas cannot be maintained in separated DAGs.

QUODLIBET [3] is a tactic-based inductive theorem proving system for first-order clauses. It does not pursue the push-bottom technology for inductive theorem proving, but it manages more complicated proofs by an effective interplay between interaction and automation. Basically, the system does all the routine work and asks the user as early as possible if intelligence or semantic knowledge is needed. QUODLIBET has been applied mostly successfully to nontrivial mathematical research, e.g. the comparison of different formalizations of the lexicographic path ordering and their properties.

The difference compared to the new PDS of the following sections is that the proof forests of QUODLIBET consist of real trees instead of dags and there are no means for changing granularity. Although the user interface admits powerful tactic programming, the proofs are always represented on the calculus level. A decade ago, this seemed reasonable: The calculus was carefully developed over years of practical evaluation to meet the requirement of being as human-oriented as possible, some of its inference steps would take ten to a hundred steps of other calculi implemented for inductive theorem proving, and the system programmer's interface admits the addition of new inference rules for further coarse grain inference steps, such as computation and decision procedures. In the current improvement phase, however, it became obvious that the system's restriction to the finest grain is a problem growing with the power of the system, and that we need the possibility for vast changes in granularity in the proof data structure.

3   Generic Proof Data Structure

The described features of the proof data structures successful in ΩMEGA and QUODLIBET are obviously orthogonal. Their combination and further generalization, including a relaxation of the granularity restrictions—following the guidelines of Section 1—result in a new proof data structure whose features exceed the features of its origins. In particular, the new data structure supports:

- the representation of alternative proof steps for *both* the reduction of a goal as well as for the expansion of a complex proof step to lower granularity
- the structuring of proof parts (i.e. lemmatization) into separate but connected parts of the data structure
- the generic representation of proof statements and justifications, biased neither to any specific calculus nor to any specific formalism for representing abstract proof plans.

In the remainder of the section, we give a formal definition of the new generic proof data structure. We start with a formal definition of the basic PDS. We then formally define PDS-views and finally we extend a single PDS to conglomerates of PDSs, so-called forests. Note that the major technical challenge to devise a mathematically sound formulation was to consistently integrate alternative proof steps for both alternatives for the reduction of goals as well as alternatives for the expansion of a complex proof step to a lower granularity.

3.1   The PDS

Our basic PDS essentially is a directed acyclic graph (dag) whose nodes contain the proof statements. The representation to be chosen for the latter is by no means constrained in our framework. The PDS has two sorts of links: *justification hyper-links* describe a relation of goal nodes to their sub-goal nodes, and *hierarchical edges* point from justifications to other justifications they refine.

Definition 1 (PDS). A PDS is composed of nodes, justifications and hierarchical edges. Each such component x of a PDS is labeled with a pair $\mathsf{label}(x) = (c, t)$, where c maintains arbitrary content and $t \in \mathbb{N}$ is a timestamp. The time information enables us to define an order in which the objects have been created. The content of the labels can be freely instantiated, for instance, with proof statements in the case of proof nodes or with names of proof rules, tactics, and methods in the case of justifications. That is, our approach is parameterized over this sort of information that is typically very specific to different proof assistants. Formally, a PDS is defined as a triple $P := \langle \mathcal{N}, \mathcal{J}, \mathcal{H} \rangle$ where

- $\mathcal{N}$ is a nonempty finite set of *nodes*. Each node $n \in \mathcal{N}$ has a label l, denoted as label(n).
- $\mathcal{J}$ is a finite set of *justifications*. Each justification $j \in \mathcal{J}$ is a triple (s, T, l). $s \in \mathcal{N}$, $T \subseteq \mathcal{N}$, and l specify the *source*, the *targets*, and the *label* of j. They are denoted as source(j), targets(j), and label(j), respectively. We will also denote justifications as $s \xrightarrow{l} T$. Generally, a justification $s \xrightarrow{l} T$ represents a *proof step* in which proof node s is reduced to the nodes T by application of the operator l. For each node $n \in \mathcal{N}$, we define the set of *incoming justifications* by $I_n := \{j \in \mathcal{J} | n \in \text{targets}(j)\}$, and the set of *outgoing justifications* by $O_n := \{j \in \mathcal{J} | \text{source}(j) = n\}$. The *graph* of $\mathcal{J}$ is $\{(\text{source}(j), n) | j \in \mathcal{J} \wedge n \in \text{targets}(j)\}$; we require it to be acyclic.
- We require that there exists exactly one node $n_r \in \mathcal{N}$ with $I_{n_r} = \emptyset$, called the *root node*.
- $\mathcal{H}$ is a finite set of *hierarchical edges* on $\mathcal{J}$. Each hierarchical edge $h \in \mathcal{H}$ is a triple (j_1, j_2, l). $j_1 \in \mathcal{J}$, $j_2 \in \mathcal{J}$, and l specify the *source*, the *target* and the *label* of h. They are denoted as source(h), target(h), and label(h), respectively. We will denote hierarchical edges also as $j_1 \xrightarrow{h}_l j_2$. The *graph* of $\mathcal{H}$ is defined as the set of pairs $\{(\text{source}(h), \text{target}(h)) | h \in \mathcal{H}\}$; we require it to be acyclic. For all hierarchical edges $j_1 \xrightarrow{h}_l j_2$ we require:

  - source(j_1) = source(j_2) (i.e. hierarchical edges may only connect justifications sharing the same source node), and
  - for each $n_2 \in \text{targets}(j_2)$ there exists an $n_1 \in \text{targets}(j_1)$ such that (n_1, n_2) is in the reflexive and transitive closure of the graph of $\mathcal{J}$ (i.e. j_1 is the first proof step of a derivation that refines the proof step characterized by j_2).

As opposed to ΩMEGA's old PDS, this definition supports alternative justifications and alternative hierarchical edges. In particular, several outgoing justifications of a node n, which are not connected by hierarchical edges, are OR-alternatives. That is, to prove a node n, only the targets of one of these justifications have to be solved. Hence they represent alternative ways to tackle the same problem n. This describes the horizontal structure of a proof. Note further that we allow sharing of refinements; i.e., two abstract justifications may be refined by one and the same justification at lower levels. Sharing justifications in refinements is motivated, for instance, as follows: Consider a justification j which represents the call to an external system that generates a set of n different solutions, all represented in a single successor node of j with outgoing alternative subproofs starting with $j_1 \ldots j_n$, one for each solution. Then, for any $i \in \{1, \ldots, n\}$, we may abstract a coarse-grain justification a_i corresponding to a path starting with $\langle j, j_i \rangle$, represented by a hierarchical edge from j to a_i. Not supporting the sharing of justifications would in this scenario require the duplication of the justification j, which is both cumbersome and not adequate.

From the problem-solving point of view we need to know if a problem—including all its related subproblems—has already been solved or which subproblems still need to be solved. We introduce the following terminology to distinguish the different situations:

Definition 2 (Open/Closed Nodes). *Let $P = \langle \mathcal{N}, \mathcal{J}, \mathcal{H} \rangle$ be a PDS and $n \in \mathcal{N}$ be a node of P. n is called* locally closed *if and only if there exists a $j \in J$ with* source$(j) = n$ *and* target$(j) = \emptyset$; *i.e. n is justified without reducing it to new subproblems. n is called* tree-wide closed *if it is locally closed or if there is a $j \in O_n$ such that all $m \in$ targets(j) are tree-wide closed. The latter says that n is justified by a reduction to subproblems $m \in$ targets(j) which are all already (recursively tree-wide) closed. A node is called* locally/tree-wide open *if it is not locally/tree-wide closed.*   □

3.2 PDS-View

Hierarchical edges construct the vertical structure of a proof. They distinguish between upper layer proof steps and related derivations which refine them at a more granular layer. This mechanism supports both recursive expansion and abstraction of proofs. A proof may be conceived at a high level of abstraction and then *expanded* to a finer grain. As opposed thereto, *abstraction* means the process of successively contracting fine-grain proof steps to more abstract proof steps.[1] Furthermore, the PDS generally supports alternative and mutually incomparable refinements of one and the same upper layer proof step. This horizontal structuring mechanism—together with the possibility to represent OR-alternatives at the vertical level—provides very rich and powerful means to represent and maintain proof attempts. In fact, such multidimensional proof attempts may easily become too complex for humans to keep an overview as a whole. In particular, since a human does not have to work simultaneously on different granularities of a proof, elaborate functionalities to access only selected parts of a PDS are useful. They are required, for instance, for user-oriented presentation of a PDS, in which the user should be able to focus on the parts of the PDS he is currently working at, while being always able to choose whether he wants to see more details for some proof step or, on the contrary, needs to be shown a coarse structure when he is lost in the details.

We define in this subsection the notion of a *PDS-view*. A PDS-view extracts from a given PDS only a horizontal structure of the represented proof attempt at chosen granularities, but with all its OR-alternatives. As an example consider the PDS fragments in Fig. 2. In the fragment on the left-hand side, the node n_1 has two alternative proof attempts and each at alternative granularities. The

---

[1] An application of recursive expansion in the ΩMEGA system is, for instance, proof planning [14]. Proof planning first establishes a proof at an abstract level. Afterwards, to be proof checked, this proof plan may have to be expanded to a (very granular) underlying calculus. An application of recursive abstraction in ΩMEGA is, for instance, the abstraction of Natural Deduction proofs to assertion level proofs which are better suited for presentation [9].

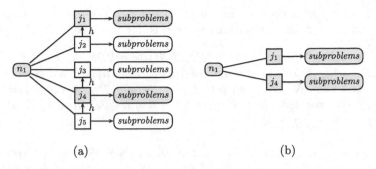

Fig. 2. (a) PDS-node with all outgoing partially hierarchically ordered justifications, and j_1, j_4 in the set of alternatives. Justifications are depicted as boxes. (b) PDS-node in the PDS-view obtained for the selected set of alternatives j_1, j_4.

fragment on the right-hand side gives a PDS-view which results by selecting a certain granularity for each alternative proof attempt, respectively. The sets of alternatives may be selected by the user and define the granularity on which he currently wants to inspect the proof. The resulting PDS-view is a slice plane through the hierarchical PDS and is—from a technical point of view—also a PDS, but without hierarchies, i.e. without hierarchical edges.

In the remainder of this subsection, we give a formal definition of a PDS-view. First, we introduce some technical prerequisites.

Definition 3 ($\mathcal{H}$-Induced Orderings $<$ and $\leq$). *Given a PDS $S = \langle \mathcal{N}, \mathcal{J}, \mathcal{H} \rangle$ we define $<$ to be the transitive closure of the graph of $\mathcal{H}$ and $\leq$ to be the reflexive closure of $<$.*

Note that $<$ and $>$ are well-founded orderings because the graph of $\mathcal{H}$ is acyclic and finite.

A PDS-node can have multiple outgoing justifications, representing alternative proof attempts or proofs at different granularity. During the proof construction or presentation, we want to restrict this set of justifications to get a complete set of alternatives at some specific granularity:

Definition 4 (Set of Alternatives). *Let $\langle \mathcal{N}, \mathcal{J}, \mathcal{H} \rangle$ be a PDS, $n \in \mathcal{N}$, and $A \subseteq O_n$ a set of justifications for n.*

- *A is adequate if there are no $k, k' \in A$ such that $k < k'$.*
- *A is complete if for all $k \in O_n$ there is a $k' \in A$ such that $k \leq k'$ or $k' \leq k$.*

A is a set of alternatives for n if it is adequate and complete. Given $j_1, j_2 \in O_n$, j_1 and j_2 are comparable, if $j_1 \leq j_2$ or $j_2 \leq j_1$; otherwise they are not comparable.

The adequacy property ensures that at most one descendant is selected for each alternative, whereas the completeness property says that there must be at least one.

For instance, the node n_1 on the left-hand side of Fig. 2 has five outgoing justifications. $\leq$ splits these justifications into 2 classes: $\{j_1, j_2\}$ and $\{j_3, j_4, j_5\}$, where the elements of a class represent the same proof alternative but at a different granularity. A set of alternatives for n is, for instance, $\{j_1, j_4\}$.

Definition 5 (PDS-View). *Let $P := \langle \mathcal{N}, \mathcal{J}, \mathcal{H} \rangle$ be a PDS, $\mathcal{N} = \{n_1, \ldots, n_m\}$ and let $A_{n_1}, \ldots, A_{n_m}$ be sets of alternatives for the nodes $n_1, \ldots, n_m$ respectively. A PDS-view is a PDS $\langle \mathcal{N}', \mathcal{J}', \emptyset \rangle$ such that $\mathcal{N}' \subset \mathcal{N}$ and $\mathcal{J}' \subset \mathcal{J}$ are the smallest sets with: (1) The root node n_r of P is in $\mathcal{N}'$, (2) if $n \in \mathcal{N}'$ then $A_n \subseteq \mathcal{J}'$, and (3) if $s \xrightarrow{l} T \in \mathcal{J}'$ then $T \subseteq \mathcal{N}'$.*

From a procedural point of view the computation of a PDS-view is a recursive process that starts at the root node n_r of a given PDS. For each node contained in the PDS-view, its set of alternatives are introduced as justifications of the PDS-view. The target nodes of these introduced justifications are then added to the nodes of the PDS-view etc. As an example consider the PDS-node and justifications on the right-hand side of Fig. 2 which are the corresponding part of the PDS-node on the left-hand side in the PDS-view.

Note that this definition of a PDS-view is not the only possible one. An alternative would be, for instance, to choose not only among hierarchical alternatives but also among the vertical alternatives. The result would be a PDS-view that contains neither hierarchical nor vertical alternatives.

3.3   Forests

We now extend the structure of a single PDS to a so-called *PDS-forest*, i.e. a set of PDSs which can be interdependent, indicated by special links between their graphs. The intuition is as follows: to prove a conjecture, further axioms, lemmas and theorems—uniformly called *lemmas*—can be used. The lemmas are either already proved or have been synthesized during proof search and are not yet proved. To accommodate either situation, new PDSs which live in the same forest as the PDS of the conjecture are introduced for the lemmas. A proof step in some PDS can then be justified by a lemma by linking the justification link to the root node the PDS for that lemma.

Definition 6 (Forest). *Let $\mathcal{I}$ be an index set. A forest is a pair $\langle (PDS_i)_{i \in \mathcal{I}}, \mathcal{F} \rangle$ where*

- *$(PDS_i)_{i \in \mathcal{I}} = (\langle \mathcal{N}_i, \mathcal{J}_i, \mathcal{H}_i \rangle)_{i \in \mathcal{I}}$ is a family of disjoint PDSs.*
- *$\mathcal{F}$ is a finite set of forest edges between PDSs. Each forest edge $f \in \mathcal{F}$ is a pair (j, n_r) consisting of the source $\mathsf{source}(f) = j$, which is a justification from some $\mathcal{J}_i$, $i \in \mathcal{I}$, and the target $\mathsf{target}(f) = n_r$, which is the root node of some PDS $PDS_{i'}$, $n_r \in \mathcal{N}_{i'}$, $i' \in \mathcal{I}$. We denote a forest link (j, n_r) by $j \dashrightarrow n_r$.*

Given a justification $j \in \mathcal{J}_i$ for some $i \in \mathcal{I}$, the set of outgoing forest links for that justification is denoted by $F_j := \{f \in \mathcal{F} | \mathsf{source}(f) = j\}$. □

Applying a new lemma on some justification j in a PDS p results in introducing a new PDS p' with root node n_r and a forest edge that connects j to n_r. Although we intuitively apply lemmas to a goal stored in a node, a forest edge starts at a justification of this node. This is necessary to determine in which alternative the lemma is to be applied. The node n is eventually *tree-wide closed* by the justification j, but n remains *forest-wide open* until the n_r in the PDS p' (just as the target nodes $\mathsf{targets}(j)$) are (forest-wide) closed.

Note that forest edges can produce cycles. This allows us to apply a lemma to itself, which is needed to represent induction in the form of *descente infinie* [18]. Moreover, due to our AND-OR proof trees these cycles may refer to different choices of AND proofs.

We now extend the notion of a PDS-view to the notion of a forest.

Definition 7 (Forest-View). *Let $\langle \mathcal{PDS}, \mathcal{F} \rangle$ be a forest. A forest-view with respect to some $p \in \mathcal{PDS}$ is a forest $\langle \mathcal{PDS}', \mathcal{F}' \rangle$, such that $\mathcal{PDS}'$ and $\mathcal{F}'$ are the smallest sets with:*

- *The PDS-view for p is in $\mathcal{PDS}'$.*
- *For all justifications j in some PDS-view from $\mathcal{PDS}'$ and for all forest edges*
  $j \dashrightarrow n_r \in \mathcal{F}$, $n_r \in p'$:
  $j \dashrightarrow n_r$ and the PDS-View for p' are contained in $\mathcal{F}'$ and $\mathcal{PDS}'$, respectively.

4   Sample Application

Our application of the proof data structure presented in this paper within the ΩMEGA project instantiates the framework with so-called *proof tasks*; i.e. they become the nodes of our proof data structure. Tasks were developed originally to represents proof situations in ΩMEGA's proof planner MULTI [13]. We extended tasks as a general technique for "natural" reasoning with abstract steps [10]. That is, the task framework allows for all kinds of steps with tasks ranging from formal steps like rewrite steps or definition expansion/contraction steps to abstract steps involving computations of external systems or merely sketching proof ideas and their flexible combination.

Proof tasks can be seen as sequents $\varphi_1, \ldots \varphi_n \vdash \psi_1, \ldots, \psi_m$ where there is always one formula—the so-called *focus*—annotated as the currently active one. The focus may be an antecedent or a succedent formula. For example, $\varphi_1, \ldots \varphi_n \vdash \underline{\psi_1}, \psi_2, \ldots, \psi_m$ describes a task where we have the context $\varphi_1, \ldots \varphi_n \vdash \psi_2, \ldots, \psi_m$ available for showing the focus $\vdash \psi_1$. In a user-interface we may want to present tasks as

$$\begin{array}{|l} \varphi_1, \ldots \varphi_n \\ \hline \underline{\psi_1}, \ldots, \psi_m \end{array}$$

and use colors to further distinguish antecedent and succedent formulas, e.g. the negative formulas in red and the positive ones in black.

In the remainder of this section, we discuss the construction of a PDS with tasks for the example theorem "$\sqrt{12}$ is irrational". The general proof technique we shall apply to this problem works as follows: Given is the conjecture "$\sqrt[j]{l}$ is irrational". Assume that $\sqrt[j]{l}$ is rational. Then there are integers n, m, which have

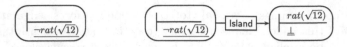

Fig. 3. PDS trees respectively after step 0 and step 1

no common divisor and for which holds that $\sqrt[i]{l} = \frac{n}{m}$. Derive a contradiction to the assumption by showing that, indeed, n, m have a common divisor. Potential candidates for the common divisor are the prime factors of l.

4.1   Proof Construction in a PDS

In a first step, we construct a PDS in the way a human mathematician would like to prove the given conjecture; see the proof sketch above. This is supported by so-called interactive island planning (see [17, 16] for details), a technique that expects an outline of the proof and has the user provide main subgoals, called *islands*. The details of the proof, eventually down to the logic level, are postponed. Hence, the user can write down *his* proof idea in a natural way with as many gaps as there are open at this first stage of the proof. Technically, in our framework the islands are tasks and all justifications between islands state island, i.e., they just indicate the intention that an island should follow from several other islands.

Step 0. The proof starts with the initial task $\vdash \neg rat(\sqrt{12})$ and the initial PDS show on the left of Fig. 3.

Step 1. In the first step we introduce the indirect argument and reduce the initial task to $rat(\sqrt{12}) \vdash \bot$ in which we assume that $rat(\sqrt{12})$ holds and the basic contradiction $\bot$ is to be proved. This action extends the PDS to one viewed on the right of Fig. 3 where the justification Island states that the action introduces a new island node.

Step 2. In the second step we derive from the assumption $rat(\sqrt{12})$ that there exist two integers n, m, which have no common divisors and for which $\sqrt{12} = \frac{n}{m}$ holds. This action further refines the PDS to

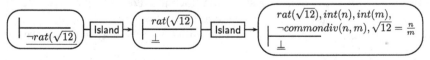

with the new task $rat(\sqrt{12}), int(n), int(m), \neg commondiv(n, m), \sqrt{12} = \frac{n}{m} \vdash \bot$.

Step 3 + 4. To complete the proof a common divisor is needed. Since 12 has the prime factors 2 and 3 there are two potential candidates. Moreover, for each candidate we have to show that both n and m are divided by it. This results in the PDS (shown below) with OR-branches (outgoing edges of PDS-nodes, e.g. $rat(\sqrt{12}), int(n), int(m), \neg commondiv(n, m), \sqrt{12} = \frac{n}{m} \vdash \bot)$) and AND-branching (outgoing links of justification nodes, e.g. Island), where Σ abbreviates all so far available

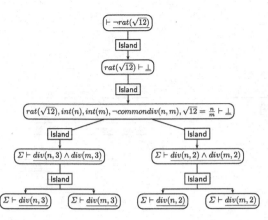

assumptions: $rat(\sqrt{12})$, $int(n)$, $int(m)$, $\neg commondiv(n, m)$, $\sqrt{12} = \frac{n}{m}$. To accomplish a proof we have now 2 possibilities: either we solve the two tasks $\Sigma \vdash div(n, 3)$ and $\Sigma \vdash div(m, 3)$, which demand to prove that both n and m have divisor 3, or we solve the two tasks $\Sigma \vdash div(n, 2)$ and $\Sigma \vdash div(m, 2)$, which demand to prove that both n and m have divisor 2.

Step 5 + ... We omit the further construction of the PDS in detail and just sketch the missing steps to derive a proof. We cannot show that both n and m have divisor 2 in the given context. Hence, the right branch of the PDS does not represent any progress. However, both n and m have divisor 3. From $\sqrt{12} = \frac{n}{m}$ follows that $m^2 * 12 = n^2$. Hence, n^2 has divisor 12 and thus also divisor 3. Then n also has divisor 3, since 3 is a prime number. This implies that $n = 3 * k$ for an integer k. Substituting n by $3 * k$ in the equation $m^2 * 12 = n^2$ results in $m^2 * 12 = 9 * k^2$. This equation can be simplified to $m^2 * 4 = 3 * k^2$. This implies that m^2 has divisor 3, from which follows that m has divisor 3 since 3 is a prime number. The introduction of all these steps closes the left branch, i.e. one alternative, of the last PDS and results in a closed PDS.

We want to remark that a proof along this idea can also be automatically proof planned in ΩMEGA; for further details we refer to [16].

4.2   Proof Expansion in a PDS

So far, our proof has been developed and sketched only at an intuitive, abstract level and logical details have been neglected. Verification of this proof requires expanding it to a logic-calculus layer. How much "effort" this expansion causes and whether it succeeds depends on the island steps and the gaps they represent. In general, an island step can be arbitrarily difficult, so that each island step may again represent a proof problem in its own right. Nevertheless, the expansion can be supported by automated tools. For instance, automated theorem provers can try to solve subproblems, computer algebra systems can perform computations, and model generators can create counterexamples, which can point out missing facts in the proof. We omit a detailed discussion of automated expansion support here and refer the interested reader to [17] and [16]. Rather, we briefly discuss the expansion of two steps in our current example PDS and sketch the resulting extended and refined PDS.

Expansion 1. Consider the first step in the current PDS, which reduces the task $\vdash \underline{\neg rat(\sqrt{12})}$ to the task $rat(\sqrt{12}) \vdash \perp$:

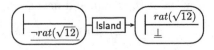

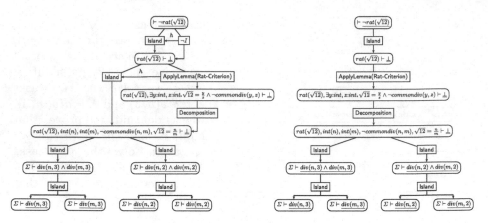

Fig. 4. (Left) Complete PDS for the running example with alternative proof attempts and different layers of granularities. (Right) A possible PDS-View determined by selection of a set of alternatives for each PDS-node in the complete PDS.

This step is already an instance of a proof step on calculus level. Indeed, it is a negation introduction step ($\neg I$). Hence, an expansion of this step simply results in a justification with $\neg I$ deriving $\vdash \neg rat(\sqrt{12})$ from task $rat(\sqrt{12}) \vdash \bot$ by a calculus step. The resulting PDS fragment is shown above.

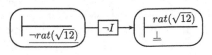

Expansion 2. The second step in the proof reduced the task $rat(\sqrt{12}) \vdash \bot$ to the task $rat(\sqrt{12}), int(n),$ $int(m), \neg commondiv(n, m), \sqrt{12} = \frac{n}{m} \vdash \bot$, which is represented by the PDS fragment above. This step implicitly encapsulates the application of the theorem that each rational number equals the fraction of two integers that have no common divisor. In the database of ΩMEGA this theorem is called *Rat-Criterion*:

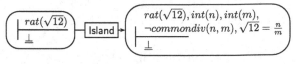

$$Rat - Criterion ::= \forall x : Rat. \; \exists y, z : int. \; \left(x = \frac{y}{z} \wedge \neg commondiv(y, z) \right)$$

It says that for all rational x there exists integers y, z, which have no common divisor and furthermore $x = \frac{y}{z}$.

The expansion of the abstract step makes the application of the Rat-Criterion theorem explicit. This works as follows: The application of Rat-Criterion to the assumption $rat(\sqrt{12})$ in the task $rat(\sqrt{12}) \vdash \bot$ derives the new assumption $\exists y, z : int. \; \left(\sqrt{12} = \frac{y}{z} \wedge \neg commondiv(y, z) \right)$, which results in the corresponding new task $rat(\sqrt{12}), \; \exists y, z : int. \; \left(\sqrt{12} = \frac{y}{z} \wedge \neg commondiv(y, z) \right) \vdash \bot$. Decomposition of the composed new assumption then derives the task

$$rat(\sqrt{12}), \; int(n), \; int(m), \; \neg commondiv(n, m), \; \sqrt{12} = \frac{n}{m} \vdash \bot$$

Altogether the resulting expanded PDS fragment at a lower,
more granular level has the form
viewed on the right. Depending
on the underlying basic calculus
these steps either present already
calculus steps or they can be further expanded. For instance, in
the CORE system lemma appli
cation is already a basic step.

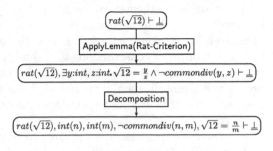

As opposed thereto, in the old ΩMEGA system lemma applications have to be
further expanded to derive Natural Deduction proofs.

The complete PDS maintaining simultaneously the initial abstract, less granular proof sketch and the lower, more granular verification of it is shown on the
left-hand side in Fig. 4. It also contains the hierarchical edges $.\overset{h}{\rightarrow}.$ which connect
the different vertical layers. It supports four different PDS-views which result
from alternative levels of granularity of the outgoing justifications of the initial
node $(\vdash \neg rat(\sqrt{12}))$ and of the outgoing justifications of the node $(rat(\sqrt{12}) \vdash \bot)$. Selecting
the upper justification for the set of alternatives for the first node and the lower
justification for the latter node results in the PDS-view shown on the right-hand
side in Fig 4.

5   Fundamental Alternatives for Modeling a PDS?

On a first glance, our modeling of a PDS may seem arbitrarily chosen among
fundamentally different possibilities and dual or isomorphic structures. After
deeper consideration, however, we believe that this is not quite the case:

Just as tasks are structured into proof attempts by recursive reduction to subtasks, alternative proof attempts result from multiple reduction of the same task.
These can be considered as a hierarchy (Fig. 5): (1) multiple proof attempts, (2)
task reduction to subtasks within one proof attempt, and (3) task composition
from formulas.

Contrary to forms of political or juristic argumentation where total evidence
is the sum of the evidences of alternative "proofs", in our area of application it
typically suffices to establish a task only once, simply because a second proof
does not give more evidence (under the current set of axioms) than a single one.

level	subject	subunits	connection of subunits	reason for the modus of the logical connection
1st	**Alternative Proof**	parallel reductions	disjunctive	area of application
2nd	**Reduction**	subtasks	conjunctive	disjunctive normal form together with level 1
3rd	**Task**	[signed] formulas	disjunctive	conjunctive normal form together with level 2

Fig. 5. Why the Logical Connections are not Arbitrary in Duality

As multiple mathematical proofs (1st level, Fig. 5) are thus connected disjunctively, it is advantageous to connect the subtasks resulting from a reduction (2nd level, Fig. 5) conjunctively, because the two levels together further the normalization of proof constructions to disjunctive normal form, resulting in a certain style. On the one hand, this style helps human beings to understand foreign proofs and maintain their own ones, and, on the other hand, makes it easier for automatic proof heuristics to recognize the triggering structures and applicable lemmas.

For the same reasons, it is advantageous to connect formulas inside a task (3rd level, Fig. 5) disjunctively. Indeed, we do not know of the dual choice of a conjunctive instead of a disjunctive connection of the formulas of a task in the literature. By tradition, both in informal human mathematical practice (starting form Aristotle's syllogisms and ending with lemmas in a modern textbook) and in formal logic calculi (Hilbert, resolution, Natural Deduction, tableau, sequent, and matrix calculi), tasks have a disjunctive structure.

6   Conclusion

This paper describes the PDS, a new generic proof data structure, which originates from and extends the successful data structures of the ΩMEGA and QUODLIBET systems. Among its key features are:

- the representation of alternative proof steps for *both* the reduction of a goal as well as for the expansion of a complex proof step to lower granularity
- the structuring of proof parts (i.e. lemmatization) into separate but connected parts of the data structure
- the generic representation of proof statements and justifications, biased neither to any specific calculus nor to any specific formalism for representing abstract proof plans.

The explicit introduction of hierarchical levels within one data structure supports the bridging between intuitive, abstract level proof development, proof explanation and proof verification. Whereas proofs are typically developed and presented at an abstract and intuitive level, proof verification typically requires some underlying calculus at a very low granularity. The PDS provides, for instance, the flexibility to perform alternative expansions of some abstract proof steps to represent the same proof idea in different underlying calculi. Maintaining simultaneously the proof at different levels of granularity accommodates, for instance, proof explanation systems, which can start with a presentation of the high-level proof, and on-demand generate presentations for expansions of *some* chosen proof steps [7]. Furthermore, the hierarchies represent the parts of the search space taken by automatic proof techniques, like for instance proof planning methods, tactics, and methodicals. Representing the search space as well as explored alternatives to represent the branches of the search space is well suited for debugging new proof techniques [6].

The PDS provides a flexible and general framework for storing and representing proofs under construction. However, the proof manipulations and refinements manipulating the PDS have to be determined and controlled by the proof system making use of the PDS. This proof system has to handle and control operations such as backtracking, instantiation of variables, collection of constraints etc. Moreover, it has to decide about whether to allow and how to realize features such as local definitions and cyclic structures in the PDS.[2]

Many proof assistants actually provide proof data structures, e.g. COQ [4], INKA [11], ISABELLE [15], NUPRL [12], TPS [1], and VSE [2] to mentioned only a few. However, to the best of our knowledge none of them has been designed to support such a horizontal and vertical representation mechanism for proofs as presented in this paper.

We implemented the generic PDS described in this paper in Allegro Lisp and defined a content independent XML format for exporting and importing forests, trees, or parts of them. Furthermore, we are able to visualize our three-dimensional graphs. For the interaction with the user, however, PDS-views are essential.

References

1. P.B. Andrews, M. Bishop, and C.E. Brown. System description: TPS: A theorem proving system for type theory. In *Proc. of the 17th Int. Conf. on Automated Deduction (CADE-17)*, LNCS, pages 164–169. Springer, 2000.
2. S. Autexier, D. Hutter, B. Langenstein, H. Mantel, G. Rock, A. Schairer, W. Stephan, R. Vogt, and A. Wolpers. VSE: Formal methods meet industrial needs. *International Journal on Software Tools for Technology Transfer, Special issue on Mechanized Theorem Proving for Technology, Springer*, september 1998.
3. J. Avenhaus, U. Kühler, T. Schmidt-Samoa, and C.-P. Wirth. How to prove inductive theorems? QUODLIBET! In *Proc. of the 19th Int. Conf. on Automated Deduction (CADE-19)*, number 2741 in LNAI, pages 328–333. Springer, 2003.
4. Y. Bertot and P. Castéran. *Interactive Theorem Proving and Program Development — Coq'Art: The Calculus of Inductive Constructions.* Texts in Theoretical Computer Science, An EATCS Series. Springer, 2004.
5. L. Cheikhrouhou and V. Sorge. PDS — A Three-Dimensional Data Structure for Proof Plans. In *Proc. of the Int. Conf. on Artificial and Computational Intelligence for Decision, Control and Automation in Engineering and Industrial Applications (ACIDCA'2000)*, 2000.
6. L. Dixon. Interactive and hierarchical tracing of techniques in IsaPlanner. In *Proc. of UITP'05*, 2005.
7. A. Fiedler. Dialog-driven adaptation of explanations of proofs. In *Proc. of the 17th International Joint Conference on Artificial Intelligence (IJCAI)*, pages 1295–1300, Seattle, WA, 2001. Morgan Kaufmann.
8. The OMEGA Group. Proof development with ΩMEGA. In *Proc. of the 18th Int. Conf. on Automated Deduction (CADE-18)*, number 2392 in LNAI, pages 143–148. Springer, 2002.
9. X. Huang. Reconstructing proofs at the assertion level. In *Proc. of the 12th Int. Conf. on Automated Deduction (CADE-12)*, LNAI, pages 738–752. Springer, 1994.

---

[2] Technically, cyclic structures can be realized by cyclic lemmas in PDS forests, [18].

10. M. Hübner, S. Autexier, C. Benzmüller, and A. Meier. Interactive theorem proving with tasks. *Electronic Notes in Theoretical Computer Science*, 103(C):161–181, November 2004.
11. D. Hutter and C. Sengler. INKA - The Next Generation. In *Proc. of the 13th International Conference on Automated Deduction (CADE-13)*, LNAI. Springer, 1996.
12. C. Kreitz, L. Lorigo, R. Eaton, R.L. Constable, and S.F. Allen. The nuprl open logical environment, 2000.
13. Andreas Meier. *Proof Planning with Multiple Strategies*. PhD thesis, Saarland Univ., 2004.
14. E. Melis and J. Siekmann. Knowledge-Based Proof Planning. *Artificial Intelligence*, 115(1):65–105, 1999.
15. L.C. Paulson. *Isabelle: A Generic Theorem Prover*. LNCS. Springer, 1994.
16. J. Siekmann, C. Benzmüller, A. Fiedler, A Meier, I. Normann, and M. Pollet. *Proof Development in OMEGA: The Irrationality of Square Root of 2*, pages 271–314. Kluwer Academic Publishers, 2003.
17. J. Siekmann, C. Benzmüller, A. Fiedler, A. Meier, and M. Pollet. Proof development with OMEGA: Sqrt(2) is irrational. In *Logic for Programming, Artificial Intelligence, and Reasoning, 9th Int. Conf., LPAR 2002*, number 2514 in LNAI, pages 367–387. Springer, 2002.
18. C.-P. Wirth. Descente infinie + Deduction. *Logic J. of the IGPL*, 12(1):1–96, 2004.

Impasse-Driven Reasoning in Proof Planning

Andreas Meier and Erica Melis

German Research Center for Artificial Intelligence (DFKI),
Saarbrücken, Germany
{ameier, melis}@dfki.de

Abstract. In a problem solving process, a step may not result in the expected progress or may not be applicable as expected. Hence, knowledge how to overcome and react to impasses and other failures is an important ingredient of successful mathematical problem solving. To employ such knowledge in a proving system requires a variety of behaviors and a flexible control. Multi-strategy proof planning is a knowledge-based theorem proving approach that provides a variety of strategies and knowledge-based guidance for search at different levels. This paper introduces reasoning about impasses as a natural ingredient of meta-reasoning at a strategic level and illustrates the use of knowledge about failure handling in the proof planner MULTI.

1 Introduction

The typical proof search in automated theorem provers relies upon local search criteria mostly referring to syntactic features of the current goal and assumptions rather than analyzing a proof situation more globally. However, in order to find a mathematical proof, more often than not the global context has to be observed, e.g., the theory, the proof history, and different proof strategies.

Humans are able to employ such information in their theorem proving. When an expected progress does not occur or when the proof process gets stuck, then an intelligent problem solver analyzes the failure and attempts a new strategy. As Schoenfeld suggests in his book on mathematical problem solving [15] *"monitoring the state of a solution as it evolves and taking appropriate action in the light of new information"* is a key skill for succeeding. This means, intelligent humans do not rely upon pre-determined control only to guide their problem solving. Instead, they draw upon a repertoire of heuristic knowledge how to deal with different situations and to dynamically guide the solution construction.

Similarly, an automated theorem proving system can monitor the solution process and can employ heuristic knowledge to reason about failed proof attempts. This requires the system to have several problem solving strategies and a flexible control, which can be guided by meta-reasoning about the overall problem solving situation.

The multi-strategy proof planner MULTI offers such a knowledge-based meta-reasoning and handles failed proof attempts as we will show in this paper. We

M. Kohlhase (Ed.): MKM 2005, LNAI 3863, pp. 143–158, 2006.

describe several examples of meta-reasoning rules that analyze and exploit failures to guide proof plan manipulations and refinements. Conceptually, the failure reasoning is supported by the architecture of MULTI, which clearly separates reasoning at different levels and provides refinement and modification strategies that can be flexibly combined. In MULTI, failure reasoning is a natural ingredient of control reasoning at a strategic level.

The paper is organized as follows. First, we briefly describe proof planning with multiple strategies and its realization in the MULTI system. Afterwards, we motivate the research on failure reasoning with an example. Section 4 introduces failure reasoning captured in general meta-reasoning rules. The role and use of failure reasoning in MULTI is illustrated for ϵ-δ-proofs in section 5. Although we use ϵ-δ-proofs for illustration throughout the paper and explain the failure reasoning with ϵ-δ-proofs, this meta-reasoning is general and applicable to other domains as well as our empirical results in section 6 evidence. Section 7 concludes the paper with a discussion of related work.

2   Background: Proof Planning with Multiple Strategies

Proof planning [4] is a theorem proving technique, which plans a proof at the abstract level of *methods*, i.e., tactics enriched by explicit pre- and postconditions. Methods result from the analysis of the common structure or common procedures of a family of proofs. They can encode not only general proof steps but also steps particular to a mathematical domain. Mathematically motivated heuristics how to proceed are encoded in the control knowledge needed to search for the sequence or hierarchy of methods that results in a solution plan. Knowledge-based proof planning [14] declaratively represents control knowledge as control rules. They are evaluated at choice points in the planning process (choice of method, choice of goal, etc.).

Knowledge-based proof planning also allows to integrate (theory-specific) constraint solving [16] and other (theory-specific) external solvers, among others for the construction/instantiation of mathematical objects (e.g., the construction of a real number that satisfies certain restrictions). For instance, proof planning for ϵ-δ-proofs delegates simple equations and inequalities containing variables to the constraint solver $\mathcal{CoSIE}$, which checks the (in)consistency of the constraints and collects consistent constraints. Thereby, the variables act as place holders for still unknown terms, and $\mathcal{CoSIE}$ can compute instantiations for the variables that satisfy the collected constraints. Such place holder variables are marked with a superscript ^{p} throughout the paper.

Simple proof planning searches at the level of methods, i.e., it searches for applicable methods and introduces the instantiated methods in the proof plan under construction until all goals are closed. Typically, the planning is monolithic in the sense that functionalities such as backtracking and instantiation of variables are part of the planning and their control is hard-coded:

> Backtrack one step in the plan, if and only if no method is applicable.
> Instantiate variables only at the end, when all goals are closed.

Multi-Strategy proof planning [13] extends proof planning by the additional hierarchical level of strategies as well as by strategic control. It allows to flexibly combine refinement and modification algorithms. The instantiation of the algorithms' parameters produces strategies, which can realize different behaviors of the algorithm. Typical strategies are those running the algorithms for method introduction, variable instantiation, and backtracking, i.e., decoupled and parameterized functionalities of the simple proof planning, realizing different kinds of method introduction, backtracking and variable instantiation.

The parameters of the algorithm for method introduction include a set of methods and a set of control rules. When such a strategy is executed, then the algorithm introduces only steps that use the methods specified in the strategy. The method-level control belongs to the strategy. That is, its choices are guided by the control rules specified in the strategy. A parameter of the instantiation algorithm is the function that determines how the instantiation for a variable is computed. A parameter of the backtrack algorithm is the function that computes a set of refinement steps that will be deleted from the partial proof plan.

Let us explain some strategies frequently occurring in ϵ-δ-proofs, i.e., in proofs of conjectures about the limit or the continuity of a function f at a point a.

The standard definitions of limit, continuity, and the derivative of a function postulate the existence of a real number δ, which may depend on an arbitrarily chosen real number ϵ:

$$\lim_{x \to a} f = l \equiv \forall \epsilon (0 < \epsilon \Rightarrow \exists \delta (0 < \delta \wedge \forall x (|x - a| > 0 \wedge |x - a| < \delta \Rightarrow |f(x) - l| < \epsilon)))$$

$$cont(f, a) \equiv \forall \epsilon (0 < \epsilon \Rightarrow \exists \delta (0 < \delta \wedge \forall x (|x - a| < \delta \Rightarrow |f(x) - f(a)| < \epsilon))).$$

A mathematical ϵ-δ-proof of such a problem constructs a real number δ depending on ϵ that satisfies certain (in)equalities.[1] A systematic procedure for discovering a suitable δ is the incremental restriction of its range. This includes the reduction of complex (in)equalities to less complex ones, the simplest of which can be propagated to range restrictions, and the determination of terms which satisfy all the restrictions.

The mathematical strategies are mirrored for proof planning. The method-introduction strategy SolveInequality corresponds to the reduction of complex inequalities to simple ones. It successively produces simpler (in)equalities until it reaches (in)equalities that are accepted by the constraint solver $CoSI\mathcal{E}$. Its methods COMPLEXESTIMATE, FACTORIALESTIMATE, and SOLVE* reduce (in)equality goals of different kinds. The connection to $CoSI\mathcal{E}$ is established by the method TELLCS, which closes inequalities and passes them to $CoSI\mathcal{E}$.

---

[1] The construction of a δ is a non-trivial task for students as well as for traditional, resolution-based automated theorem provers. Bledsoe proposed several versions of the problem LIM+ (see next section) as a challenge problem for automated theorem proving [3]. The simplest versions of this problem (problem 1 and 2 in [3]) are at the edge of the capabilities of traditional automated theorem provers but the harder versions are beyond their capabilities. More difficult problems such as Cont-If-Deriv (see next section) cannot be proved by traditional provers.

COMPLEXESTIMATE reduces inequality goals of the form $|b| < e$. It exploits the triangle inequality by rewriting $b = k * a + l$ for an a for which an assumption of the form $|a| < e'$ is in the proof context and postulates for the existence of a real number v^p serving as an auxiliary variable. The resulting simpler goals are $|l| < \frac{e}{2}$, $e' < \frac{e}{2*v^p}$, $|k| \leq v^p$, and $0 < v^p$.

The method FACTORIALESTIMATE reduces inequality goals of the form $|\frac{t}{t'}| < e$. It also postulates for the existence of a real number v^p serving as an auxiliary variable and creates three simpler goals: $0 < v^p$, $v^p < |t'|$, and $|t| < e * v^p$.

Applications of SOLVE* exploit the transitivity of $<, >, \leq, \geq$ and reduce a goal of the form $a_1 < b_1$ to a new goal $b_2 \sigma \leq b_1 \sigma$ in case an assumption of the form $a_2 < b_2$ can be used and a_1, a_2 can be unified by the substitution σ.

Two other domain-independent method introduction strategies prepare composed conjectures and proof assumptions to make relevant subformulas available for the methods of Solvelnequality.

The instantiation strategy ComputeInstFromCS corresponds to the actual construction of mathematical objects (real numbers). It instantiates variables that occur in the constraints collected by the constraint solver $\mathcal{CoSIE}$. ComputeInstFromCS is applicable, when $\mathcal{CoSIE}$ can compute an instantiation for a variable that is consistent with the constraints collected so far.

The domain-independent backtrack strategy BackTrackStepToGoal realizes a simple goal-triggered backtracking. When applied wrt. to a goal it removes the step that introduced this goal.

The Implementation in MULTI

The implementation of proof planning with multiple strategies in MULTI [11] works with two blackboards, a (object-level) proof plan blackboard and a control blackboard. During their execution, the strategies change the proof plan blackboard content and a Meta-Reasoner changes the control blackboard to guide the selection of strategies. To do so, it evaluates the strategic control knowledge, which is declaratively represented by strategic control rules. Hence, there is no need to hard-code the sequence of strategies. In a nutshell, MULTI operates according to the following cycle:

Job Offers. Applicable strategies post their applicability (for the current partial plan) as 'job offers' onto the control blackboard.

Guidance. Strategic control rules are evaluated to rank the job offers in the light of situation information.

Invocation. The strategy with the highest ranked job offer is invoked.

Execution. The strategy is executed and works on the proof plan blackboard.

3   Motivating Examples

We contrast the proof planning for two ϵ-δ-proofs: for LIM+ (without impasse) and for Cont-If-Deriv, which encounters an impasse.

LIM+ states that the limit of the sum of two functions f and g equals the sum of their limits:

if $\lim\limits_{x \to a} f(x) = l_1$ and $\lim\limits_{x \to a} g(x) = l_2$, then $\lim\limits_{x \to a} (f(x) + g(x)) = l_1 + l_2$.

When the definition of *lim* is expanded, the proof planning problem consists of two assumptions

$$\forall \epsilon_1 (0 < \epsilon_1 \Rightarrow \exists \delta_1 (0 < \delta_1 \wedge \forall x_1 (|x_1 - a| > 0 \wedge |x_1 - a| < \delta_1$$
$$\Rightarrow |f(x_1) - l_1| < \epsilon_1)))$$

and

$$\forall \epsilon_2 (0 < \epsilon_2 \Rightarrow \exists \delta_2 (0 < \delta_2 \wedge \forall x_2 (|x_2 - a| > 0 \wedge |x_2 - a| < \delta_2$$
$$\Rightarrow |g(x_2) - l_2| < \epsilon_2)))$$

and the conjecture

$$\forall \epsilon (0 < \epsilon \Rightarrow \exists \delta (0 < \delta \wedge \forall x (|x - a| > 0 \wedge |x - a| < \delta$$
$$\Rightarrow |(f(x) + g(x)) - (l_1 + l_2)| < \epsilon))).$$

Proof planning LIM+ first decomposes the conjecture and assumptions. This results, among others, in the two new assumptions $|f(x_1{}^p) - l_1| < \epsilon_1{}^p$ and $|g(x_2{}^p) - l_2| < \epsilon_2{}^p$ and the new goals $0 < \delta^p$ and $|(f(c_x) + g(c_x)) - (l_1 + l_2)| < c_\epsilon$, where c_x and c_ϵ are constants that replace x and ϵ, respectively.[2] ϵ_1, x_1, ϵ_2, x_2, and δ become place holder variables (labeled with the ^{p} superscript) for which $\mathcal{CoSIE}$ can collect constraints. Both goals are tackled by the Solvelnequality strategy. The goal $0 < \delta^p$ is closed by TELLCS and passed to $\mathcal{CoSIE}$. The second goal $|(f(c_x) + g(c_x)) - (l_1 + l_2)| < c_\epsilon$ requires further decomposition by COMPLEXESTIMATE, which employs the new assumption $|f(x_1{}^p) - l_1| < \epsilon_1{}^p$ and yields five new goals:

$$\epsilon_1 < \frac{c_\epsilon}{2 * v^p} \quad (1)$$
$$|1| \leq v^p \quad (2)$$
$$0 < v^p \quad (3)$$

$$|g(c_x) - l_2| < \frac{c_\epsilon}{2} \quad (4)$$
$$x_1{}^p = c_x. \quad (5)$$

(1), (2), (3), and (5) can be closed by TELLCS and are passed to $\mathcal{CoSIE}$. Goal (4) is reduced by SOLVE* wrt. the assumption $|g(x_2{}^p) - l_2| < \epsilon_2{}^p$ to the goals $\epsilon_2{}^p \leq \frac{c_\epsilon}{2}$ and $x_2{}^p = c_x$, which can both be closed by TELLCS. The decomposition of the assumptions on f and g results in some further goals, which are all solved by Solvelnequality.

When all goals are closed, the constraint solver $\mathcal{CoSIE}$ computes instantiations for the variables that are consistent with the collected constraints.

---

[2] During the decomposition of the assumptions further goals are created and during the decomposition of the conjecture further assumptions are derived. However, in order to illustrate the basic proof planning approach we ignore these additional goals and assumptions.

Cont-If-Deriv states that, if a function f has a derivative f' at point a, then f is continuous at a:

$$\text{if } \lim_{x \to a} \frac{f(x)-f(a)}{x-a} = f', \text{ then } cont(f, a).$$

When the definitions of *lim* and *cont* are expanded, the proof planning problem consists of the assumption

$$\forall \epsilon_1 (0 < \epsilon_1 \Rightarrow$$
$$\exists \delta_1 (0 < \delta_1 \wedge \forall x_1 (|x_1 - a| < \delta_1 \wedge |x_1 - a| > 0 \Rightarrow |\frac{f(x_1)-f(a)}{x_1-a} - f'| < \epsilon_1)))$$

and the conjecture

$$\forall \epsilon (0 < \epsilon \Rightarrow \exists \delta (0 < \delta \wedge \forall x (|x - a| < \delta \Rightarrow |f(x) - f(a)| < \epsilon))).$$

Proof planning for Cont-If-Deriv fails because of a typical exception. More detailed, the proof planning goes as follows: As for LIM+ the initial conjecture and assumption are decomposed. The main resulting goal is

$$|f(c_x) - f(a)| < c_\epsilon \tag{6}$$

and a new assumption is

$$|\frac{f(x_1{}^p) - f(a)}{x_1{}^p - a} - f'| < \epsilon_1{}^p. \tag{7}$$

Using this assumption the goal (6) can be proved in several steps.[3] However, when decomposing the initial assumption, the goal $|c_x - a| > 0$ was created as a side goal and cannot be proved. This gives rise to an impasse because the goal-triggered backtracking with strategy BackTrackStepToGoal does not lead to a solution. This impasse is not a problem of missing methods. Rather, since the condition $|c_x - a| > 0$ is not always true, mathematically it is necessary to consider the cases $|c_x - a| > 0$ and $|c_x - a| \leq 0$. That is, $|f(c_x) - f(a)| < c_\epsilon$ has to be proved twice, once under the condition $|c_x - a| > 0$ and once under the condition $|c_x - a| \leq 0$. Because the impasse information is surfaced only at a later stage of the proof planning, its analysis should lead to a conclusion on how to modify (i.e., not just refine) the overall proof plan in order to circumvent the exception.

4   Meta-reasoning About Failed Proof Attempts

This and many other examples show that proof planning may encounter impasses, i.e., situations in which an open goal cannot be closed because there are no applicable methods or strategies or in which no instantiation for a variable can be found.

Impasses in multi-strategy proof planning may occur at two levels, inside strategies or at strategy choice points. When an impasse occurs during the

---

[3] That is, COMPLEXESTIMATE is applied to (6) with assumption (7) and all resulting goals can again be solved by proof planning.

execution of a strategy, the strategy interrupts and the failure is recorded. When the next strategy has to be selected for execution, the strategic control rules can reason about the failures. When an impasse occurs at strategy choice, then the failing applicability of a strategy is recorded together with the failure reasons and the strategic control rules can reason upon.

Most systems have a default behavior that is called, when such errors occur. So has MULTI. For instance, the default behavior to deal with failures in the method introduction algorithm is the goal-triggered backtracking with the strategy BackTrackStepToGoal (i.e., to backtrack the step that introduced the goal for which no applicable method can be found). However, the default reaction is only one of a variety of possible reactions to failures that is evaluated in MULTI. An evaluation of a repertoire of alternative reactions to failures is useful for many reasons, two of which are:

(1) Theorem proving may require refinements, modifications, or additional information, which are hard to predict from common proof patterns since they are exceptions. Some heuristics can exploit failures since in some cases these failures hold information that is necessary in order to discover a solution plan at all.
(2) A (knowledge-based) search and backtracking procedure that is generally suitable for different mathematical domains including domains that require higher-order formalizations is difficult to devise. Rather, knowledge of suitable failure handling and backtrack points in different domain-dependent proof situations can be used to organize the search.

Hence, in MULTI failure handling is not hard-coded once and forever. Rather, the analysis of frequent failures and possible reactions results in general (and informal) meta-reasoning rules, which can be formalized in control rules that MULTI can use. These control rules analyze information about the failure situation, the current proof plan, the history, etc. and suggest suitable proof plan modifications and refinements. In the remainder of this section, we shall introduce three general meta-reasoning rules. In the subsequent section, we shall illustrate their encoding in control rules in MULTI as well as their application to ϵ-δ-proofs.

4.1   Case Split Introduction

Case split is well-known in mathematics. More often than not, it is not obvious in advance, when it is useful to apply a case split and which cases to consider. The following rule describes the need for a case split at an abstract level.

A main goal can be solved by some methods which introduces side goals, called 'conditions'. In a situation where one of these conditions cannot be proved (while the main goal is solved), an impasse is reached. The impasse can be removed by introducing a case split on the failing condition and its negation earlier in the plan. The modification requires to prove the main goal in each of the cases.

This description is represented by the general meta-reasoning rule

Case-Split Introduction:
IF   failing condition C while some methods solve main goal
THEN introduce case split $C \vee \neg C$ before application of methods

In section 5, we shall explain how this meta-reasoning rule is encoded into control rules and how the main goal and the side goals are determined.

4.2   Analysis of Variable Dependencies

Goals sharing the same variable but belonging to different proof plan branches are dependent. The instantiations and constraints of those variables may cause failures. Take, e.g., two goals g and g' that both contain a variable v^p. Lets assume that a partial proof plan for g is created, which binds v^p in such a way that g' cannot be proved anymore. The default reaction – standard goal-triggered backtracking – would remove g'. However, if the problem for the failure is not g' but the selection of an appropriate instantiation for v^p, then this backtracking will not lead to a solution proof plan. Rather, part of the subplan for g has to be removed to introduce another subplan that constrains v^p differently.

This heuristic gives rise to the general meta-reasoning rule:

Analyze VarDependencies:
IF   failure on goal caused by variable instantiation/constraints
THEN backtrack variable instantiation/constraints

Again, this rule can be represented by declarative control rules. Its application to ϵ-δ-proofs is illustrated in section 5.

4.3   Unblock Desirable Steps

Often classes of proofs exhibit a common proof pattern, which suggests a particular hierarchy or combination of proof steps. This implies that during the solution process particular proof steps become 'desirable', i.e., the proof pattern suggests to apply these steps next. If such a desirable step is blocked, then meta-reasoning can analyze how to unblock the application of the desirable step. In its most general form, this meta-reasoning rule can be formulated as

Unblock Desirable Steps:
IF   step S is desirable but blocked
THEN perform other steps to enable S

In section 5, we shall discuss two instances of this general rule, which both rely on the common proof pattern for ϵ-δ-proofs introduced in section 2 (i.e., their determination of desirable steps is wrt. to the proof pattern for ϵ-δ-proofs).

The first instance analyzes a blockage of the instantiation strategy Compute-InstFromCS caused by insufficiently constrained variables. To overcome this impasse the meta-reasoning suggests actions to enable the collection of further constraints. The second instance analyzes failing method applications and suggests the speculation of a lemma that would make a desirable method applicable.

5   Examples from the Domain of ϵ-δ-Proofs

To illustrate the failure reasoning we detail some of it for ϵ-δ-proofs – a domain of mathematics for which up to now the proofs are not trivially automated. As the empirical results in section 6 show, the same meta-reasoning is also applicable to other domains.

5.1   Guiding the Introduction of Case Splits

The proof of Cont-If-Deriv is an example in which the introduction of a case split is necessary. As described in section 3 proof planning fails to prove the condition $|c_x - a| > 0$ that was created as a side goal when decomposing the initial assumption. At this point the meta-reasoning rule *Case Split Introduction* suggests the introduction of a case split earlier in the proof plan.

Technically, the meta-reasoning rule *Case Split Introduction* is formalized in two control rules in MULTI that guide suitable backtracking and the introduction of the case split. This works as follows: if SolveInequality fails to prove a condition of an assumption that was used to prove the main goal, then a strategic control rule triggers the backtracking of all steps following the introduction of the failing condition. Afterwards, another control rule introduces the case split with the failing condition and its negation.

When proof planning encounters the impasse for Cont-If-Deriv (see section 3), the case split is introduced before the main goal $|f(c_x) - f(a)| < c_\epsilon$ is tackled. Then, SolveInequality continues and has to prove $|f(c_x) - f(a)| < c_\epsilon$ once enriching the context with $|c_x - a| > 0$ and once enriching the context with $\neg(|c_x - a| > 0)$. In the first case the failing condition $|c_x - a| > 0$ follows from the context of the case. The second case is proved differently by SolveInequality: First, it simplifies the hypothesis $\neg(|c_x - a| > 0)$ to $c_x = a$. Afterwards, it uses this equation to simplify the goal $|f(c_x) - f(a)| < c_\epsilon$ to $0 < c_\epsilon$, which follows from the context.

Other ϵ-δ-proofs also require this kind of failure reasoning and the same failure handling can also be used to introduce case splits in other mathematical domains (see section 6).

5.2   Analyzing Variable Dependencies

The meta-reasoning rule *Analyze VarDependencies* analyzes dependencies of goals that share some variables – either directly in their formulas or in the assumptions in their contexts. The meta-reasoning rule is encoded in the strategic control rule `analyze-varfailure`. When a goal with shared variables cannot be closed, `analyze-varfailure` analyses whether sub-plans for other goals with these variables introduce constraints on the variables. If this is the case, it guides the backtracking of those steps that introduced such constraints rather than to employ the standard goal-triggered backtracking, which would backtrack the goal that cannot be solved. Afterwards, re-opened goals can be solved differently.

As example for an ϵ-δ-proof that needs the analysis of variable dependencies consider the following problem:

$$\text{If } \lim_{x_1 \to 0} f(x_1) = l \text{ and } a > 0, \text{ then } \lim_{x \to 0} f(a * x) = l.$$

The decomposition of the initial goal and the initial assumption yield the goal $|f(a * c_x) - l| < c_\epsilon$ and the new assumption $|f(x_1{}^p) - l| < \epsilon_1{}^p$. SolveInequality solves the goal with this assumption by an application of the method SOLVE*.

The decomposition of the initial assumption also results in the two goals

$$|a * c_x| > 0 \tag{8}$$

$$|a * c_x| < c_{\delta_1} \tag{9}$$

These two goals can be solved with two assumptions from their context

$$|c_x| > 0 \tag{10}$$

$$|c_x| < \delta^p \tag{11}$$

which were created during the decomposition of the initial theorem.

When tackling these two goals SolveInequality first proceeds as follows: It applies SOLVE* to goal (8) wrt. the assumption (11). This is possible since $|c_x| < \delta^p$ equals $\delta^p > |c_x|$ and δ^p can be trivially unified with $|a * c_x|$. The application of SOLVE*, however, introduces the constraint $\delta^p \mapsto |a * c_x|$ on δ^p, which affects the assumption (11) in the context of goal (9). Next, SolveInequality tackles goal (9) but fails, since with the constraint on δ^p no solution is possible (i.e., assumption (11) cannot be used as necessary, see below).

Guided by `analyze-varfailure` MULTI first backtracks the application of SOLVE* to goal (8). Afterwards, SolveInequality solves this goal differently: It applies the method COMPLEXESTIMATE with the assumption (10) to the goal and passes the resulting inequality goals with TELLCS to $\mathcal{CoSIE}$. With this solution proof plan for goal (8) SolveInequality can also solve goal (9) by applying the method COMPLEXESTIMATE with assumption (11) and passing the resulting inequality goals with TELLCS to $\mathcal{CoSIE}$.

More ϵ-δ-proofs as well as problems from other domains require this failure reasoning to analyze and overcome variable dependencies (see section 6).

5.3   Meta-reasoning for Insufficiently Determined Constraints

Remember the *Unblock Desirable Steps* rule from section 4.3. One of its instances unblocks the instantiation strategy ComputeInstFromCS.

To illustrate this failure reasoning we detail the proof planning process for the problem Lim-Div, which states that the limit of the function $\frac{1}{x}$ at point $c \neq 0$ is $\frac{1}{c}$, i.e., for $c \neq 0$

$$\forall \epsilon (0 < \epsilon \Rightarrow \exists \delta (0 < \delta \wedge \forall x (x \neq 0 \wedge |x - c| < \delta \wedge |x - c| > 0 \Rightarrow |\tfrac{1}{x} - \tfrac{1}{c}| < \epsilon))).$$

The proof planning works as follows: Decomposition of the initial goal results in the two goals $0 < \delta^p$ and $|\tfrac{1}{c_x} - \tfrac{1}{c}| < c_\epsilon$. SolveInequality closes the goal $0 < \delta^p$ by TELLCS and simplifies the second goal to $|\tfrac{c - c_x}{c_x * c}| < c_\epsilon$. It continues with the application of FACTORIALESTIMATE, which reduces this goal to three simpler subgoals $0 < v^p$, $|c_x * c| > v^p$, and $|c - c_x| < v^p * c_\epsilon$ with a new variable v^p. SolveInequality closes these three goals with TELLCS. Since all goals are closed the strategy ComputeInstFromCS becomes a highly desirable strategy that should return instantiations for the variables δ^p and v^p computed by $\mathcal{CoSIE}$.

Now, $\mathcal{CoSIE}$ fails to determine instantiations because the constraints collected so far

$$\begin{array}{ccc} \dfrac{|c_x - c|}{c_\epsilon} < v^p & 0 < v^p & v^p < |c_x * c| \\ 0 < \delta^p & c \neq 0 & 0 < c_\epsilon \end{array}$$

are insufficient to compute a solution.[4] Hence, also the application of the desirable strategy ComputeInstFromCS is blocked.

A possibility to overcome this problem is to create further constraints by further proof planning. An instance of the meta-reasoning rule *Unblock Desirable Steps* suggests this "repair":

IF	a constraint solver fails to provide instantiations because of insufficient constraints while all goals are closed
THEN	consider actions to create and pass further constraints

Technically, the idea to overcome highly desirable but blocked variable instantiations by a constraint solver is encoded in the strategic control rule `unblock-constr`. When all goals are closed, but instantiation strategies are not applicable since connected constraint solvers fail to compute instantiations, then `unblock-constr` analyzes the current proof plan for possible further constraints.

One possibility to derive further constraints is the refinement of existing constraints closed by applications of TELLCS. If `unblock-constr` detects such constraints that likely can be refined to simpler constraints, it triggers the backtracking of the corresponding TELLCS applications (only these selected applications).[5] Afterwards, the re-opened goals can be tackled again and can be refined. Note that this backtracking serves the applicability of methods uncovering further constraints rather than the traversal of the search space.

---

[4] This is a common situation in constraint solving: A set of constraints has been accepted since no inconsistency could be detected, so far. Nevertheless, the collected constraints are not sufficient to compute a solution for the constrained variables. The critical constraints here are the constraints on v^p, which state that $\tfrac{|c_x - c|}{c_\epsilon}$ has to be less than v^p, which has to be less than $|c_x * c|$. These constraints are not inconsistent, but a solution for v^p exists only, if $\tfrac{|c_x - c|}{c_\epsilon} < |c_x * c|$ holds. This, however, does not follow from the constraints collected so far.

[5] Currently, the critical constraints are chosen by heuristics encoded in the control rule. We are enhancing $\mathcal{CoSIE}$ to return the critical constraints directly.

In proof planning of Lim-Div, `unblock-constr` (successively) triggers the backtracking of the applications of TELLCS that close $|c - c_x| < v^p * c_\epsilon$ and $|c_x * c| > v^p$. Solvelnequality reduces the re-opened goals with applications of the method COMPLEXESTIMATE and passes the resulting constraints by applications of TELLCS to $\mathcal{CoSIE}$. This leads to the following constraint store (the variables $v_1{}^p$ and $v_2{}^p$ are introduced by the applications of COMPLEXESTIMATE):

$$
\begin{array}{llll}
c_\epsilon > 0 & c \neq 0 & v^p \geq v_1{}^p * \delta^p & v_1{}^p > c \\
v^p > 0 & v_2{}^p > 1 & \frac{c_\epsilon * v^p}{2} > 0 & \delta^p > 0 \\
\delta^p \leq \frac{c_\epsilon * v^p}{2 * v_2{}^p} & v^p * 2 \leq c^2 &
\end{array}
$$

Now the following instantiations consistent with these constraints can be computed: $v_2{}^p \rightarrow 2$, $v_1{}^p \rightarrow c + 1$, $v^p \rightarrow \frac{c^2}{2}$, and $\delta^p \rightarrow min(\frac{c_\epsilon * c^2}{8}, \frac{c^2}{2*(c+1)})$.

This example is not an isolated one. More ϵ-δ-proofs require this failure reasoning (see section 6). Moreover, the meta-reasoning rule is generally applicable also to other domains in which constraints solvers are used.

5.4   Lemma Speculation

Another typical instance of the *Unblock Desirable Steps* rule in section 4.3 unblocks method applications whose matching with the proof situation requires additional information. It reads as follows

IF	the application of a desirable method fails because of a unification residuum and the residuum is likely to be provable in the current context
THEN	speculate residuum as lemma and apply it to unblock the desirable method

Since lemma speculation may open a Pandora's box, the restriction *the residuum is likely to be provable in the current context* needs to be defined. For instance, in proof planning that uses a constraint solver the constraint solver can be exploited to decide whether a residuum is a promising lemma. For ϵ-δ-proofs, the meta-reasoning queries $\mathcal{CoSIE}$ whether it accepts the residuum and only then meta-reasoning suggests the speculation of the lemma. This way, we combine the domain-independent unification and matching with the domain knowledge in $\mathcal{CoSIE}$.[6] Technically, this lemma speculation is encoded by the control rule `unblock-method`.

To illustrate this failure reasoning we detail planning for the problem

$$
\text{If } \lim_{x \to 0} f(x + c) = l, \text{ then } \lim_{x_1 \to c} f(x_1) = l.
$$

The decomposition of the initial goal results, among others, in the goal $|f(c_{x_1}) - l| < c_{\epsilon_1}$. Decomposition of the initial assumption yields the new assumption $|f(x^p + c) - l| < \epsilon^p$. Solvelnequality should apply the method SOLVE*

---

[6] Alternatively, theory unification incorporates domain-specific axioms and theorems into the unification procedures. However, the decidability of theory unification is difficult to determine and depends on the concrete set of domain equations (e.g., see [2]). Undecidable unification and matching, however, could block the complete proof planning process.

to tackle the new goal with the new assumption. However, this fails since the application-conditions of SOLVE* request the unification of $|f(x^p + c) - l|$ and $|f(c_{x_1}) - l|$, which fails. Since no other applicable method or assumption are available, MULTI's default control would backtrack, which would not lead to a solution proof plan.

When SolveInequality fails to tackle $|f(c_{x_1}) - l| < c_{\epsilon_1}$ with the assumption $|f(x^p + c) - l| < \epsilon^p$, then the analysis of the failure by unblock-method yields the residuum $x^p + c = c_{x_1}$, which is accepted by $\mathcal{CoSIE}$. Hence, the control rule fires and introduces $x^p + c = c_{x_1}$ as lemma and guides the rewriting of the goal $|f(c_{x_1}) - l| < c_{\epsilon_1}$ with this equation. This results in the goal $|f(x^p + c) - l| < c_{\epsilon_1}$. The application of SOLVE* to this goal and the assumption $|f(x^p + c) - l| < \epsilon^p$ is now possible, and SolveInequality can solve all resulting goals.

For other ϵ-δ-proofs the same meta-reasoning can overcome blocked unifications and matchings. For the application in other domains the only prerequisite is that a means exists that can decide whether a lemma is promising.

6   Empirical Results

The meta-reasoning rules in section 4 describe general situations in mathematical proof processes. Although our contribution is fundamentally conceptual and architectural, we had to show whether it is empirically relevant as well. Therefore, we tested the benefit in three domains, the ϵ-δ-proofs from the analysis textbook [1], the residue class domain, and inductive proofs. Table 1 gives sample problems from all three domains and the failure-reasoning they require. The numbered colons denote (i) case split introduction, (ii) unblock constraint solving, (iii) unblock by lemma speculation, (iv) analyze variable dependencies. Note that $x \to a^-$ and $x \to a^+$ denote the left-hand limit and the right-hand limit, respectively.

The relevance of failure reasoning is not only demonstrated by Table 1. Its figures alone are underestimating because many similar problems can be formulated. Moreover, the relative frequency of failure reasoning is also important. Therefore, the fact that 25 out of 70 ϵ-δ-proofs constructed by MULTI from the systematically explored testbed [1] involve failure reasoning evidences the crucial role of failure reasoning.

Residue Class Problems. The residue class conjectures classify given residue class structures wrt. their algebraic category. An example theorem is "the residue class structure $(\mathbb{Z}_5, \bar{+})$ is associative". Other problems from this domain concern the isomorphy of two algebraic structures. An example is "the residue class structures $(\mathbb{Z}_5, \bar{+})$ and $(\mathbb{Z}_5, (x\bar{+}y)\bar{+}\bar{1}_5)$ are isomorphic".

To tackle residue class problems we developed several techniques encoded in four different method-introduction strategies in MULTI. In one of these strategies, the TryAndError strategy (see [12]), the *Analyze VarDependencies* rule is crucial since MULTI has to deal with nested existential quantifiers, which result in 'nested' variables shared by several goals. Hence, dependencies among the variables and the goals have to be analyzed.

Table 1. Sample proofs whose solution requires meta-reasoning about failures

Conjecture	(i)	(ii)	(iii)	(iv)
ϵ-δ-Proofs				
$\lim\limits_{x\to 0}(f(a+x)-f(a))=0 \Rightarrow cont(f,a)$	x		x	x
$\lim\limits_{x\to a^-} f(x)=l \wedge \lim\limits_{x\to a^+} f(x)=l \Rightarrow \lim\limits_{x\to a} f(x)=l$	x			
$\lim\limits_{x\to a^-} f(x)=f(a) \wedge \lim\limits_{x\to a^+} f(x)=f(a) \Rightarrow cont(f,a)$	x			
$\lim\limits_{x\to 2} \frac{1}{1-x} = -1$		x		
$\lim\limits_{x\to a} f(x)=l_f \wedge \lim\limits_{x\to a} g(x)=l_g \wedge \forall x\, g(x)\neq 0 \Rightarrow \lim\limits_{x\to a} \frac{f(x)}{g(x)} = \frac{l_f}{l_g}$		x		
$\lim\limits_{x\to\infty} f(x)=l \Rightarrow \lim\limits_{x\to\infty} \frac{f(x)}{x} = 0$		x		
$\lim\limits_{x\to 0} f(x+a)=l \Rightarrow \lim\limits_{x\to a} f(x)=l$			x	
$\lim\limits_{x\to 0+} f(\frac{1}{x})=l \Rightarrow \lim\limits_{x\to\infty} f(x)=l$			x	
$\lim\limits_{x\to 0} f(x)=l \wedge a>0 \Rightarrow \lim\limits_{x\to 0} f(a*x)=l$				x
$\lim\limits_{x\to a} f(x)=l \Rightarrow \lim\limits_{x\to 0} f(x+a)=l$				x
Residue Class Problems				
$closed(\mathbb{Z}_3\backslash\{\bar{0}_3\}, \bar{*})$				x
$\neg closed(\mathbb{Z}_3\backslash\{\bar{0}_3\}, \bar{+})$				x
$\neg\exists e{:}z_9\, unit(\mathbb{Z}_9, \bar{-})$				x
$\neg inverses(\mathbb{Z}_6, \bar{*}, \bar{1}_6)$				x
$\neg divisors(\mathbb{Z}_6, \bar{*})$				x
$\neg commutative(\mathbb{Z}_8, \bar{-})$				x
$\neg distibutive(\mathbb{Z}_4, \bar{-}, \bar{-})$				x
Inductive Proofs				
$\forall x{:}item\forall y,z{:}list\ x\in y \Rightarrow x\in concatenate(y,z)$	x			
$\forall x{:}item\forall y,z{:}list\ (x\in y \vee x\in z) \Rightarrow x\in concatenate(y,z)$	x			
$\forall x{:}item\forall y{:}list\ x\in insert(x,y)$	x			
$\forall y{:}list\ length(y)=length(isort(y))$	x	x		
$\forall x{:}item\forall y{:}list\ x\in isort(y) \Rightarrow x\in y$	x	x		
$\forall x{:}item\forall y{:}list\ count(x,isort(y))=count(x,y)$	x	x		
$\forall y{:}list\ reverse(reverse(y))=y$			x	
$\forall x{:}nat\ even(x+x)$			x	

We proved about 19.000 residue class conjectures with MULTI. About half of these theorems, in particular, theorems refutating a property, could be proved with the TryAndError strategy only (see [12] for the detailed description of the experiments). Some representative examples occur in Table 1.

Inductive Proofs. So far, we did not apply MULTI to inductive proofs. The inductive theorems in Table 1 are taken from [9], which describes failure reasoning by so-called critics in the proof planner CLAM. Since the critics employed in CLAM are a special case bound to a particular method (see related work in section 7), our general failure reasoning rules for case-split introduction and lemma speculation are applicable for inductive proofs as well. For a more complete list of inductive proofs that require failure reasoning see [9].

7   Conclusions and Related Work

We presented a novel conceptual and architectural contribution to knowledge-based automated theorem proving. We demonstrated how MULTI's novel theorem proving architecture supports failure reasoning and the automatic discovery of heureka steps such as introduction of case splits and speculation of lemmas. As evidenced by empirical results, the discussed failure reasoning is generally applicable rather than overly specific.

Proof planning may fail because of an exception from proof structures and procedures captured by proof planning methods, strategies and heuristic control. Such failures can be analyzed automatically and this analysis gives rise to applying proof planning strategies, which can revise the proof plan or invent new knowledge that is needed to complete a proof, e.g., case splits or new conjectures. That is, often failures hold the key to the construction of a solution proof plan.

The failure reasoning and the subsequent proof plan modifications are possible in MULTI, since MULTI's architecture does not enforce a pre-defined backtracking or other pre-defined control. Rather, when a failure occurs, then strategic control rules, which declaratively encode failure handling heuristics, can analyze the failure and dynamically guide promising refinements and modifications of the proof plan. Further meta-reasoning that contributes to this flexible control in MULTI is discussed in [10].

Related Work
Unblocking desirable steps in MULTI is related to the control reasoning in elaborate blackboard systems, see [5] and [6]. When a highly desirable knowledge source is not applicable, then reasoning on the failure can suggest the invocation of knowledge sources that unblock the desired knowledge source.

The speculation of residuum lemmas is related to constrained resolution [7], which intertwines resolution with unification. We also intertwine unification with the main proof process by speculating unification residues as lemmas. As opposed to constrained unification, our meta-reasoning controls the speculation of lemmas since it suggests only lemmas that are directly accepted by $CoSIE$.

In their interesting work described in [8, 9] Andrew Ireland and Alan Bundy extended proof planning by so-called critics as a means to patch failed proof attempts in proof planning inductive proofs. Their proof planner CIAM is specialized for proving theorems by mathematical induction and employs the so-called rippling technique that is mainly encoded in the **wave** method. The critics in CIAM are associated with the **wave** method and capture patchable exceptions to the application of this method. A critic can, e.g., introduce a case split directly preceding the wave method, in order to make a conditional wave rule applicable, in case the **wave** method is blocked because of this condition.

The failure reasoning in MULTI considerably differs from the critics mechanism in CIAM in its conceptual design. Critics are method-like entities that are bound to failing preconditions of a particular method. Moreover, the critic's patch is a special procedure that changes the proof plan. In contrast, failure reasoning in MULTI is represented by declarative control rules. These control rules are not associated with a particular method but can reason about the current proof plan and about other information such as the proof planning history. The patch of a failure is not implemented into special procedures but is carried out by methods and strategies whose application is suggested by the control rules.

To summarize, because of its flexible proof construction at the strategy level, MULTI can realize a more general failure reasoning approach not bound to particular methods.

References

1. R.G. Bartle and D.R. Sherbert. *Introduction to Real Analysis*. John Wiley& Sons, New York, 1982.
2. K.H. Bläsius and H.J. Bürckert, editors. *Deduktionssysteme*. Oldenbourg, 1992.
3. W.W. Bledsoe. Challenge Problems in Elementary Analysis. *Journal of Automated Reasoning*, 6:341–359, 1990.
4. A. Bundy. The Use of Explicit Plans to Guide Inductive Proofs. In *Proceedings of CADE-9*, volume 310 of *LNCS*, pages 111–120. Springer, 1988.
5. D.D. Corkill, V.R. Lesser, and E. Hudlicka. Unifying Data-Directed and Goal-Directed Control. In *Proceedings of AAAI-82*, pages 143 – 147. AAAI Press, 1982.
6. E.H. Durfee and V.R. Lesser. Incremental Planning to Control a Blackboard-Based Problem Solver. In *Proceedings of AAAI-86*, pages 58 – 64. AAAI Press, 1986.
7. G.P. Huet. *Constrained Resolution: A Complete Method for Higher Order Logic*. PhD thesis, Case Western Reverse University, 1972.
8. A. Ireland. The Use of Planning Critics in Mechanizing Inductive Proofs. In *Proceedings of LPAR'92*, volume 624 of *LNAI*, pages 178–189. Springer, 1992.
9. A. Ireland and A. Bundy. Productive Use of Failure in Inductive Proof. *Journal of Automated Reasoning*, 16(1-2):79–111, 1996.
10. A. Meier. MULTI – *Proof Planning with Multiple Strategies*. PhD thesis, Fachbereich Informatik, Universität des Saarlandes, Saarbrücken, 2004.
11. A. Meier and E. Melis. MULTI: A Multi-Strategy Proof Planner. In *Proceedings CADE-20*, Springer, 2005.
12. A. Meier, M. Pollet, and V. Sorge. Comparing Approaches to Explore the Domain of Residue Classes. *Journal of Symbolic Computation*, 34(4):287–306, 2002.
13. E. Melis and A. Meier. Proof Planning with Multiple Strategies. In *Proceedings CL-2000*, volume 1861 of *LNAI*, pages 644–659. Springer, 2000.
14. E. Melis and J. Siekmann. Knowledge-Based Proof Planning. *Artificial Intelligence*, 115(1):65–105, 1999.
15. A.H. Schoenfeld. *Mathematical Problem Solving*. Academic Press, New York, 1985.
16. J. Zimmer and E. Melis. Constraint solving for proof planning. *Journal of Automated Reasoning*, 2004. accepted.

Literate Proving: Presenting and Documenting Formal Proofs

Paul Cairns and Jeremy Gow

UCL Intereaction Centre,
University College London,
London, WC1E 7DP, UK
{p.cairns, j.gow}@ucl.ac.uk

Abstract. Literate proving is the analogue for literate programming in the mathematical realm. That is, the goal of literate proving is for humans to produce clear expositions of formal mathematics that could even be enjoyable for people to read whilst remaining faithful representations of the actual proofs. This paper describes maze, a generic literate proving system. Authors markup formal proof files, such as Mizar files, with arbitary XML and use maze to obtain the selected extracts and transform them for presentation, e.g. as LATEX. To aid its use, maze has built in transformations that include pretty printing and proof sketching for inclusion in LATEX documents. These transformations challenge the concept of faithfulness in literate proving but it is argued that this should be a distinguishing feature of literate proving from literate programming.

1 Introduction

Whilst formal languages, such as those used in formalised mathematics and programming, are ideal for communicating with a computer, they are far removed from the natural discourses that take place between humans in natural languages. Indeed, it can be argued that overcoming the distinction between human-human discourse and human-computer discourse is the entire basis for the discipline of human-computer interaction.

In formal mathematical proofs, humans may rapidly become lost in the detail and fail to understand the proof [19] or even to recognise them as a proof [4]. The corresponding difficulty of understanding programming languages has long been recognised. Documentation is considered to be a valuable resource for communicating the purpose and concepts embodied in the formal code [9]. However, it is also recognised that coding and documenting code are quite different activities and that the documentation of code is rarely of a high, readable standard.

To help overcome the division between program and documentation, Knuth devised the notion of *literate programming* [8]. Here the code and its explanation are combined together in a single document. Knuth's hope was that in doing so, programs would be faithfully described because when changes were made to the code, corresponding changes could be made to the documentation located

M. Kohlhase (Ed.): MKM 2005, LNAI 3863, pp. 159–173, 2006.

in the same place. Of course, this is what people normally do with code comments. However, a commented program does not make for easy reading for a human. Knuth's further step was to transform the single document in two different ways — one way to produce the compilable version of the code, the other to produce a human readable document. In this way, documentation would not only be faithful to the code but might even become an enjoyable literary work in its own right.

Formal mathematics shares many of the features of programming languages. In recent years, large repositories of formal mathematics such as the Mizar [10] and Coq libraries are being developed (and more are planned with on-going MKM). Whilst these libraries are not expected to change in the same manner as program code, there is nonetheless a need to document the formal language to help guide a reader through ideas, notation and even the trickier parts of formalisation [7]. In fact, Cruz-Filipe et al. (*ibid.*) recognise that some form of *literate proving*, that is, the weaving together of formal proofs and documentation, would be valuable.

This paper describes maze, a working system that implements literate proving. It deviates from Knuth's original implementation, instead following more modern versions of literate programming [14], as will be discussed in the next section. The way maze provides literate proving functionality is based on three key principles:

1. No changes made to the proof tool being used.
2. Minimal interference with the proof source files.
3. Minimal restrictions on the form of the proof or its presentation.

This is achieved by allowing the author to use arbitary XML markup to structure their proof source files and using comment syntax to 'hide' the markup from the proof tool. For instance, using maze an author of a Mizar article can extract parts of the article for ready inclusion in a LATEX document, whilst still writing a normal, checkable Mizar article. The principles embodied in maze are quite general, making it widely applicable. We have so far used it to produce literate versions of proofs from the Mizar, Isabelle and Phox systems. The key point is that regardless of the underlying formal system, maze provides a good level of presentation with very little effort on the part of the author.

In order to aid discussion, the maze system will be described before we expand on the principles behind the system and literate proving in general. Most of the examples provided are taken from Mizar, specifically [18], as Mizar represents a substantial library of formal mathematics that could conceivably benefit from being presented via maze. Examples are also given for other proving systems to demonstrate the flexibility of this approach. Also, we have included as an appendix a short article presenting the proof of the irrationality of e as presented in the Mizar library. The maths in the article was generated entirely automatically from the original Mizar article [15][1] using maze.

---

[1] http://www.mizar.org/JFM/

2   From Programming to Proving

In Knuth's conception of literate programming, the source code for a program and the documentation of a program would be written jointly in a single document. This document could then be transformed one of two ways to give either the executable version of the program code or the typeset version of the documentation, as depicted in Figure 1. In this way, the source code as described in the documentation and the source code as compiled were one and the same — there was no duplication through cut-and-paste and no slightly different versions between that presented and that compiled. Morever, the original document did not need to follow the logical structure of the code at all but instead the documentation and the code were interleaved and interlinked as best fitted a literary exposition. For this reason, Knuth called his system web.

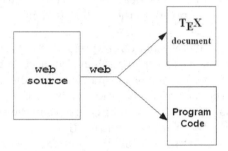

Fig. 1. The structure of Knuth's web

The drawback with this was that the program was actually written through a series of macros and that macro language needed to be learned. In addition, web and subsequent literate programming systems, such as cweb [13], tend to impose their own formatting. Though these native formats could no doubt be changed, they would require some intimate knowledge of the workings of the system.

Thus, the tightly coupled documentation and code of web imposes its own limitations. With regard to say the Mizar library, a literate proving system of this sort would also require refactoring significant parts of the Mizar library to be able to generate the original articles and their corresponding documentation.

Also, it must be noted that though the motivations of literate proving are similar to those of literate programming, the goals of literate proving are not. For example, the same piece of mathematics may be used in different ways. For instance, a description of the real line may strongly vary depending on whether it is to be used in developing the theory of fields or as a prototypical metric space. It is unlikely that similar concerns would arise in program description.

This suggests that a literate proving system would require rather looser coupling between the proofs and their documentation. A more recent literate programming system, Warp, shows a possible way forward [14]. With warp, the literate programming system is essentially a way of extracting code from a program and into manageable chunks. These chunks can be incorporated, in any

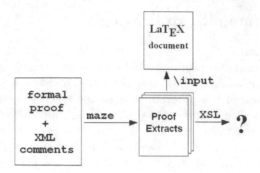

Fig. 2. The structure of maze

order, into a larger document that describes what they are and also typesets them appropriately. This is precisely the model used in maze and is shown in Figure 2.

With maze then, the principles of literate programming are maintained. That is, there is only one version of the formal mathematics and there is a freely flexible relationship between the description of the mathematics and its logical structure. This clearly comes at the cost of a separation between the formal mathematics and the documentation. The risk then is of course that the source mathematics could change independently of its description. However, the author who cared about the description would still be obliged to update the documentation if it were to make sense.

The next section gives details of how maze extracts chunks of Mizar articles. The interested reader is refered to Thimbleby's paper on warp for a deeper discussion of the various types of literate programming systems and their relative merits and drawbacks [14].

3   maze

For maze to extract the required parts of a formal proof script, it is necessary to insert markers in the proof. Of course, any markers will necessarily stand out from the formal language. To prevent the markers from interfering in the validation of the proof, the comment tags for the article are used to hide the maze instructions from the formal system. The instruction to maze, within a comment, is in the form of an XML tag. For example, in Mizar, the comment tag is :: and so the tags marking the beginning and end of a portion to extract would be:

```
::<maze id="extract">
...Mizar article...
::</maze>
```

As you might expect, when maze encounters this pair of tags, everything between (and including) the tags is written to the file extract.xml.

Note that apart from the comment tags for the particular formal language, maze requires no other knowledge of the underlying language. This makes adapting it to other formal languages relatively straightforward.

The XML style, following Warp [14], seems a natural choice. A model of maze is that it extracts any portion of an article between such tags and in the process drops the comment markers from any such tags. Any non-XML comments are marked with <comment> tags. From this simple process, to use the extracts in documents such as LATEX or HTML, further transformations of the extracted parts are optionally performed using XSL. For example from the marked up section of Mizar as follows:

```
::<maze id="demo">
::<theorem><statement>
theorem Th21:
 -K = -Q implies K = Q
::</statement><proof>
  proof
    ::Uses the natural property of inverses
  --K = K & --Q = Q
   ::<ref>
   by PRE_TOPC:20;
   ::</ref>
  hence thesis;
  end;
::</proof></theorem></maze>
```

maze produces the following:

```
<maze id="demo">
<theorem>
<statement>
theorem Th21:
 -K = -Q implies K = Q
</statement>
<proof>
  proof
<comment>
Uses the natural property of inverses
</comment>
  --K = K & --Q = Q
<ref>
   by PRE_TOPC:20;
</ref>
  hence thesis;
  end;
</proof>
</theorem>
</maze>
```

However, rather than require an author to also have knowledge of XSL or to spend time on developing an appropriate stylesheet, `maze` has certain transformations built-in. These are described in the following subsections.

3.1   maze for LaTeX

The simplest method for getting text into LaTeX is to have `maze` produce plain text extracts within the LaTeX `verbatim` environment. If the target, though, is truly readable LaTeX then of course a `verbatim` version of text falls short of nicely presented mathematics in several ways. Specifically concentrating on Mizar, the Mizar language is entirely in standard ASCII text and so lacks the finesse of the rich character sets and symbols that LaTeX has. For this reason, `maze` is also able to produce a pretty version of the output from Mizar articles suitable for LaTeX. For example, here is a theorem and its proof that has been pretty printed from [18]:

THEOREM Th_{20} :
$Kc = Q$ iff K misses $-Q$
PROOF
$A_1 : -Q = Q'$ by STRUCT$_0$:def 5;
hereby assume $Kc = Q$; then
$K \setminus Q =$ by XBOOLE$_{-1}$: 37; then
$K/ \setminus Q' =$ by SUBSET$_{-1}$: 32;
hence K misses $-Q$ by A_1, XBOOLE$_0$:def 7;
END;
assume K misses $-Q$; then
$K/ \setminus -Q =$ by XBOOLE$_0$:def 7; then
$K \setminus Q =$ by A_1, SUBSET$_{-1}$: 32;
hence thesis by XBOOLE$_{-1}$: 37;
END;

The pretty printing follows some simple heuristics to produce this output. These are as follows. First, single characters have been interpreted as variables and so are put into the math environment. Trailing numerals after a string of letters are assumed to be subscripts. Thirdly, keywords such as "theorem" have been put into the small caps font. Finally, characters that would normally correspond to LaTeX control codes have been appropriately converted to appear correctly in the final document.

Though the output is obviously more "latex-y", it could not really be called pretty. Symbols that are a natural feature of Mizar such as `c=` for subset, appear peculiar in LaTeX as $c =$. Though heuristics could also be developed to handle these, they would begin to require more detailed parsing of the Mizar source and this is not without hazard or even always possible [5]. Instead, a lighter method has been chosen that places the control in the hands of the author. When extracting text for pretty LaTeX, `maze` is also able to consult what we have called a match file. In this, strings from the Mizar article appearing in the match file have replacement LaTeX code that is used instead of the string. The match file takes priority over any of the other heuristics. The result appears like this:

THEOREM Th$_{20}$:
$K \subseteq Q$ iff K misses $-Q$
PROOF
$A_1 : -Q = Q'$ by `struct_0` :def 5;
hereby assume $K \subseteq Q$; then
$K \setminus Q = \emptyset$ by `xboole_1` : 37; then
$K \cap Q' = \emptyset$ by `subset_1` : 32;
hence K misses $-Q$ by A_1, `xboole_0` :def 7;
END;
assume K misses $-Q$; then
$K \cap -Q = \emptyset$ by `xboole_0` :def 7; then
$K \setminus Q = \emptyset$ by A_1, `subset_1` : 32;
hence thesis by `xboole_1` : 37;
END;

In this example, the ASCII representations of some symbols have been replaced with the corresponding LaTeX symbols. Other symbols have been replaced with a more usual one such as the replacement of {} with $\emptyset$. Also, the match file need not be confined to mathematical symbols and here has been used to make the references to other articles appear in a different font. The match file that was used here consisted of only eight lines of text, one line per match and its replacement, and so could easily be produced by an author.

The example in appendix A uses the match file extensively to help produce a version of the Mizar proof that uses the standard summation notation for series. The appendix also uses `skip` tags to help present the proof — these work as described in the next section but for pretty printing insert ellipsis to show omitted material.

3.2   Proof Sketching with maze

Literate proving, like literate programming, should strive to be about producing a document that not only elucidates the proofs presented but also makes that experience enjoyable to the human reader. However, as is well known, formal mathematics of the sort found in the Mizar library has a tendency to be verbose and for key insights to be lost in the detail [16, 6]. To aid in exposition and enjoyment, some form of simplification is often necessary.

For this reason, maze is able to automatically generate proof sketches. The form chosen was that proposed by Wiedijk as a way to aid constructing formal proofs in Mizar [17]. The two things that distinguish a formal proof sketch from a completed Mizar proof are: references in support of proofs steps are always omitted; some proof steps are also omitted. The sketch should therefore be a summary of the most salient proof steps of the formal proof.

The omission of references is easily achieved automatically in maze since all references are indicated by the keyword by. In contrast, the assessment of which proof steps are most salient can only be made by the author. For this reason, a second maze specific tag was introduced, namely skip. Where skip tags are placed within maze tags, all of the text except the text marked up to be skipped is placed in the file specified by the maze tag. The result is put in the verbatim environment. Here is an example of a complete Mizar proof that has been sketched to appear in this article:

```
theorem
 P is dense & Q is dense & Q is open implies P /\ Q is dense
 proof
  assume that A1:P is dense and A2:Q is dense and A3:Q is open;
    [#] TS c= Cl(P /\ Q)
    proof
     now let C be Subset of TS; assume A7: C is open;
      assume x in C;
      then  Q meets C   then
      A8:Q /\ C <> {}
      Q /\ C is open
      then P meets (Q /\ C)   then
            hence (P /\ Q) meets C
    end;
  hence thesis
  end;
   then Cl(P /\ Q) = [#] TS
  hence thesis
 end;
```

Of course, this output could also be pretty printed if required. Any automation of the proof sketching necessarily needs to know something about the structure of the formal language. Currently, therefore, maze only does proof sketching for Mizar.

3.3   Tactic-Style Proof Assistants

Having illustrated maze's functionality using Mizar, we now briefly show the versatility of the system with example based on the Phox system [12]. Phox is a proof assistant with a tactic-style form of interaction: the user specifies how the system should transform the proof state, in a similar fashion to Isabelle, HOL and many other proof tools. This is a very different form of interaction to the declarative proof style of Mizar — nonetheless, maze can still be used for literate proving.

As with Mizar, a section of Phox proof file can be marked up freely by the author, as follows:

```
(* <maze id="square"><theorem> *)
(* Product of squares equal to square of the product. *)
(* <statement> *)
fact square.mult /\x,y (square x * square y = square (x * y)).
(* </statement><proof><skip> *)
```

```
intros.
(* </skip><rewrite> *)
unfold square.
(* </rewrite> *)
(* (x.x).(y.y) = (x.y).(x.y) *)
(* <calc> *)
rewrite -p 1 mult.assc.R.
rewrite -p 4 mult.comm.R.
(* </calc> *)
(* = (x.(x.y)).y *)
(* <final> *)
rewrite mult.assc.R.
(* <skip> *)
trivial.
(* </skip></final></proof></theorem></maze> *)
```

Here the author has chosen to mark up the short six-line proof as three steps:
two `<rewrite>` steps (one line each) and a `<calc>` step (two lines), with the
trivial first and last lines skipped over. They have also annotated the script with
comments that declarative describe some of the proof states. This is transformed
into the following XML:

```
<maze id="square">
<theorem>
<comment>
Product of squares equal to square of the product.
</comment>
<statement>
fact square.mult /\x,y (square x * square y = square (x * y)).
</statement>
<proof>
<rewrite>
unfold square.
</rewrite>
<comment>(x.x).(y.y) = (x.y).(x.y)</comment>
<calc>
rewrite -p 1 mult.assc.R.
rewrite -p 4 mult.comm.R.
</calc>
<comment>= (x.(x.y)).y</comment>
<final>
rewrite mult.assc.R.
</final>
</proof>
</theorem>
</maze>
```

This example illusrates how maze allows the author to freely annotate and
structure their proofs, and consequently to present them in any way they like.
This freedom requires the author to design their own presentations (or to use

others designs), but allows them to design and present literate formal proofs, where the structure is tailored to suit their presentation needs.

3.4 Implementation Details

maze is implemented in Java with the formal source being either passed as a command line parameter or through the standard input. A single source file may have multiple `<maze>` sections extracted, each going to an individual file as specified in the id attribute. If id has value foo the output file is foo.xml, if it is omitted the output is sent to the standard output.

The system can configured for a particular proof tool by providing it with a simple description of the system's comment syntax. The different transforms described in the paper are produced by engaging various modes via command line options. These modes are:

Raw: Extracts data without any changes.
Text: XML tags are not shown in output.
All: Data within of skip tags is extracted.
Verb: Places a LATEX verbatim environment around the output.
Suffix: changes the filename suffix of the output file.
Help: produces a summary of the flags and match file structure.

maze is freely available along with all the source code used in the production of this article from the author or from the web-site: www.uclic.ucl.ac.uk/paul.

4   The Principles of Literate Proving

First and foremost, it is worth noting that maze embodies a generic set of principles, like the Warp literate programming system [14], that are not specific to this particular implementation. These are:

1. The use of "commenting out" for maze instructions so that they are ignored by a proof checker
2. Proof extracts being XML marked up for further transformations
3. The match file concept to aid pretty printing without deep semantic knowledge of the formal language
4. The use of skip to aid proof sketching
5. A small and simple instruction set (currently two commands).

Of course, there are specific features such as the pretty printing of keywords in small caps and how proof sketching is performed that are specific to Mizar. Even then, it should be noted that Mizar, as a large body of formal mathematics, is being used to demonstrate literate proving but maze is not deeply entangled in that particularly library. maze can be trivially reconfigured to work with other proof tools — we have so far also used it with Phox and Isabelle. Where literate proving differs across these systems is not in the nature of the tool but actually what it means to make the different sorts of formal proofs literate:

contrast the use of maze with Mizar's declarative proofs against Phox's procedural proof scripts. Literate proving allows us to attach some structured declarative information to selected procedural steps, greatly improving proof presentation for these systems.

The general concept of literate proving though does seem to have separate concerns from literate programming. Literate programming is concerned with producing programmes that are enjoyable for human readers to read. Likewise, literate proving is about producing formal proofs that are enjoyable for humans to read. Literate programming has the goal also to be faithful to the actual program, hence the close coupling of the code and the documentation. In maze, it is possible to have entirely faithful presentations of proofs but in fact it produces more readable documents if they are pretty printed and also sketched. Thus, the presentation of the proof can differ significantly from the original. This could have consequences where a person reading both literate proofs and the original proofs is unable to easily make a connection between the two. The impact of the superficial differences may be not be so superficial and only time will tell.

Tools such as maze provide a guarantee that faithfulness is preserved — the presented proof is automatically generated from the original. Of course, the author defines any match file used in pretty printing but it is to be hoped that an author would make logical or at least acceptable choices for ensuring that Mizar symbols are replaced with suitable LaTeX symbols. Where unusual choices of matching symbol have been used, the author should have some responsibility to explain their choice.

Another difference between literate programmming and proving is that the description of a program is deliberately quite specific to the tasks of that particular program. The documentation is therefore aimed at explaining how the program meets the overall tasks or how subtasks lead to the completion of the tasks. Standard algorithms such as quick sort would not be the most interesting ones to explain in a program's documentation. In contrast, literate proving could have a number of goals.

A literate proof could be along the lines of a literate program to explain how a particular formal proof represents a more traditional proof or how particular features of the formal language were used. Unlike programs though, proofs are objects of interest to mathematicians in and of themselves not just the tasks they achieve [19]. They are communication acts that lead to other ideas and mathematicians may wish to explain a proof in different ways depending on how that proof is used in a particular domain. For example, a diagonalisation proof may be explained very differently for a mathematician than for a computer scientist. Moreover, there can be no sense in which a given explanation that is not merely about language specifics could be sufficient for all time. Fortunately, maze's implementation is such that there is no constraint to have only one explanation but it provokes interesting questions as to the role of literate proving in the wider realm of mathematical knowledge management.

5   Related Work

Several proof tools already provide some form of document generation. With the
Coq system [3] comes the CoqDoc tool to transform Coq proof files into LaTeX
and HTML. CoqDoc allows pretty printing of formal objects, explanatory text
in comments (including LaTeX/HTML commands) and hiding of subparts of the
document. maze reproduces this functionality, and with an appropriate stylesheet
could produce identical presentations. However, although some presentation details may be specified, CoqDoc produces documents of a roughly fixed structure
and style. In contrast, with maze the author is free to choose both the structure
and style: any form of XML markup may be used and transformed to a range
of presentation formats and styles. While CoqDoc is well suited to producing
more readable versions of proof files in a standard LaTeX or HTML format, maze
allows much more flexible document generation.

We believe literate proving to be better supported by systems that give the
author freedom to markup and present formal proofs as she sees fit, rather than
imposing a standard structure that is bound to the underlying proof files. Apart
from CoqDoc, all other systems we are aware of suffer from similar drawbacks.
For example, Isabelle/Isar allows generation of LaTeX and HTML, but of a prede-
termined structure and style [11]. The HELM project [2] has developed an XML
generation tool for Coq, but the structure is that of the underlying proof terms,
and does not provide support of a more free-form development of literate proofs.

6   Conclusion

Literate proving is the analogue of literate programming for formal proof languages and maze is a system that implements literate proving as demonstrated
on the Mizar library. It is clear from maze that some straightforward principles
can be used to do literate proving for any formal mathematical language but
that some form of pretty printing and sketching is likely to be desirable. It is
also worth noting that literate proving is not entirely a mathematical version of
literate programming but rather literate descriptions of proofs could be manifold
and need updating as more mathematics is learned. Quite how literate proofs
may fit into the wider body of mathematical knowledge remains to be seen.

Acknowledgements

Thanks to Harold Thimbleby for his comments on both the paper and the maze
system.

References

1. A. Asperti, B. Buchberger & J.H. Davenport (2003), editors, Mathematical Knowledge Management, Proceedings of MKM 2003. LNCS **2594**, Springer.
2. A. Asperti et al (2003), Mathematical Knowledge Management in HELM. *Annals
   of Mathematics and Artificial Intelligence*, **38**(1):27–46.

3. Y. Bertot & P. Castéran (2004), *Coq'Art: The Calculus of Inductive Constructions.* Springer. See also `http://coq.inria.fr/`
4. P. Cairns & J. Gow (2003), A theoretical analysis of hierarchical proofs. In [1], pp175–187.
5. P. Cairns & J. Gow (2004), Using and parsing Mizar. *Electronic Notes in Theoretical Computer Sci.*, **93**:60-69.
6. P. Cairns, J. Gow & P. Collins (2003), On dynamically presenting a topology course. *Annals of Mathematics and Artificial Intelligence*, **38**:91–104.
7. L. Cruz-Filipe, H. Geuvers & F. Wiedijk (2004), C-CoRN, the constructive Coq repository at Nijmegen. In A. Asperti, G. Bancerek, A. Trybulec (eds), Mathematical Knowledge Management, Proceedings of MKM 2004. LNCS **3119**, Springer, pp88–103.
8. D. E. Knuth (1984), Literate programming. *The Computer Journal* **27**:97-111.
9. S. McConnell (1993), *Code Complete: A practical handbook of software construction.* Microsoft Press.
10. The Mizar Mathematical Library `http://mizar.org`
11. L. Paulson (1994), *Isabelle: a generic theorem prover.* Springer.
12. C. Raffalli & R. David (2002), Computer Assisted Teaching in Mathematics. *Proc. Workshop on 35 Years of Automath*, Edinburgh.
13. H. Thimbleby (1986), Experiences of 'literate programming' using cweb (a variant of Knuth's WEB). *The Computer Journal* **29**(3):201–211.
14. H. Thimbleby (2003). Explaining code for publication. *Software — Practice and Experience* **33**:975–1001.
15. F. Wiedijk (1999), Irrationality of *e*. *Journal of Formalized Mathematics* **11**(42).
16. F. Wiedijk (2000), *The De Bruijn Factor*, Poster at TPHOL 2000.
17. F. Wiedijk (2003), Formal Proof Sketches. Types for Proofs and Programs, proceedings of TYPES 2003. LNCS **3085**, Springer, pp378–393.
18. M. Wysocki, A. Darmochwał (1989), Subsets of topological spaces. *Journal of Formalized Mathematics* **1**(28).
19. C. Zinn (2004), *Understanding Informal Mathematical Discourse.* PhD Thesis, Arbeitsberichter des Instituts für Informatik, Friedrich-Alexander-Universität, **37**(4).

A   Example: The Irrationality of *e*

As an extended example we present a short article based on Freek Wiedijk's Mizar proof of the irrationality of *e* [15]. Here maze has been used to provide pretty printed extracts from a marked-up copy of the original Mizar article. Some of the extracts are sketches of the underlying formal proofs.

A.1   Overview

THEOREM
e is irrational

In the Mizar library, *e* is defined as the usual infinite sum. More explicitly, *e* is the sum of the sequence, eseq, where:

DEFINITION
func eseq $->$ Real_Sequence means
: Def_5 : for k holds it.$k = 1/(k!)$;

Briefly, the proof mainly considers the terms of eseq multiplied by $n!$ for some appropriately chosen n. In this case, if e were rational, the sum of the final n terms of eseq$\times n!$ would have to be an integer. However, this expression can be bounded by a geometric series and hence must be a positive integer strictly between 0 and 1 — a contradiction.

A.2   Bounds on the Terms of eseq $\times n!$

First we require two lemmas on the terms of eseq. The proof of the first lemma is a straightforward induction on k and tells us that each term of the sequence eseq $\times n!$ is bounded by a corresponding term in the geometric series.

THEOREM Th_{39}
$x = 1/(n+1)$ implies $(n!)/((n+k+1)!) <= x \uparrow (k+1)$

The second lemma gives us a bound for the tail of the series for e. Note that the bound is in fact the sum of the geometric series from the previous lemma.

THEOREM Th_{40} :
$n > 0 \& x = 1/(n+1)$ implies $n! \times \sum_{n+1}^{\infty}(\text{eseq}) <= x/(1-x)$
PROOF
$\vdots$

$A_4 : 0 < x \& x < 1$ by $A_1, A_2, \text{REAL\_2} : 127, \text{SQUARE\_1} : 2$;
deffunc $F(\text{Nat}) = x \uparrow (\$1 + 1)$;
consider seq being Real_Sequence such that
A_5 : for k holds seq.$k = F(k)$ from SEQ_1 :sch 1;
$\vdots$

then A_{10} : seq is summable $\& \sum(\text{seq}) = \text{seq}.0/(1-x)$ by $A_4, A_7, \text{SERIES\_1} : 29$;
$A_{11} : \sum(\text{seq}) = x/(1-x)$ by $A_6, A_8, A_9, \text{SERIES\_1} : 29$;
$A_{12} : (\text{eseq})_{k=n+1}^{\infty}$ is summable by $\text{Th}_{24}, \text{SERIES\_1} : 15$;
now let k;
$A_{13} : (n!(\#)((\text{eseq})_{k=n+1}^{\infty})).k = n! \times (((\text{eseq})_{k=n+1}^{\infty}).k)$ by SEQ_1 : 13
$. = n! \times \text{eseq}.(n+1+k)$ by SEQM_3 :def 9
$. = n! \times (1/((n+k+1)!))$ by Def_5
$. = n!/((n+k+1)!)$ by $\text{XCMPLX\_1} : 100$;
hence $(n!(\#)((\text{eseq})_{k=n+1}^{\infty})).k >= 0$ by Th_{34};
seq.$k = x \uparrow (k+1)$ by A_5;
hence $(n!(\#)((\text{eseq})_{k=n+1}^{\infty})).k <= \text{seq}.k$ by $A_1, A_{13}, \text{Th}_{39}$;
END;
then $\sum(n!(\#)((\text{eseq})_{k=n+1}^{\infty})) <= \sum(\text{seq})$ by $A_{10}, \text{SERIES\_1} : 24$;
hence $n! \times \sum_{n+1}^{\infty}(\text{eseq}) <= x/(1-x)$ by $A_{11}, A_{12}, \text{SERIES\_1} : 13$;
END;

A.3   The Proof of the Irrationality of e

PROOF
assume e is rational;
then consider n such that $A_1 : n >= 2 \& n! \times e$ is integer by Th_{32};
$A_2 : n! \times e = n! \times ((\sum_1^n (\text{eseq})) + \sum_{n+1}^\infty (\text{eseq}))$
by $\text{Def}_6, \text{Th}_{24}, \text{SERIES\_1} : 18$
$. = n! \times (\sum_1^n (\text{eseq})) + n! \times \sum_{n+1}^\infty (\text{eseq})$;
reconsider $N = n! \times e$ as Integer by A_1;
reconsider $N' = n! \times \sum_1^n (\text{eseq})$ as Integer by Th_{38};
$A_3 : n! \times \sum_{n+1}^\infty (\text{eseq}) = N - N'$ by A_2;
set $x = 1/(n+1)$;
$A_4 : x/(1-x) < 1$ by A_1, Th_{41};
$n > 0$ by A_1;
then $n! \times \sum_{n+1}^\infty (\text{eseq}) <= x/(1-x)$ by Th_{40};
then $n! \times \sum_{n+1}^\infty (\text{eseq}) < 0 + 1$ by $A_4, \text{AXIOMS} : 22$;
then $n! \times \sum_{n+1}^\infty (\text{eseq}) <= 0$ by $A_3, \text{INT\_1} : 20$;
hence contradiction by Th_{36};

Semantic Matching for Mathematical Services

William Naylor and Julian Padget

Department of Computer Science, University of Bath, UK
{wn, jap}@cs.bath.ac.uk

Abstract. The amount of machine oriented data on the web as well as the deployment of agent/Web Services are simultaneously increasing. This poses a service-discovery problem for client agents wishing to discover Web Services to perform tasks. We discuss a prototype mathematical service broker and look at an approach to circumventing the ambiguities arising from alternative but equivalent mathematical representations occurring in mathematical descriptions of tasks and capabilities.

1 Introduction

The amount of machine-oriented data on the Web is increasing rapidly as semantic Web technologies achieve greater up-take. At the same time, the deployment of agent/Web Services is increasing and together they create a problem for software agents that is the analog of the human user searching for relevant HTML pages. In this paper we discuss the problem of reasoning about the semantics of the description of a mathematical query and how it relates to descriptions of services that may be suitable for solving it. This discussion takes place in the context of a broker architecture (shown in figure 1), specifically targeted at the discovery of mathematical services.

If a service is to be found and used, it must advertise itself. There are many generic aspects of a service and several specific to mathematical services, but for the purposes of this paper, we will concentrate on four, *viz.* the (i) signatures of inputs, (ii) signatures of outputs, (iii) pre-conditions and (iv) post-conditions, which specify the service's requirements and capabilities.

The role of the broker is to act as an intelligent mediator between clients and services, selecting services that match the client's problem statement (task). This task description must specify the signatures of inputs, outputs and the pre- and post-conditions that characterise the service that is sought. The broker's job is to identify service(s) or combinations of services satisfying the attributes given in the task. One of the major problems we address here is how to deal with the alternative but equivalent representations that occur in mathematical descriptions of task and capability.

The remainder of this paper is laid out as follows: In Section 2 we summarise the various XML schemas used in MONET/GENSS, focusing on Mathematical Service Description Language (MSDL), then in Sections 3 and 4 we address the question of how to canonicalise the mathematical descriptions in order to deal with the differences that arise from the many alternative but equivalent representations. Having established how to restructure the mathematical information for ease of comparison, Section 5 defines a similarity measure on which to base a ranking of matching services. Section 6 discusses the matching process.

M. Kohlhase (Ed.): MKM 2005, LNAI 3863, pp. 174–189, 2006.

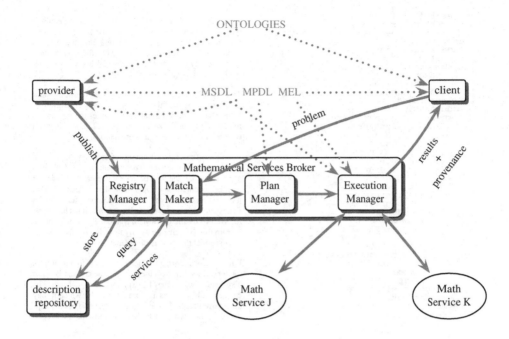

Fig. 1. MONET/GENSS – Brokerage Architecture

2 Encodings for Mathematical Services

In the MONET [15, 5] and GENSS [21] projects the objective was and is (respectively) mathematical problem solving through service discovery and composition by means of intelligent brokerage [1]. Various ontologies were defined as part of the MONET project which are utilised in our brokerage mechanism. Schemas using these ontologies include the following:

- Mathematical Service Description Language (MSDL), see [3], is an extension of WSDL that incorporates more (mathematical) information about a service, in particular pre- and post-conditions, taxonomic references *etc.*. An example MSDL document describing a factorisation service for square-free integers is shown in Figure 2[1]. It is this language which we shall be mostly concerned with in this paper, as it allows representation of all of the concepts we require for service discovery.
- Mathematical Problem Description Language (MPDL) see [4]. This schema allows a client to pose a mathematical problem by specifying an MSDL element, which embodies the problem, as part of the `output` or `post-condition` element,

---

[1] The prefix `monet` is used to denote the namespace:
  `http://monet.nag.co.uk/monet/ns`
whilst the prefix om is used to denote the namespace:
  `http://www.openmath.org/`

```
<monet:definitions
  targetNamespace=
    "http://monet.nag.co.uk/problems/">
<monet:problem
  name ="Factorisor">
  <monet:header>
    <monet:taxonomy
      taxonomy=
        "http://gams.nist.gov"
      code="GamsB"/>
  </monet:header>
  <monet:body>
    <monet:input name ="n">
      <monet:signature>
        <om:OMOBJ>
          <om:OMS cd ="setname1"
                  name ="Z"/>
        </om:OMOBJ>
      </monet:signature>
    </monet:input>
    <monet:output
      name ="list_factors">
      <monet:signature>
        <om:OMOBJ>
          <om:OMA>
            <om:OMS cd ="sts2"
                    name ="list"/>
            <om:OMS
              cd ="setname1"
              name ="Z"/>
          </om:OMA>
        </om:OMOBJ>
      </monet:signature>
    </monet:output>
    <monet:pre-condition>
      <om:OMOBJ>
        <om:OMBIND>
          <om:OMS cd ="quant1"
                  name ="forall"/>
          ...
```

**OpenMath markup for n
must be square free**

```
          ...
    </monet:pre-condition>
    <monet:post-condition>
      <om:OMOBJ>
        <om:OMA>
          <om:OMS cd ="relation1"
                  name ="eq"/>
          <om:OMV name ="n"/>
          ...
```

**OpenMath markup for the
product of the items in
list_factors**

```
          ...
  </monet:body>
</monet:problem>
</monet:definitions>
```

The name attribute to the monet:problem element may be used to identify the particular service description. In the MSDL file given in figure 2 the name of the service is Factorisor. The MSDL file may be held on a publicly (or restricted) visible Web server. So that the MSDL may be accessed via a URL.

The monet:taxonomy references an ontology via a URL, *e.g.* in figure 2 the GAMS taxonomy is referred to. It also has a code attribute which gives the position in the taxonomy of the service, in figure 2 code B is given which refers to numerical computations.

Each monet:input element refers to an input, *i.e.* a parameter to the service. The content of the element is a signature, *i.e.* the type of the parameter, this signature may be specified using OpenMath. The input element has a name attribute by which it may be referred to. In figure 2 the service takes one input which must have type referred to by the OpenMath Symbol (OMS) with name Z from the CD setname1. By referring to this CD, one may see that this type is the integers.

Each monet:output element refers to an output, *i.e.* a return value from the service. The content is again a signature, which may be expressed in OpenMath. The output element has a name attribute by which it may be referred to. In figure 2 the OpenMath signature represents a list, where each element must be an integer.

Each monet:pre-condition specifies some condition on the inputs which the service is allowed to assume is true. For example, a factorisor which can only factorise square free numbers may have a pre-condition which states that the input is a square free number.

Each monet:post-condition specifies some condition on the outputs from the service. This specifies a relationship between the inputs and the outputs of the service. For example a factorisation service might have a post-condition which says that the product of its outputs is equal to the input.

Fig. 2. MSDL description of a factorisation service

perhaps along with a directive specifying what to do with the problem (evaluate, lookup, prove, *etc.*). A service may then receive this message and perform the problem posed.

– Mathematical Explanation Language (MEL) see [7]. With this schema the service is able to return the results of a calculation to a client. Facilities also exist for returning an explanation of the results, this may be a trace of execution, the proof of a result, supporting evidence, a reference to a formulae evaluated *etc.*. A required element in this ontology is a reference to identify which problem this is replying to.

Naturally, documents using the above schemas must refer to mathematical objects at various points. The schemas are agnostic on how these objects should be described or what ontologies should be used. In practice there are two main contenders: (i) Content MathML [24] and (ii) OpenMath [22]. One of the advantages of MathML over OpenMath, is that it is more widely known, hence more tools have been written which understand it. Also it has a close association with Presentation MathML which makes the rendering of objects trivial. OpenMath has the advantage that it is an extensible language, whereas MathML only covers a fixed dialogue of concepts and requires an external definition mechanisms to define other concepts. Indeed, the MathML specification suggests OpenMath as this definition mechanism.

3   Normal Form

There are many equivalent ways to describe mathematical conditions: for example if $i, j \in \mathbb{Z}$, then $i \leq j$ and $i - 1 < j$ are equivalent. This implies that the descriptions of mathematical services may not be unique, and consequently creates a significant problem for the broker in identifying mathematical services, as the descriptions must necessarily often contain complex mathematical objects. This is the problem we currently face and we will now describe how we are tackling it. We have observed that most expressions take the form of $Q(L(R))$ where:

- Q is a quantifier block $e.g.$ $\forall x \exists y$ s.t. $\cdots$
- L is a block of logical connectives $e.g.$ $\wedge, \vee, \Rightarrow, \cdots$
- R is a block of relations. $e.g.$ $=, \leq, \geq, \neq, \in \cdots$

The block Q along with the input and output elements serve to define the scope of variables within the document. This will be relevant for steps 4 and 5 of our method. The block L consists of logical connectives which are relevant to step 1. The block R consists of relations which will restrict the value of one or more variables. Examples of relations are equalities or inequalities between expressions, or boolean valued set operators $(\subset, \subseteq, \in, \cdots)$.

Despite the fact there can be no absolute normal form [17], we can nevertheless carry out a sequence of transformations to construct a normal form suited to our brokerage task, thus:

1. The logical parts of the task and capability are rewritten in Disjunctive Normal Form (DNF) (see for example [19]), which is convenient for the calculation of similarity values (see §5). It also proves convenient for deducing dependencies allowing service composition, mentioned in Section 8. The transformations use a number of basic re-write rules, for example de Morgan's rule, distributivity of $\wedge$, over $\vee$, $etc.$.

   Example 1.
   $(a \wedge b) \Rightarrow \neg(c \vee \neg d)$ rewrites to the equivalent expression $\neg a \vee \neg b \vee (\neg c \wedge d)$

2. Associativity. Various operations are n-associative $e.g.$ $+, *, \cup, \cap, \wedge, \vee$, this means the operation takes n arguments and if the operation is denoted by $\otimes$, then: $a \otimes (b \otimes c) = (a \otimes b) \otimes c$. A natural form is to *flatten* the arguments, $i.e.$ provide each argument as an argument of the operation at the first level. $i.e.$

$$a \otimes (b \otimes c) \rightarrow \otimes(a, b, c) \text{ and } (a \otimes b) \otimes c \rightarrow \otimes(a, b, c)$$

3. The numerous domain specific mathematical equivalences that exist, *e.g.* the example mentioned in the first paragraph of Section 3, are addressed by means of a database of context sensitive rules.
   We use a number of techniques for deducing context information:
   (i) Performing a prior pass of the document and looking for constructs such as $x \in \mathbb{Z} \Rightarrow \cdots$.
   (ii) Scanning the `sts` (Small Type System [8]) files which record information about the signatures of OpenMath symbols.
   (iii) Scanning signatures from the input and output variables given in the MSDL.
   An important consideration is that cycles must not exists whilst applying these rules *i.e.* we must not have the two rules $A \rightarrow B$ and $B \rightarrow A$ concurrently in our system (or any generalisation of this case).
4. α conversion[2] addresses the situation where there are bound variables in the description as a result of quantification or a differential operator *etc.*. There is no reason for task and capability parameter (or result) names to coincide, but a straightforward transformation ensures they do, making subsequent manipulations simpler.

Example 2. Consider the situation where a capability exists, which knows how to integrate univariate functions. The description of the capability in the MSDL is likely to be expressed in terms of anonymous functions as:

$$R = \int \lambda : x \rightarrow f(x)\mathrm{dx} \tag{1}$$

here x is an anonymous variable and its name is independent of the meaning. The approach we use is to normalise the variables so that the normalised variable is dependent on the position in the document, for example if it is the 1st variable encountered in a pre-order traversal of the document tree, it might be normalised to n_1. In which case equation 1 would be normalised to:

$$R = \int \lambda : n_1 \rightarrow f(n_1)\mathrm{dn_1} \tag{2}$$

5. Commutative, like associative, operators offer opportunity for confusion since there will be a number of ways to represent an expression involving them. Our solution is to define an ordering on the elements of OpenMath objects, then when a capability is registered, the children of any commutative operations will be stored in order. A similar sorting is performed on task descriptions. As long as the sorting is structurally based this means that regardless of the ordering (as long as it is well founded), capability and task will be identical down to their leaves. There is, however, a problem with identifying variables which are direct descendants of a commutative operation; we deal with this by constructing equivalence classes of documents that are structurally identical modulo variable differences.
   We have defined a syntactic ordering relation on two OpenMath objects `o1` and `o2` as described in Algorithm 1, which may be used to enable this transformation.

---

[2] By which we mean consistent variable renaming.

Algorithm 1. OpenMath Ordering

---

if o1 and o2 have different element tags **then**
    OMI < OMF < OMSTR < OMB < OMS < OMA < OMBIND < OMV
else if if o1 and o2 are OMI elements **then**
    order on their content (an integer)
else if o1 and o2 are OMF elements **then**
    order on the numerical value of the dec or hex attribute[3]
else if o1 and o2 are OMSTR elements **then**
    order lexicographically on their content (a string)
else if o1 and o2 are OMB elements **then**
    order on the base64 (defined in reference [10]) content of the OMB elements
else if o1 and o2 are OMS elements **then**
    order lexicographically on the value of the cd attribute followed by the value of the name
    attribute
else if o1 and o2 are OMA elements **then**
    recursively order on their children (in document order)
else if o1 and o2 are both OMBIND elements **then**
    recursively order on:
       1)   first child
       2)   number of variables in the OMBVAR child
       3)   the third child
else
    o1 and o2 are OMV elements, these are treated equally.
end if

3.1   Complexity of Algorithm 1

If the result returned by algorithm 1 is true, its time complexity is $O(n)$ (worst case) where n is the size of the smallest OpenMath object. The best and average case are $O(1)$ coinciding with a false result.

An example which displays our method for overcoming the commutativity problem is the following:

Example 3. Consider a capability which can integrate piecewise functions, of the following form:

$$f(x) = \begin{cases} x < a & f(x) = g(x) \\ x > a & f(x) = h(x) \end{cases} \tag{3}$$

where the inputs are a, g and h. The restriction given by equation 3, where the right hand side is a piecewise construct, may be given as a pre-condition specified by the capability. Perhaps a task has specified that f is of the form:

$$f(x) = \begin{cases} x \geq a & f(x) = h(x) \\ x < a & f(x) = g(x) \end{cases} \tag{4}$$

This specifies the same condition, but in a different form *i.e.* piecewise is commutative. The pre-condition must be normalised, which results in the children of the

---

[3] We need not be concerned with the incomparability of floating point values as these are IEEE floating point values and thus a finite subset.

`piecewise` descendant of the `pre-condition` element being ordered using the ordering given in algorithm 1. We see that (in this ordering),

```
<om:OMS cd="relation1" name="geq"/>
```

is smaller than

```
<om:OMS cd="relation1" name="lt"/>
```

and thus the *part* of the function $f(x)$:

$$x \geq a \;\; f(x) = h(x)$$

is less than

$$x < a \;\; f(x) = g(x)$$

so the normal form chosen (before variable normalisation) for $f(x)$ is that given by equation 4.

3.2   Complexity of the Normalisation Process

The complexity of performing the above steps is clearly the sum of the complexity of performing each one of them, *viz.*

$$\text{total complexity} = C_L + C_A + C_{Meq} + C_\alpha + C_{com}$$

where:

i) C_L is the cost for performing the conversion to DNF. The worst case is $O(2^{n+1})$ [19] where n is the number of terms.

ii) C_A is the cost for flattening every associative operation

$$C_A = n_a \times |e|$$

where: n_a, e denote the number of associative operators, and the expression (the output from step 1), respectively and $|e|$ denotes the length of expression e.

iii) C_{Meq} is the cost for applying the theorems. This is independent of the number of theorems, as we use a hashing technique for applying the theorems. In fact

$$C_{Meq} = M_{th} \times C_{ap}$$

where: M_{th}, C_{ap} is the number of matched theorems and the cost for applying a specific theorem respectively. The time cost for application of a theorem will be dependent on the size of the sum of reduct and redex for this particular theorem, whilst the size cost will be dependent on the redex alone.

iv) C_α is the cost for performing the α-conversion

$$C_\alpha = n_v \times |S_T|$$

where: n_v, $|S_T|$ is the number of variables and the size of the subtree over which the α-conversion is being applied, respectively.

v) C_{com} is the time cost for resolving commutative operators

$$C_{com} = n_c(s_t + o_c \times o_t)$$

where: n_c, s_t, o_c, and o_t are the number of commutative operators, the search time (time complexity only), the number of occurrences of this particular operator and the order time for ordering the children of this operator (dependent on the number of children and the time taken for the OpenMath order).

Since n_a, M_{th}, n_v and n_c are constants and in most cases we expect $C_{ap}, |S_T|, s_t, o_c$ and o_t to be small the overall worst case complexity will be $O(2^{n+1})$.

4 Dealing with Inputs and Outputs

Users must necessarily specify inputs and outputs to any problem they wish to solve. This is so that the variables can be given names and referenced in the pre- and post-conditions. The user will know what type of inputs they have, so we can require them to provide the types. The user, however, will not know the name that the service uses or the order that the service specifies them in. This gives us a combinatorial (in the number of inputs/outputs with the same signatures) number of possible orderings for the inputs and outputs. Another way to deal with this problem is to impose a restriction on the user; that is to fix the names of the variables, or the order in which they occur. These two approaches need not be incompatible, as we may keep trying different orderings until a match has been found, or the required number of matches have been found. This may be made part of the normalisation scheme by adding a conjunction of set inclusions to the pre-conditions for the inputs, and post-conditions for the outputs. With each conjunction of set inclusions, a different order of *normalised* names must be forwarded to the α renaming process.

Example 4. If the MSDL of the query specifies the following:

inputs:	$x \in \mathbb{Z}$	**outputs:**	$R \in \mathbb{Z}$
	$y \in \mathbb{Z}$		
	$z \in \mathbb{Q}$		

pre-condition: $P_{pre}(x, y, z)$ **post-condition:** $P_{post}(x, y, z, R)$

where x, y, z are the names of the inputs, R is the name of the output and P_{pre}, P_{post} are predicates specifying the pre- and post- conditions respectively. This will be normalised to the following pair of conjunctions:

pre-condition: $n_1 \in \mathbb{Z} \wedge n_2 \in \mathbb{Z} \wedge n_3 \in \mathbb{Q} \wedge P_{pre}(n_1, n_2, n_3)$

and

post-condition: $n_1 \in \mathbb{Z} \wedge n_2 \in \mathbb{Z} \wedge n_3 \in \mathbb{Q} \wedge \tilde{n}_1 \in \mathbb{Z} \wedge P_{post}(n_1, n_2, n_3, \tilde{n}_1)$

where n_1, n_2, n_3 and $\tilde{n}_1$ are the normalised forms of x, y, z and R respectively. A second alternative may be appropriate for this query, *viz.*:

pre-condition: $n_1 \in \mathbb{Z} \wedge n_2 \in \mathbb{Z} \wedge n_3 \in \mathbb{Q} \wedge P_{pre}(n_2, n_1, n_3)$

and

post-condition: $n_1 \in \mathbb{Z} \wedge n_2 \in \mathbb{Z} \wedge n_3 \in \mathbb{Q} \wedge \tilde{n}_1 \in \mathbb{Z} \wedge P_{post}(n_2, n_1, n_3, \tilde{n}_1)$

Example 5. To consider a more concrete example, there may be a capability with MSDL:

inputs: $x \in \mathbb{Z}$	**outputs:** $d \in \mathbb{Z}$
$y \in \mathbb{Z}$	$r \in \mathbb{Z}$
post-condition: $P_{post}(x, y, r, d) \equiv d * y + r = x$	

A client issues the following task MSDL to the broker:

inputs: $x \in \mathbb{Z}$	**outputs:** $d \in \mathbb{Z}$
$y \in \mathbb{Z}$	$r \in \mathbb{Z}$
post-condition: $P_{post}(x, y, r, d) \equiv d * x + r = y$	

Now even though these MSDL documents are different, the broker must determine that the capability is relevant, and it may do this by trying different permutations of variables with the same types, during the normalisation process, until a match is found.

5   Calculating a Similarity Measure

Once the task description has been normalised, it can be compared with the capability descriptions registered with the broker, with the objective of calculating a similarity value. We denote the pre- and post-conditions of task and capability descriptions by T_{pre}, T_{post}, C_{pre} and C_{post}. We express the matching requirement between them as:

$$T_{pre} \Rightarrow C_{pre} \wedge C_{post} \Rightarrow T_{post}$$

That is to say, the pre-conditions of the capability must be satisfied by the pre-conditions of the task and the post-conditions of the task must be satisfied by the post-conditions of the capability. In all the following , we consider pre- and post- conditions in DNF, so $x \in C_{pre}$ means *x is a conjunct in the DNF for the capability pre-condition*. Superfluous capability pre-conditions (task post-conditions) do not effect whether the function may be performed. It is necessary, however, that there are no extra task pre-conditions (capability post-conditions) as this might allow the client to provide conditions incompatible with the capability pre-condition (capability post-condition). This may be formalised in the following:

$$\forall x_1 \in T_{pre} \; \exists y_1 \in C_{pre} \; \text{s.t.} \; x_1 \Rightarrow y_1 \tag{5}$$

and

$$\forall x_2 \in C_{post} \; \exists y_2 \in T_{post} \; \text{s.t.} \; x_2 \Rightarrow y_2 \tag{6}$$

One way of proceeding is to treat the pre- and post-conditions separately in order to get two similarity values S_{pre} and S_{post}. If it so happens that the pre- and post-conditions are equally important, then the average of these values will provide a good measure for the similarity value, however this will not always be the case, and other feasible measures are to weight S_{pre} and S_{post} linearly with the number of matching disjuncts in the pre-condition match as opposed to the post-condition match. This can be justified by observing that there are a linear number of different ways for the conditions to match.

We shall denote the DNF for C_{pre} (or T_{post}) by $R = R_1 \vee \cdots \vee R_n$ and for C_{post} (or T_{pre}) by $S = S_1 \vee \cdots \vee S_{\tilde{n}}$. To calculate a value $\in [0.0, 1.0]$ indicating how well equations 5, 6 are satisfied, we shall use the formula:

$$similarity(R, S) = \sum_{i=1..\tilde{n}} M_1(R, S_i) \frac{1}{\tilde{n}} \qquad (7)$$

where M_1 is a function which indicates how well the expression $S_i \Rightarrow R$ holds. This is equivalent to stating how well S_i matches with one of the conjuncts making up R. A good formula to calculate this is:

$$M_1(R, S_i) = \max_{j=1\cdots n} \{M_2(R_j, S_i)\} \qquad (8)$$

where M_2 is a similarity function for conjuncts. We may calculate a value for $M_2(R_i, S_j)$ as:

$$M_2(R_j, S_i) = \sum_{k=1\cdots\delta} m(R_j, S_{i,k}) \frac{1}{\delta} \qquad (9)$$

where δ is the number of terms in S_i, $S_{i,k}$ are terms in S_i and:

$$m(R_j, S_{i,k}) \text{ returns } \begin{array}{l} 1.0 \text{ if } S_{i,k} \text{ matches a term in } R_j, \\ 0.0 \text{ otherwise.} \end{array} \qquad (10)$$

The term matches which must be performed in order to evaluate the function given by expression 10 may be achieved in a variety of ways, some more effective than others. Two of these methods *algebraic match* and *value substitution match* are expounded in [13]. A third rather simplistic method is simply to perform a syntactic equivalence test on the XML. An important principle which must never be disregarded is that the term matches must be relatively cheap to perform, it is clearly ridiculous to perform an equivalence test once for every service (in the repository) which is as expensive as the service to be discovered!

5.1   Complexity of the Similarity Calculation

Theorem 1. *If the terms in each conjunct in the DNF are stored in order, then the average complexity for calculating the similarity value defined in Section 5 is:*

$$O(\overline{n}^2 \overline{m} \log_2(\overline{m}))$$

where $\overline{n}$, $\overline{m}$ are the average number of conjuncts in the DNF, average number of terms in each conjunct respectively.

If the terms are unordered, then the complexity is: $O(\overline{n}^2 \overline{m}^2)$

Algorithm 2. Register a capability

---

input: C_{MSDL} {MSDL of capability}
    : C_{URL} {URL of capability}

N_{MSDL} ←Normalise MSDL
{normalise the service MSDL, following Section 3}
Store $< N_{MSDL}, C_{MSDL}, C_{URL} >$ tuple in registry database
{we store the normalised form as this will save calculating it every time we have a look up, we must still store the original MSDL (for reference) and the capability URL (for access)}

The following abuse of notation shall be used. we shall say $O(f(x))$ when we mean the (time) complexity of the calculation of $f(x)$.

Proof. The complexity of calculation of $similarity(R, S) =$

$$O \left(\sum_{i=1..\tilde{n}} M_1(R, S_i) \frac{1}{\tilde{n}} \right) = O(\tilde{n}) \times O(M_1(R, S_i)) \tag{11}$$

$$= O(\tilde{n}) \times O \left(\max_{j=1 \cdots n} \{M_2(R_j, S_i)\} \right) \tag{12}$$

$$= O(\tilde{n} \times n) \times O(M_2(R_j, S_i)) \tag{13}$$

where $n, \tilde{n}$ are the number of conjuncts in the DNF for R, S respectively. Then:

$$O(M_2(R_j, S_i)) = O \left(\sum_{k=1 \cdots \delta} m(R_j, S_{i,k}) \frac{1}{\delta} \right) = O(\delta) \times O(m(R_j, S_{i,k})) \tag{14}$$

We see that if the terms are stored in order (using perhaps the ordering defined in Algorithm 1) a *binary chop* technique may be used to determine if $S_{i,k}$ occurs in R_j, its complexity is $O(\log_2 \tilde{m})$ where $\tilde{m}$ is the number of terms in R_j. To get a general complexity value, fix $n = \tilde{n} = \overline{n}$, the average number of conjuncts in each DNF and $\delta = \tilde{m} = \overline{m}$, the average number of disjuncts in each DNF respectively. Then:

average complexity $= O(\overline{n}^2 \overline{m} \log_2(\overline{m}))$

If the terms are not stored in order, then the search will take $O(\tilde{m})$ time, so:

average complexity $= O(\overline{n}^2 \overline{m}^2)$

6   Overall Matching Algorithm

Our matchmaking architecture is based around two main algorithms: The first is for registering capabilities in the database, this is detailed in algorithm 2.

The second takes the description of a task, it then returns an ordered list of the capabilities in the database ordered on their similarity as defined in section 5.

6.1   Scalability

The problem with the approach taken in algorithms 2 and 3 is that algorithm 2 is relatively cheap, the main cost being conversion to normal form, while algorithm 3 is not. It must execute the following steps:

Algorithm 3. Task Capability comparison

---

   input: T_{MSDL} {MSDL of task}
  output: Collection of triples, {consisting of:
           MSDL and URL of capability,
           the similarity value of T_{MSDL} to the capabilities MSDL}

$T_{N(\mathrm{MSDL})}$ ←normalised MSDL of the task
ret ←new Collection()
{This loop will accumulate the values to be returned}
for each entry in the registry database **do**
    C_{MSDL} ←this capabilities MSDL
    $C_{N(MSDL)}$ ←this capabilities normalised MSDL
    C_{URL} ←this capabilities URL
    S_{Val} ←similarity($T_{N(MSDL)}$, $C_{N(MSDL)}$) {The similarity value of the normalised task
    MSDL and the normalised capability MSDL}
    add($< C_{MSDL}, C_{URL}, S_{Val} >$ to ret {add the capabilities MSDL, URL and its similarity
    value to the task, to the collection to be returned}
end for
sort ret by the similarity values
return ret

---

1 Convert the query to normal form
2 Compare the query with every item in the registry
3 Sort the similarity values

This algorithm does not scale well because as the registry becomes large steps 2 and 3 become the overriding factors, as they depend on the size of the registry. Also the ordered list returned will be of length equal to the size of the registry, most of which will be irrelevant.

A better approach would be only to return the best match (or best few matches). An efficient implementation of this approach is not possible using the present data structures. A promising approach is centred on storing the terms in some well-founded order on OpenMath objects (perhaps that defined in Algorithm 1). The major cost then becomes registering a service, this is only done once. Looking up a service will then have the advantage that fast $O(log_2(n))$ algorithms may be utilised to look up terms. This is not a straightforward look up of the service, however, due to the non-bijective correspondences between task and capability descriptions.

7 Approaches to Matchmaking

Conventional service matching techniques [20], which mostly seem to rely on subsumption reasoning, view services in terms of pre-conditions and effects. Although this could be applicable to the domain we are exploring, it appears to lack the necessary precision and for the present we are investigating the use of pre- and post-conditions that define a *functional relationship* between the inputs and outputs.

It is quite natural to develop a specification of a mathematical service in terms of inputs, outputs, pre-conditions and post-conditions, where the post-condition may express quite complex mathematical relationships between the inputs and outputs. Although we have yet to explore fully the practicality of the conventional service matching approach, it seems that it might be possible to cast each such type as a concept in an ontological hierarchy—in effect not unlike hashing—assuming one could reason sufficiently accurately about inclusion relationships over the type information to locate the type correctly in the hierarchy. Consequently a problem description could be recast in the same manner and Description Logic (DL) reasoning applied to determine the applicability of a service or otherwise. This, however seems questionable, for two reasons:

(i) The richness of mathematical expression means that two equivalent descriptions may appear very different so that it is not clear how readily two equivalent descriptions (task and capability) might result in the identification of the correct nodes in the ontological hierarchy.

(ii) The work involved in determining equivalence—essentially using mathematical reasoning—to identify the above nodes effectively appears to offer the means to resolve the issue anyway if carried through to conclusion, such that using a DL reasoning step may be unhelpful and even misleading.

These two observations may appear to conflict, in that in each case apparently opposing arguments rest on the work involved in comparing the descriptions of the task and the capability. A resolution of this point is that in case (i) mathematical reasoning is only used so far as it enables DL reasoning to be applied, while in case (ii) mathematical reasoning is carried through to the conclusion of the process.

We also note that an approach in which each set of input, output, pre- and postconditions has the potential to induce a new node in the ontological hierarchy may result in a very large number of nodes, the benefit of which is not yet clear.

7.1   Review of Related Work

A variety of matchmaking systems have been reported in the literature, and we review some related systems below.

The SHADE (SHAred Dependency Engineering) matchmaker [12] operates over logic-based and structured text languages. The aim is to dynamically connect information sources. The matchmaking process is based on KQML (Knowledge Query and Manipulation Language) communication [9]. Content languages of SHADE are a subset of KIF (Knowledge Interchange Format) [11] as well as a structured logic representation called MAX (Meta-reasoning Architecture for "X"). Matchmaking is carried out solely by matching the content of advertisements and requests. There is no knowledge base and no inference performed.

COINS (COmmon INterest Seeker) [12] is a matchmaker which operates over free text. The motivation for the COINS is the need for matchmaking over large volumes of unstructured text on the Web or other Wide Area Networks and the impracticality of using traditional matchmakers in such an application domain. Initially the free text matchmaker was implemented as the central part of the COINS system but it turned out that it was also useful as a general purpose facility. As in SHADE the access language is KQML. The System for the Mechanical Analysis and Retrieval of Text (SMART) [18] information retrieval system is used to process free text.

LARKS (Language for Advertisement and Request for Knowledge Sharing) [20] was developed to enable interoperability between heterogeneous software agents and had a strong influence on the DAML-S specification. The system uses ontologies defined by a concept language ITL (Information Terminology Language). The technique used to calculate the similarity of ontological concepts involves the construction of a weighted associative network, where the weights indicate the belief in relationships. While it is argued that the weights can be set automatically by default, it is clear that the construction of realistically weighted relationships requires human involvement, which becomes a hard task when thousands of agents are available.

InfoSleuth [16] is a system for discovery and retrieval of information in open and dynamically changing environments. The brokering function provides reasoning over the advertised syntax and the semantics. InfoSleuth aims to support cooperation among several software agents for information discovery, where agents have roles as core, resource or ontology agents. A central service is the broker agent which is equipped with a matchmaker which matches agents that require services with agents that can provide those services.

The GRAPPA [23] (Generic Request Architecture for Passive Provider Agents) system allows multiple types of matchmaking mechanisms to be employed within a system. It is based on receiving arbitrary matchmaking offers and requests, where each offer and request consist of multiple criteria. Matching is achieved by applying distance functions which compute the similarities between the individual dimensions of an offer and a request. Using particular aggregate functions, the similarities are condensed to a single value and reported to the user.

MathBroker ([14] and [2]) is a project at RISC-Linz with some elements in common with those described here, including providing semantic descriptions of mathematical services. Caprotti *et al.*[6] note that MathBroker uses ebXML for service registration and discovery, while [2] suggests that most of the matchmaking is achieved through traversing taxonomies and states that actual understanding of the pre- and post-conditions is still an open problem.

The matchmaking mechanism demonstrated in the MONET project, as described in [6], was based on description logic reasoning over OWL service descriptions, where the query term is a concept or an expression over concepts. Some illustrative examples in [6] involve terms such as the GAMS class of the service, the name of the kind of algorithm, properties of the service platform, etc. Thus, the MONET broker represents a significant step forward in searching for services, by moving from syntax (keywords) to semantics (concepts), but the user must still then be in a position to select between the matches, which will typically require comparison of pre- and post-conditions. Indeed, as [6] state in their conclusions:

> Matching on pre- and post-conditions would be a very powerful technique, partly because it would give the user much more freedom in formulating problems but also because it would help automate service orchestration. Services whose post- and pre-conditions matched could be plugged-into each other. However it is a well-known fact that proving that two mathematical expressions are identical is in theory undecidable [17] (although there are many mechanisms which will solve a large class of cases in practice). Given that we have access to general purpose computer algebra systems such as Maple within the MONET framework it would not be too difficult for a broker to use them as oracles to decide whether two statements were equivalent.

That is the task that we are examining here.

Most of the projects above have focused on providing a generic matchmaker, capable of being adapted for a particular application. The motivation, however, for many such projects has primarily been e-commerce (as a means to match buyers with sellers, for instance). Some projects are also focused on the use of a particular multi-agent interaction language (such as KQML), to enable communication between the matchmaker and other agents. Our approach, however, is centred on the implementation of a matchmaker that is specific to mathematical relations. Similar to GRAPPA, our matchmaker can support multiple comparison techniques.

8   Future Directions

The above method for semantic matching of mathematical capabilities may be extended in a natural way to discover a composition of capabilities that may be used to perform some task. The algorithm required for calculating the similarity value for conjuncts given by expression (9) in section 5 requires determining a match between individual terms, this allows us to determine which conditions have and which conditions have not been satisfied (by the capability) or required (by the task). Future work involves designing effective methods for determining compositions of services where overall the task post-conditions are met and where the pre-conditions may be satisfied by a conjunct of the task pre-conditions and post-conditions of other services (whose pre-conditions have been met).

9   Conclusion

In this paper, we have considered some of the issues involved in mathematical service matching. We point out the ambiguities occurring in mathematics and suggest a way in which these may be circumvented by converting expressions occurring in the descriptive pre- and post-conditions into a normal form. We suggest a similarity value which measures how similar two services are to each other and analyse how the pre- and post-conditions of the task and the capability contribute to this similarity value. Finally we take a brief look at composition of services and how this ties into the above work.

Acknowledgements. The work reported here is partially supported by the Engineering and Physical Sciences Research Council of the United Kingdom under the Semantic Grids call of the e-Science program (grant reference GR/S44723/01).

References

1. M-L. Aird, W. Barbera Medina, and J. Padget. Brokerage for Mathematical Services in MONET. In L. Cavedon, Z. Maamar, D. Martin, and B. Benatallah, editors, *Extending Web Services Technologies*, volume 13 of *Multiagent Systems, Artificial Societies, and Simulated Organizations*. Springer, May 2005. ISBN: 0-387-23343-1.
2. R. Baraka, O. Caprotti, and W. Schreiner. A web registry for publishing and discovering mathematical services. In *2005 IEEE International Conference on e-Technology, e-Commerce, and e-Services (EEE 2005), 29 March–1 April 2005, Hong Kong, China*, pages 190–193. IEEE Computer Society, 2005.

3. S. Buswell, O. Caprotti, and M. Dewar. Mathematical service description language. Technical report, The MONET Project, 2004.
4. O. Caprotti, D. Carlisle, A.M. Cohen, and M. Dewar. Mathematical problem ontology. Technical report, The MONET Project, 2004.
5. O. Caprotti, M. Dewar, J.H. Davenport, and J. Padget. Mathematics on the (Semantic) Net. In Christoph Bussler, John Davies, Dieter Fensel, and Rudi Studer, editors, *Proceedings of the European Symposium on the Semantic Web*, volume 3053 of *LNCS*, pages 213–224. Springer Verlag, 2004. ISBN 3-540-21999-4.
6. O. Caprotti, Mike Dewar, and Daniele Turi. Mathematical service matching using description logic and owl. In *Mathematical Knowledge Management*, pages 73 – 87, 2004.
7. Y. Chicha, J.H. Davenport, and D. Roberts. Mathematical explanation ontology. Technical report, The MONET Project, 2004.
8. J.H. Davenport. A Small OpenMath Type System. *SIGSAM Bulletin*, 34(2):16–21, June 2000.
9. T. Finin, R. Fritzson, D. McKay, and R. McEntire. KQML as an agent communication language. In *Proceedings of 3rd International Conference on Information and Knowledge Management, pp. 456-463, 1994*.
10. N. Freed and N. Borenstein. Base64 Encoding. http://www.mhonarc.org/~ehood/MIME/2045/rfc2045.html#6.8.
11. M. Genesereth and R. Fikes. Knowledge interchange format, version 3.0 reference manual. Technical report, Computer Science Department, Stanford University, 1992. Available from http://www-ksl.stanford.edu/knowledge-sharing/papers/kif.ps.
12. D. Kuokka and L. Harada. Integrating information via matchmaking. *Intelligent Information Systems 6(2-3), pp. 261-279, 1996*.
13. S. Ludwig, O. Rana, W. Naylor, and J. Padget. Agent-based matchmaking of mathematical web services. *AAMAS*, 2005. To appear.
14. Mathbroker project. http://poseidon.risc.uni-linz.ac.at:8080.
15. MONET Consortium. MONET Home Page, www. Available from http://monet.nag.co.uk, last accessed May 2005.
16. M. Nodine, W. Bohrer, and A.H. Ngu. Semantic brokering over dynamic heterogenous data sources in infosleuth. In *Proceedings of the 15th International Conference on Data Engineering, pp. 358-365, 1999*.
17. D. Richardson. Some unsolvable problems involving elementary functions of a real variable. *Journal of Computational Logic*, 33:514–520, 1968.
18. G. Salton. *Automatic Text Processing*. Addison-Wesley, 1989.
19. K. Sutner. Disjunctive normal form. http://www-2.cs.cmu.edu/afs/andrew.cmu.edu/course/15/354/www/Lectures/PLogic.pdf.
20. K. Sycara, S. Widoff, M. Klusch, and J. Lu. *Larks: Dynamic matchmaking among heterogeneous software agents in cyberspace*. Kluwer, 2002.
21. The GENSS Project. GENSS Home Page, www. Available from http://genss.cs.bath.ac.uk, last accessed May 2005.
22. The OpenMath Society. The OpenMath Standard, October 2002. Available from http://www.openmath.org/standard/om11/omstd11.xml.
23. D. Veit. *Matchmaking in Electronic Markets*, volume 2882 of *LNCS*. Springer, 2003. Hot Topics.
24. W3C Recommendation. Mathematical Markup Language (MathML) Version 2.0, October 2003. Available from http://www.w3.org/TR/MathML2/.

Mathematical Knowledge Browser
with Automatic Hyperlink Detection*

Koji Nakagawa and Masakazu Suzuki

Faculty of Mathematics, Kyushu University,
Kyushu Univ. 36, Fukuoka 812-8581, Japan
{nakagawa, suzuki}@math.kyushu-u.ac.jp

Abstract. Mathematical OCR (Optical Character Recognition) systems retrieve character sequences and the structure of mathematical formulae from raster images scanned from mathematical documents. In this paper a method for detecting hyperlinks, e.g. formula links, from mathematical OCR output is described. We also experimentally demonstrated the effectiveness of the method. By using the method we implemented a prototype system of a mathematical knowledge browser that helps people read mathematical articles.

1 Introduction

An important activity in mathematics is the reading of articles or books. Recently mathematical knowledge has started to be stored and browsed in computers, but most mathematical knowledge is still stored and browsed in printed media. Computer assistance can help the reading activity by effective functionalities, e.g. navigation with hyperlinks, which are not possible in printed media. In [5] we presented the idea of a 'mathematical knowledge browser' that helps people read mathematical articles.

For the implementation of a mathematical knowledge browser we need a method of extracting the logical structure from mathematical articles and a method of detecting hyperlinks. The method of extracting logical structure was shown to be achieved in [5], and in this paper we propose a method to automatically detect hyperlinks, which then enable effective browsing of mathematical articles. Detection of hyperlinks from OCR output for general documents was achieved in [4]. However an attempt to achieve this in mathematical documents has not been realized. Mathematical documents have more intricate structures and more types of hyperlinks, e.g. formula links, than other documents.

In Section 2 we describe the functionalities of the mathematical knowledge browser and discuss hyperlink types. By using the hyperlink detection and the automatic logical structure extraction methods we implement a prototype mathematical knowledge browser. The implemented prototype is explained in

* This work is (partially) supported by Kyushu University 21st Century COE Program, Development of Dynamic Mathematics with High Functionality, of the Ministry of Education, Culture, Sports, Science and Technology of Japan.

M. Kohlhase (Ed.): MKM 2005, LNAI 3863, pp. 190–202, 2006.

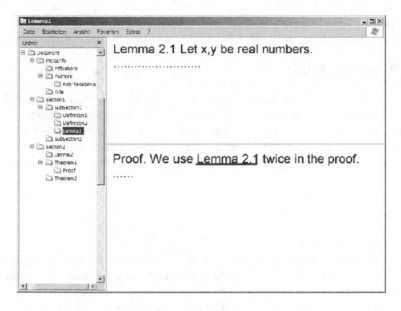

Fig. 1. Mathematical Knowledge Browser (Sketch)

Section 3. In Section 4 the hyperlink detection method is presented. An experimental result of the proposed method is described in Section 5. Finally, we conclude in Section 6.

2   Mathematical Knowledge Browser

The mathematical knowledge browser helps people read mathematical articles. One of the inputs for this mathematical knowledge browser is the printed mathematical document. Initially, the printed mathematical document can be scanned and processed by OCR. Then the logical structure and some hyper links are automatically extracted and shown to users.

2.1   User Interface

The mathematical knowledge browser consists of three panes: structure, reference and browser panes (Fig. 1). In the structure pane located on the left side, structural information is shown as a tree that shows the logical structure and links to mathematical components such as theorems or propositions. The browser pane at the right bottom and the reference pane at the right top show the same mathematical text, but can show different positions of the text.

While reading an article one is often tempted to view different parts of the article at the same time, e.g. by looking back at definitions, propositions, or formulae. By clicking on a source of a hyperlink in the browser pane, the text pointed to by the hyperlink will be shown in the reference pane. By browsing this way while reading one does not lose ones attention and so can better focus on the content. For example in Fig. 1, by clicking the hyperlink 'Lemma 2.1' of the browser pane the content of 'Lemma 2.1' appears in the reference pane.

2.2 Hyperlink Types

Hyperlinks facilitate browsing activities and enhance the readability of a document by effective navigation. There are two types of hyperlinks, internal and external. In an article an internal link points to a position within the article, while an external link points to a position in another information source. Fig. 2 shows some examples of hyperlinks. Here are possible internal and external hyperlinks.

Internal Hyperlinks

- **formulae number**
  In mathematical papers formulae are often numbered for reference purposes. A formula number is located at the left or right of a formula. For example in Fig. 2, '(0.1)' is an example of a formula number. Hyperlinks to the formula should be made in places where the string sequence '(0.1)' appears.
- **citation**
  An article cites other documents usually by bracketed strings, e.g. '[12]' or '[BR2]'. Detailed information of the cited documents is shown in the reference list at the end of the article. A hyperlink can be made from the place where the bracketed string is to the corresponding entry in the reference list.
- **mathematical components**
  One of distinct characteristics of mathematical articles is that there are mathematical components (e.g. Definition, Lemma and Theorem). Also in an article these mathematical components are often mainly referred to in proofs. For example, in text "By Lemma 2.4 it suffices to prove ..." the string 'Lemma 2.4' should link to the place where the description of 'Lemma 2.4' is.
- **headings (e.g. chapter, section, subsection)**
  In text, chapters or sections are sometimes referred to. For example, the sentence "This concept will be described in Section 2." can appear in the text. Then the string 'Section 2' should have a link to the description of 'Section 2'.
- **technical terms**
  If some new notions are introduced, they are named by special keywords. It is also convenient to have a hyperlink from the place where such a keyword appears to the place where the corresponding notion is introduced. However, it is difficult to recognize automatically those items. It should be solved in a different way.
- **footnote**
  A footnote identifier is usually written in the upper-script of a word at the end of a line of text. A link can be made from the footnote identifier to the footnote within the same page. See the example in Fig 2.

Ark. Mat., 35 (1997), 185-199
[PageHeader]

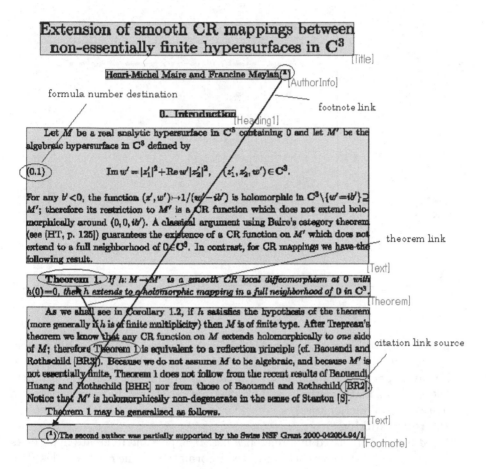

Extension of smooth CR mappings between non-essentially finite hypersurfaces in $\mathbf{C}^3$ [Title]

Henri-Michel Maire and Francine Meylan[*] [AuthorInfo]

formula number destination

0. Introduction [Heading1]

footnote link

Let M be a real analytic hypersurface in $\mathbf{C}^3$ containing 0 and let M' be the algebraic hypersurface in $\mathbf{C}^3$ defined by

$$(0.1) \qquad \operatorname{Im} w' = |z_1'|^2 + \operatorname{Re} w'|z_2'|^2, \quad (z_1', z_2', w') \in \mathbf{C}^3.$$

For any $b' < 0$, the function $(z', w') \mapsto 1/(w' - ib')$ is holomorphic in $\mathbf{C}^3 \setminus \{w' = ib'\} \supseteq M'$; therefore its restriction to M' is a CR function which does not extend holomorphically around $(0, 0, ib')$. A classical argument using Baire's category theorem (see [HT, p. 125]) guarantees the existence of a CR function on M which does not extend to a full neighborhood of $0 \in \mathbf{C}^3$. In contrast, for CR mappings we have the following result. [Text]

theorem link

Theorem 1. *If $h: M \to M'$ is a smooth CR local diffeomorphism at 0 with $h(0)=0$, then h extends to a holomorphic mapping in a full neighborhood of 0 in $\mathbf{C}^3$.* [Theorem]

As we shall see in Corollary 1.2, if h satisfies the hypothesis of the theorem (more generally if h is of finite multiplicity) then M is of finite type. After Treprean's theorem we know that any CR function on M extends holomorphically to *one* side of M; therefore Theorem 1 is equivalent to a reflection principle (cf. Baouendi and Rothschild [BR3]). Because we do not assume M to be algebraic, and because M' is not essentially finite, Theorem 1 does not follow from the recent results of Baouendi, Huang and Rothschild [BHR] nor from those of Baouendi and Rothschild [BR2]. Notice that M' is holomorphically non-degenerate in the sense of Stanton [S].

citation link source

Theorem 1 may be generalized as follows. [Text]

(1) The second author was partially supported by the Swiss NSF Grant 2000-042054.94/1 [Footnote]

Fig. 2. Examples of Internal Links

- **figure, table**

  A figure or a table can be identified by numbers separated by dots. Then a link should be made from a keyword such as 'Figure 2.3' or 'Table 1.2' to the place where the figure or the table is.

External Hyperlinks

- **common mathematical technical terms**

  There are common mathematical technical terms such as 'real number', 'group' or 'ring'. It would be useful to have links from these terms to online

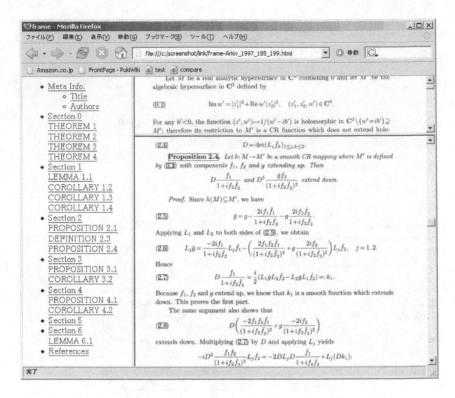

Fig. 3. Screenshot of Prototype Implementation

mathematical dictionaries, e.g. MathWorld[1], so that one can easily understand or recall the notions without having to physically look up books. Since it may happen that a common term can mean different concepts in different areas of mathematics, the link destination should be search pages of mathematical dictionaries.

– **reference linking**

Cited articles are listed in a reference list. It is possible to create hyperlinks from articles of the reference list to the information of the cited articles. The technology to identify articles is called 'reference linking'[1, 3]. The destination of such a hyperlink can be an entry of a mathematical review site[2] or the place where the article is.

3   Prototype Implementation

We implemented a prototype of our mathematical knowledge browser using ordinary functionalities of standard web browsers (Fig. 3). Here the article is shown as a sequence of bitmap images. Sources and destinations of hyperlinks are shown

---

[1] http://www.mathworld.com

[2] http://www.ams.org/mathscinet

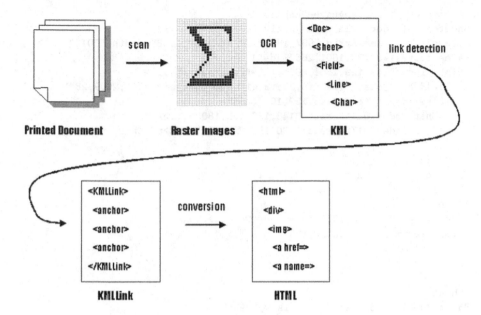

Fig. 4. Process Flow of Prototype Implementation

as surrounding boxes overlapping the bitmap images. The source of a hyperlink is colored red, and the destination green.

The process flow of the prototype implementation is shown in Fig. 4. At first, printed materials are scanned and then converted into raster image files. Then these images are processed by an OCR engine. We use an integrated OCR system for mathematical documents called INFTY[3][6]. INFTY reads the scanned page images of a mathematical document and provides character recognition results. One of the important characteristics of INFTY is that it can recognize two-dimensional mathematical expressions. The recognition result can be saved in a XML format called KML, which includes the results of logical structure analysis. A KML file is analyzed by a link detection program that produces the result in KMLLink format. From KMLLink and KML files some HTML files can be produced by a conversion program and browsed by ordinary web browsers.

The contributions this paper makes are the link detection program and the program for conversion into HTML. These programs are written in Python[8], which is a script language that conveniently handles XML.

3.1 KML: An OCR Result Format with Meta-information and Logical Structure

INFTY produces output in a XML format called KML. For example, Fig. 5 shows the output results in KML for the scanned image shown in Fig. 2. The top element is 'Doc' which contains some 'Sheet' elements representing pages. A 'Sheet'

---

[3] INFTY is freely available from http://www.inftyproject.org/en/

```
<Doc version="1.1" language="English" ...>
 <Sheet id="1" doc_file_name="Arkiv_1997.kml"
   image_file_name="Arkiv_1997_185.tif" height="4438" width="3015" ...>
  <Area rect="148,129,1801,266" id="1" ...>
   <Text rect="148,129,1801,266" tag="PageHeader" ...>
    <Field base_char_size="16,30,13,41" sub_char_size="11,20,9,28">
     <Line id="1" rect="148,129,1086,195">
      <Char code="0141" rect="148,133,195,180" ...>A</Char>
      <Char code="0172" rect="200,151,223,180" ...>r</Char>
      ...
     </Line>
     <Line id="2" rect="149,200,1801,266">
      ...
  </Area>
  <Area rect="279,948,3022,1239" id="2" ...>
   <Text rect="279,948,3022,1239" tag="Title" ...>
    ...
  </Area>
  ...
 </Sheet>
 <Sheet id="2" doc_file_name="Arkiv_1997.kml"
   image_file_name="Arkiv_1997_186.tif" height="4432" width="3002" ...>
  <Area rect="229,169,326,215" id="1">
   <Text rect="229,169,326,215" tag="PageNumber" ...>
    <Field base_char_size="16,28,12,39" sub_char_size="11,19,8,26">
     <Line id="1" rect="229,169,326,215">
    ...
   </Text>
  </Area>
  <Area rect="1088,168,2358,228" id="2">
   <Text rect="1088,168,2358,228" tag="PageHeader" ...>
    <Field base_char_size="16,29,12,40" sub_char_size="11,20,8,27">
     <Line id="1" rect="1088,168,2358,228">
   ...
  </Area>
  <Area rect="231,405,3224,701">
   <Text tag="Theorem">
    <Field>
     <Line id="1" rect="392,405,3224,486">
      <Char code="2154" rect="392,410,452,467" bold="1"...>T</Char>
      <Char code="2168" rect="458,409,506,467" bold="1"...>h</Char>
      ...
  </Area>
 <CharInfo>... </CharInfo>
</Doc>
```

Fig. 5. Example of KML Output from an INFTY OCR Engine

```
<KMLLink verion="1.0" page="23">
<anchor type="destination" kind="mathcomp" label="THEOREM 2.1"
        page="3" rect="106,3290,137,3340" />
...
<anchor type="source" kind="mathcomp" label="THEOREM 2.1"
        page="10" rect="2462,1734,2495,1789" />
...
</KMLLink>
```

Fig. 6. Example of KMLLink Format

element contains some 'Area' elements whose positions and sizes are indicated by 'rect' attributes. The value of the 'rect' attribute "*left,up,right,down*" indicates the positions of left, up, right, down borders, respectively, of the rectangle. An 'Area' element contains a 'Text' element having a 'Field' element. A 'Field' element has several 'Line' elements that again have several 'Char' elements.

To satisfy the need to put additional information for meta-information and logical structure, the 'tag' attribute for the 'Text' element exists to represent the type of the text field. The values of the 'tag' attribute are 'PageHeader', 'PageNumber', 'Caption', 'Title', 'AuthorInfo', 'AbstractHeader', 'Abstract', 'Keywords', 'Heading1', 'Heading2', 'Heading3', 'Heading4', 'Heading5', 'Text', 'Bibitem', 'Definition', 'Axiom', 'Theorem', 'MainTheorem', 'Proposition', 'Corollary', 'Lemma', and 'Footnote'.

3.2   KMLLink: Link Description Language for KML

The link detection program takes a KML file as input and produces results in the KMLLink format. Fig. 6 shows an example in KMLLink. The top element is 'KMLLink' that contains only 'anchor' elements. There are two types of 'anchor': 'source' and 'destination' specified by the 'type' attribute. The 'kind' attribute takes one of three values: 'citation', 'formula', or 'mathcomp' (mathematical component). The 'page' and 'rect' attributes specify a rectangle in a page. The 'label' attribute specifies the identifier of a link. A pair of 'source' and 'destination' anchors that have the same label indicate a link. For example, in Fig. 6 a pair of two anchors indicate a 'Theorem 2.1' link from a place in page '10' to a place in page '3'.

3.3   Conversion from KMLLink to HTML

The conversion program takes a KMLLink file and a KML file as input, and produces three HTML files:

- frame file,
- navigation file,
- content file.

The frame file forms the outline that contains three panes by using the 'FRAME' element of HTML. The content of the navigation pane is described

```
<div style="position:relative;top: 0px; left:0px;...">
 <img src="/images/InvM_1970_121_134-0.jpg">
 <a href="#Theorem 2.4" style="position:absolute;
                    left:397px; top:262px; width:5px; height:9px;
                    border: 1px solid red;" target="reference"></a>
 ...
 <a name="Theorem 2.4" style="position:absolute;
                     left:27px; top:528px; width:386px; height:80px;
                     border: 1px solid green;"> ...</a>
</div>
```

Fig. 7. HTML Realization of Hyperlinks over a Raster Image for a Page

in the navigation file. Production of the navigation file needs the result of the logical structure analysis that is stored in the KML file. Both the reference and the browse panes show the same content file in which scanned images are vertically allocated and browsed by scroll bars. (As images used for OCR are large for browsing, the size of scanned images is decreased by 15%.)

It is also possible to show the content in MathML(+HTML), since the INFTY OCR engine can recognize mathematical formulae. However, we chose raster images for showing pages because there are some miss-recognitions by OCR and it is not always the case that web browsers can display MathML properly.

In a content file, destinations and sources of hyperlinks are indicated by surrounding boxes that are realized by specifying the 'style' attribute of HTML. For example, a page is represented in Fig. 7. An image is shown by the 'img' element. The source of a hyperlink is realized by the 'a' element with the 'href' attribute. The 'target' attribute specifies the target window "reference", which represents the reference pane. The destination of a hyper link is realized by the 'a' element with the 'name' attribute. In the 'style' attribute, the positions and sizes of surrounding boxes are specified by 'left','top', 'width', and 'height' with 'border'.

4　Automatic Link Detection Method

In this paper we focus on the detection of three internal link types: formula, citation, and mathematical components; because these are especially useful in browsing. Other links will be the subject of future work. Automatic link detection can be achieved by looking for specific string patterns. A link is specified by its source and its destination. In most cases, the string pattern of the source and the destination of a link are the same.

Although we can not expect string patterns that will work for all articles, in this paper we use fixed patterns that should work in most cases. Fig. 8 shows the fixed patterns (regular expressions) used for detecting destinations and sources of links. Basically the algorithm looks for these fixed patterns line by line, and decides whether what it finds is a destination or a source.

For there to be more accurate recognition of links, there needs to be some mechanisms by which one can specify the string patterns of links. For example, in

kind	regular expression	example
formula	`\([0-9]+(\. ?[0-9]+)*\'?\)`	(2) (1.2) (3')
citation	`\[([^\[^\]]*)\]`	[2] [Mar80,Buc99]
mathcomp	`Theorem( [0-9]+(\. ?[0-9]+)*'?| [a-zA-Z]+|)` `Lemma( [0-9]+(\. ?[0-9]+)*'?| [a-zA-Z]+|)` `...`	Theorem 3.2 Lemma II

Fig. 8. Used Regular Expressions

[4] they define a link specification language called LITHP (Link Type description language for HyperText Processing) by which one can define link patterns.

Here the detailed algorithm is explained for each link.

4.1 Formula Link

Link Destination Detection. Sometimes a formula has a label written in a parenthesized number, or numbers separated by dots, e.g. '(2)' or '(2.1)' or '(2.2.3)', at the left or right of the formula. However all such labels do not necessarily become formula link destinations. For example:

> tion $e^{i\varphi}$ may be regarded as being rapid. In fact, even though an upper bound for α is not attained numerically (as in the case of Theorem 3), an upper bound furnished by (19)

Here '(19)' must not be recognized as a formula link destination. To avoid this problem, only the first occurrences of such labels are considered to be destinations of formula links, because in most cases these labels that are not destinations come after the formula is labeled.

Link Source Detection. All strings that match the regular expression for 'formula' in Fig. 8, and are the same as link destinations' labels are link sources.

4.2 Citation Link

Link Destination Detection. Destinations of citation links can be detected from the reference section. Usually an reference entry starts with either a bracket string (e.g. '[Buc2004]') or numbers with a dot (e.g. '12.').

Link Source Detection. A citation link source is usually written in the form of '[$str_1, \cdots, str_n$]'. However all $str_1, \cdots, str_n$ do not always indicate the source of citation links. For example, '[8, Theorem 3]' indicates 'Theorem 3' of the paper that is indicated by the citation number '8'. The label '[7, pp. 38]' indicates that it refers to the 38th page of the article cited by the number '7'. Another example is an interval notation '[a,b]'.

Here, the way to distinguish is that strings occurring in the reference list (link destinations) are considered to be citation sources. Then the examples 'Theorem 3' and 'pp. 38' are not strings of a citation source. Additionally after the first occurrence of a non-citation string, all such strings are considered to be non-citation strings. Namely suppose we have '$[str_1, \cdots, str_{i-1}, str_i, \cdots, str_n]$' and the strings from str_1 to str_{i-1} appear in the reference list, but str_i does not appear in the reference list, the strings from str_i to str_n are not considered to be citation sources. For example, let us consider the case '[8, Theorems 3, 4]'. Suppose '4' and '8' appear in the reference list. In this case '8' is considered to be a citation label, but '4' is not because a non-citation string 'Theorems 3' appears before '4'.

4.3   Mathematical Component Link

Link Destination Detection. Destinations can easily be detected after logical structure extraction, because in KML the 'Text' elements are tagged by keywords, e.g. 'Theorem'. The beginnings of such 'Text' elements are mathematical component link destinations. For example, in Fig. 2 'Theorem 1' is a mathematical component link destination.

Table 1. Experimental Result of Detecting Hyperlinks

paper ID	formula		citation		math. comp.	
	source	dest.	source	dest.	source	dest.
ActaM_1970_37_63	92/92[2]	54/55[1]	18/18[1]	7/7[0]	54/54[2]	16/16[0]
ActaM_1998_283_305	0/0[23]	0/0[5]	33/33[0]	12/12[0]	38/38[3]	30/34[4]
AIF_1970_493_498	0/0[0]	0/0[1]	6/6[0]	2/2[0]	4/7[0]	1/1[0]
AIF_1999_375_404	4/4[18]	1/4[3]	18/18[0]	12/12[0]	44/46[0]	34/34[0]
AnnMS_1971_157_173	2/2[3]	0/2[3]	17/18[0]	6/6[0]	6/12[3]	11/11[0]
AnnM_1970_550_569	55/55[0]	29/29[0]	24/24[0]	20/20[0]	40/46[0]	6/6[0]
Arkiv_1971_141_163	0/0[0]	3/3[0]	24/24[0]	7/7[0]	41/42[0]	24/24[0]
Arkiv_1997_185_199	53/53[4]	42/42[2]	24/24[0]	12/12[0]	30/32[2]	16/16[0]
ASENS_1970_273_284	0/0[0]	0/0[0]	32/32[0]	14/14[0]	9/9[2]	7/7[0]
ASENS_1997_367_384	0/0[13]	1/1[2]	33/33[0]	15/15[0]	34/41[3]	18/18[0]
BAMS_1971_157_159	7/7[0]	9/9[0]	7/7[0]	6/6[0]	6/6[0]	3/3[0]
BAMS_1971_160_163	0/5[0]	0/3[0]	6/6[0]	6/6[0]	1/1[0]	6/6[0]
BAMS_1974_1219_1222	0/0[0]	0/0[0]	6/6[0]	2/2[0]	0/4[0]	9/9[0]
BAMS_1998_123_143	0/0[0]	0/0[0]	113/113[0]	48/48[0]	33/35[0]	34/34[0]
BSMF_1970_165_192	18/18[0]	15/15[0]	71/71[0]	8/8[0]	41/50[0]	16/16[0]
BSMF_1998_245_271	50/50[0]	34/34[0]	41/41[0]	21/21[0]	37/48[0]	19/20[0]
InvM_1970_121_134	46/46[1]	19/19[1]	30/30[0]	7/7[0]	0/0[2]	2/2[0]
InvM_1999_163_181	31/31[26]	18/18[5]	0/18[0]	0/6[0]	0/3[21]	17/19[0]
JMKU_1971_181_194	16/16[8]	14/14[0]	10/11[0]	9/9[0]	16/20[10]	8/12[3]
JMKU_1971_373_375	0/0[4]	0/0[6]	6/6[0]	4/4[0]	1/1[1]	3/3[0]
JMS_1975_281_288	10/10[0]	6/6[0]	18/18[0]	11/11[0]	19/19[0]	16/16[0]
JMS_1975_289_293	3/3[0]	6/6[0]	6/6[0]	3/3[0]	1/2[0]	3/3[0]
JMS_1975_497_506	43/43[0]	43/43[0]	19/19[0]	10/10[0]	4/5[1]	7/7[0]
KJM_1999_17_36	42/42[2]	27/27[5]	22/22[0]	7/7[0]	40/40[5]	21/24[1]
MA_1977_275_292	38/39[0]	29/30[0]	24/24[0]	14/14[0]	11/13[2]	12/12[0]
MA_1999_175_196	39/39[1]	37/37[2]	36/36[0]	27/27[0]	13/15[1]	6/6[0]
TMJ_1973_317_331	0/0[0]	0/0[0]	19/22[0]	12/12[0]	17/17[0]	11/11[0]
TMJ_1973_333_338	0/0[0]	0/0[0]	6/9[0]	6/6[0]	6/6[0]	5/5[0]
TMJ_1990_163_193	109/116[53]	41/44[6]	57/65[0]	31/31[0]	51/64[0]	25/26[0]
Sum	658/671[158]	428/441[42]	726/760[1]	339/345[0]	597/676[58]	386/401[8]
	98.1%	97.1%	95.5%	98.3%	88.3%	96.3%

Link Source Detection. Strings that are the same as the strings of mathematical component link destinations are mathematical component link sources.

5 Experiment

To show the effectiveness of the link detection method we set up an experiment. A large-scale database of mathematical articles [6, 7] stored in KML was utilized. We randomly chose 29 English articles on pure mathematics (issued in 1970 - 1999) from different journals. Basically, an old and a new paper are chosen for each journal.

From the database in KML we made a correct KMLLink database and initiated the experiment (Table 1). A table entry is in the form '*success/all[excess]*'. '*all*' is the number of all correct elements. '*success*' is the number of elements successfully detected by the method presented in this paper. '*excess*' is the number of elements excessively detected by the method. Note that the decrease of the number of '*excess*' means better result. In total, the result achieved a 95.1% success rate with 267 excessively detected elements, which was 8.1% of all correct elements.

6 Conclusion

A method to detect several types of hyperlinks from printed mathematical documents was proposed. Using the method, we implemented a prototype mathematical knowledge browser. The authors believe that automatically detected hyperlinks make browsing of mathematical articles more effective. We intend to improve our hyperlink detection method and apply the improved version to larger scale databases.

In general, the style assumptions described in this paper do not work for mathematical documents whose styles are completely different. For example for citations some articles use a parenthesized form, e.g. '(Buc 2000)'. To adapt the system to such cases, a mechanism by which one can specify string patterns by regular expressions is needed. With such a mechanism, the system will work for exceptional cases.

For the prototype implementation of the mathematical knowledge browser we used standard web browsers, but for greater functionality we will need to implement standalone software. The following improvements can be considered for our mathematical knowledge browser:

- Elaborate Search
  Automatic detection of internal hyperlinks of technical keywords is a difficult task. A practical solution would be to provide an elaborate search functionality in the browser. By selecting a keyword in the browser and pressing a button, all words that are the same as the keyword in the paper will be marked and they can then be browsed sequentially.

- Showing Overview
  An article can be shown in an overview mode. For example, it is possible to show only the numbered mathematical formulae that appear in an article. In this way, one may be able to get the general idea of the paper.

Mathematical knowledge needs to be stored in a content-base format rather than a presentation-base format so that it can be used for various purposes. Mathematical knowledge should be store in a higher level format, because at this higher level practical usage is enhanced. However, currently most mathematical knowledge is stored in printed media and the situation will not change much without some action been undertaken. The technologies presented here support converting lower level formatted knowledge into higher level formatted knowledge. We hope that in the future people will store mathematical knowledge in a content-base format such as OMDoc[2].

References

1. Donna Bergmark. Automatic extraction of reference linking information from online documents. Technical report, 2000. CSTR 2000-1821.
2. M. Kohlhase. OMDoc: An Infrastructure for OpenMath Content Dictionary Information. *SIGSAM Bulletin (ACM Special Interest Group on Symbolic and Algebraic Manipulation)*, 34(2):43–48, 2000.
3. Steve Lawrence, C. Lee Giles, and Kurt Bollacker. Digital libraries and Autonomous Citation Indexing. *IEEE Computer*, 32(6):67–71, 1999.
4. A. Myka and U. Güntzer. Automatic Hypertext Conversion of Paper Document Collections. In *Digital Libraries: Current Issues, Digital Libraries Workshop, Newark, NJ, USA, May 19-20, 1994, Selected Papers*, volume 916 of *Lecture Notes in Computer Science*, pages 65–90. Springer, 1995.
5. K. Nakagawa, A. Nomura, and M. Suzuki. Extraction of Logical Structure from Articles in Mathematics. In A. Trybulec A. Asperti, G. Bancerek, editor, *Mathematical Knowledge Management, Third International Conference, MKM 2004, Bialowieza, Poland, September 19-21*, volume 3119 of *Lecture Notes in Computer Science*, pages 276–289. Springer, 2004.
6. M. Suzuki, F. Tamari, R. Fukuda, S. Uchida, and T. Kanahori. INFTY — An Integrated OCR System for Mathematical Documents. In *ACM Symposium on Document Engineering (DocEng '03), Grenoble, France, Nov. 20-22*, 2003.
7. S. Uchida, A. Nomura, and M. Suzuki. Quantitative analysis of mathematical documents. *International Journal on Document Analysis and Recognition*, 2005. ISSN: 1433-2833 (Paper) 1433-2825 (Online).
8. Guido van Rossum. *Python Reference Manual*. Python Software Foundation, release 2.4.1 edition, March 2005.

A Database of Glyphs for OCR of Mathematical Documents

Alan Sexton and Volker Sorge

School of Computer Science, University of Birmingham, UK
{A.P.Sexton, V.Sorge}@cs.bham.ac.uk
http://www.cs.bham.ac.uk/~aps|~vxs

Abstract. Automatic document analysis tools for mathematical texts are necessary to enlarge the pool of mathematical knowledge available in electronic form. However, development of such tools is currently hindered by the weakness of optical character recognition systems in dealing with the large range of mathematical symbols and the often subtle but important distinctions in font usage in mathematical texts. Research on developing better systems for mathematical optical character recognition crucially depends on having an extensive, high quality database of glyphs used in mathematical texts for training and test purposes. We present such a database of symbols constructed from a large set of characters available in the LaTeX document preparation system that can serve as a basis mathematical text recognition. We describe its integration into a prototypical system optical character recognition system for mathematics that enables the construction of LaTeX source documents from mathematical documents available as images. From the lessons learned in this work we derive a road map for further research into the area of mathematical text analysis.

1   Introduction

There is a conspicuous need to translate the knowledge locked in the existing large body of printed scientific documents into a more accessible, searchable and versatile electronic form. A critical component in the document analysis technology required to fill this need is effective and accurate optical character recognition for type-set mathematics. However, optical character recognition (OCR) of mathematical texts poses some special problems. Unlike non-scientific text, there is a very large range of symbols commonly used. The Comprehensive Latex Symbol List [5], for example, presents 2,826 different symbols available in LaTeX. More than that, while it is not uncommon to meet a mixture of only upright, italic and bold font faces in a non-mathematical text, the situation in mathematics is very different with a large number of font variants not only being common, but crucial to an understanding of the text because of significant semantic information being carried by often relatively subtle distinctions in font faces. Possibly because of the low current demand in the mass market for such functionality and the high cost in processing and memory overheads to provide it, commercial suppliers have not, to date, devoted a great deal of effort to

M. Kohlhase (Ed.): MKM 2005, LNAI 3863, pp. 203–216, 2006.
© Springer-Verlag Berlin Heidelberg 2006

the problems of these areas. Nevertheless, the automatic processing of scientific texts is highly desirable as their availability in electronic form would make their content more widely accessible by specialist audiences or by users with special needs.

Academic work on mathematical OCR (MOCR) and document analysis has developed since Anderson's initial work in 1968 [1], but the INFTY system [4], which is possibly the most advanced mathematical document analysis system currently extant, still relies on a mixture of a standard commercial OCR system for normal text and a custom mathematical symbol OCR system that caters for only 564 symbols (including the various alphabets) in mathematical sections. One of the problems that hinders development in MOCR is the lack of a suitable database of glyphs and symbols to use to train MOCR systems and upon which to base experimental development of scientific document analysis systems.

We have developed such a database by extracting symbols from a specially fabricated document containing approximately 5,300 different mathematical, scientific and textual symbols. This document is originally based on [5] and has been extended to cover most of the mathematical and textual alphabets and symbols currently freely available in LaTeX. We have integrated the database into our own experimental MOCR system, which derives from a metric based technique for font recognition [7]. In spite of the fact that it is still only in prototype form, the system already enables us already to process mathematical documents given as images and use the database to construct LaTeX source files that reproduce the input documents.

Our work could be seen as approaching the same goal as [8] but from the opposite direction. [8] aims at building a ground truth set of mathematical symbols by compiling a database of characters from a selected set of mathematical articles. While our database may not enjoy the same authority as a full ground truth set, it has more breadth in that it contains most supported LaTeX symbols rather than just the necessarily limited set contained in the publications considered. In particular, we can deal with the rapidly growing number of symbols used in diverse scientific disciplines such as computer science, logics, and chemistry.

We continually use the terms *character*, *symbol* and *glyph* in this document. We use the first two interchangeably to refer to the most elementary indivisible graphical element from a font directly available to a writer. A glyph, however, is more elementary in that it is a single connected component of a graphical element, one or more of which are required to make a symbol. Thus the three symbols "a", "%" and "≡" can be seen to be composed of 1, 2 and 3 glyphs, respectively.

In Sect. 2 we discuss some of the issues that must be addressed in the design of an MOCR system and their consequences for a glyph database. We describe the actual database and its contents in in Sect. 3. Section 4 discusses, using an example, the process that our experimental MOCR system applies in using the database to analyse the image of a document and reconstruct it from the database. We address some lessons learned and future work in Sect. 5 and summarise the final conclusions in Sect. 6.

2   Design Issues

The aim of our work is to design a database of symbols that can be used as a basis for a wide range of research in the field of MOCR. In particular, we wish to facilitate research on MOCR without prejudicing possible approaches to the problems involved. Hence we would like our database to be as neutral as possible with respect to the pattern recognition technologies that may be employed on its contents. Furthermore, we would like to make the system such that writing code to manipulate, analyse and process the database is relatively simple so that even inexperienced users can use it for worthwhile OCR and MOCR projects. In this way, we believe, we lower the barrier to entry to research in the field.

For this reason, we have designed a simple, if rather wasteful, LaTeX file, where symbols are listed with identification and calibration markings. The layout of the file is designed for easy processing and our code base provides a tool to extract all the glyphs from the symbols in the file and save them as separate tiff images in a directory structure that allows easy access. An index file is also generated that, for each symbol, identifies its component glyphs and relevant information about them.

There are a number of issues that must be taken into consideration both directly for the design of the database and indirectly because of the effect they have on consequent MOCR pattern matching strategies.

Base points: The *base point* of a character is a distinguished position relative to the image of the symbol (but not necessarily within the bounding box of the image) that is used to align it with other symbols in a text. A document processing system aligns symbols on a line by placing them so that their base point lies on a common base line and that two neighbouring symbols are placed so the base point of the second is placed in a horizontal position relative to a point calculated as the horizontal component of the base point of the first plus the width of the first. This horizontal position can be adjusted depending on the needs for filling lines or adjusting spacing.

The base point of a symbol bears no automatic relationship to the bounding box of the symbol. For example, a symbol with a descender such as "g", "j", "p", "q" or "y", will have its base point positioned above the lower limit of its bounding box. Many symbols such as "-" or "=", have their base point positioned below the lower limit of their bounding boxes.

Thus a symbol has to be identified before its base point can be found, although the knowledge of where the baseline is together with a database of symbol baselines can be used to constrain the choice of possible matching symbols. Our database must therefore relate the base point for each symbol, and indeed, of each glyph of each symbol, to the bounding box of the symbol.

Multi-glyph symbols: Some symbols are composed of multiple glyphs. For non-mathematical OCR, this does not pose significant problems because there are relatively few such cases (mostly accents, a few punctuation symbols and an occasional symbol such as "=") and there are obvious approaches that can be used to deal with such exceptions. For MOCR, however, there are many more cases and, frequently, cases where different symbols share common components e.g., $\cup, \uplus, \uplus$ or $\int, \iint, \iiint, \iiiint$ or $\precsim, \succapprox, \precsim, \prec, \precsim, \preceq$.

This presents a choice for a MOCR system: should it try to directly recognise symbols as a whole or should it try to recognise individual glyphs and reconstruct symbols from glyphs at a different level in the system? Since it is quick and easy to reconstruct a symbol from the component glyphs, but not to do the reverse, the database should store glyphs rather than symbols, but with the necessary information so that symbols can be conveniently reconstructed.

Duplicate glyphs: In LATEX, there are often multiple different fonts which have symbols which render to precisely the same graphical object at some resolution. For example there are many different instances of simple squares or circles in different fonts which are graphically indistinguishable when rendered (although they may have different base points). From the pure graphical pattern matching point of view, such objects are representatives of equivalence classes and it is the class that matters. Nonetheless, it can be significant to higher levels of mathematical document analysis as to precisely which representative it is, so the information about all such representatives must be maintained in the database with the actual final choice being made on the basis of, possibly, further contextual evidence.

Scanning resolution: Commercial, mass market OCR systems tend to be tuned to work best on relatively low resolution (200-400 dpi) images. This is possibly because of the time, memory and processing power demands that handling higher resolutions images would place on customer's resources. For normal OCR, this appears to be adequate given the relatively small symbol sets that need to be recognised and the relatively low demands for type face discrimination that current customers appear willing to put up with.

For MOCR, given the large symbol sets and the large number of fonts and typefaces that must be reliably distinguished, it seems unlikely that such resolutions will suffice. Our preliminary studies indicate that 600dpi may be adequate for MOCR and Suzuki has reported in [4] excellent recognition accuracy at this resolution. This is fortunate as mid to high volume scanning devices are currently readily available at this resolution but become extremely expensive at higher resolutions.

Many publishers, libraries and other organisations are currently working on digitising their collections and back catalogues. As with JSTOR [3], there seems to be a consensus on scanning at 600dpi. Therefore, for at least the foreseeable future, MOCR systems will need to be able to handle documents at this level of resolution. Nonetheless, we would like our database to be adaptable to researchers who wish to experiment with different, or even multiple resolutions.

Whatever the resolution, there will be issues of aliasing due to the limits of both the printing and the scanning technologies.

Point size: Merely scaling a symbol from one point size to match the appropriate dimensions of the glyph at a different point size can have the effect of upsetting its weight[1]. For this reason, the same symbol rendered at different point sizes often have different shapes, and different approaches have been taken in OCR systems to deal with it. Some systems try to use features invariant in the type of shape changes that occur between different point sizes. Others train their systems on collections of symbols that include glyphs at a few, or even at many different point sizes. All such options should be available to users of our glyph database.

Scaling: Irrespective of the point size the symbols of the target documents were originally rendered in, the document may have been scaled via photocopying or the printing process before being scanned. This complicates the issue of correctly identifying glyphs if different point sizes are being explicitly catered for.

Unknown symbols: Whatever the extent of a glyph database, it can never be complete. Not only will obscure symbols arise from the past that have not been included, but new symbols will be designed — sometimes by an author deliberately positioning multiple symbols onto overlapping locations, sometimes by design of new fonts and symbols. A MOCR glyph database has to be easily extensible and must provide support for allowing an MOCR system to decide correctly that there is no glyph in the database that is a significant match to a target glyph in a document.

3   A Database of Glyphs

Our database currently consists of a set of LATEX formatted documents (one per point size for 8, 9, 10, 11, 12, 14, 17 and 20 points) and rendered to tiff format (multi-page, 1 bit/sample, CCITT group 4 compression) at 600dpi, and an annotation text file, automatically generated from the LATEX sources during formatting, containing one line for each symbol described in the LATEX documents which associates the identifier of the symbol with the LATEX code necessary to generate the symbol together with the information on what extra LATEX packages or fonts, if any, are required to process the code and whether the symbol is available in *math* or *text* mode. Together with the documents, we provide Java programs to process them to extract the glyphs from the documents and store them in a suitable directory structure with one tiff file per glyph, and no more than 100 symbols per directory, and an index file containing the requisite extra information such as bounding box to base point offsets, identification of sibling glyphs in a symbol etc. We render the tiff images from the formatted LATEX documents using ghostview to obtain the cleanest possible images devoid of the kind of noise, skew, distortion, scaling and machine dependent incidental problems that printing and scanning would introduce. We do not currently use a relational database to store the images and related information. However that is

---

[1] The weight of a symbol refers to the thickness of the strokes used to draw it. If the weights of the symbols in a font are badly matched, as, for example, you might get in a poorly designed font, the overall effect of a block of text in that font is unaesthetic and looks "unbalanced".

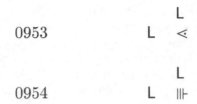

Fig. 1. Sample image from the database document

```
953,\ABXprecdot,ABX,math
954,\ABXVvdash,ABX,math
```

Fig. 2. Sample entries from the database annotation file

merely for ease of experimental development and switching to such an approach presents no difficulties.

Implementation of an MOCR program requires implementation of a glyph matching algorithm which normally requires analysing the glyphs in the database to construct a configuration set for the matcher. Use of the matcher normally requires access to the configuration set (usually quite small) but not to the actual glyphs. For our own metric based matcher [6], we generate a feature vector set from the database which can be quickly and efficiently loaded. It is intended that researchers could choose to use their own matcher and compile their own configuration set from the glyphs or, if their focus is on different levels of the mathematical document analysis process, simply use our provided matcher.

The LaTeX documents of the database enumerates all the symbols and homogenises their relative positions and sizes with the help of horizontal and vertical calibrators. The single symbols are then extracted by recognising all the glyphs a symbol consists of as well as their relative position to each other and to the calibrators. Each entry in the database thus consists of a collection of one or more glyphs together with the relative positions and the code for the actual LaTeX symbol they comprise. A sample of two symbol entries from a LaTeX database document is shown in Fig. 1 and the corresponding excerpt from the annotation file in Fig. 2. The sans-serif "L" symbols above and to the left of the target symbols serve to identify the horizontal and vertical components respectively of the base point of the symbol. The four digit number to the left is the identification number of the target symbol which is used to relate the symbol to the appropriate line in the annotation file.

The first page of the LaTeX document contains only the 10 digits and the calibration "L" symbol as registration data for a simple mini OCR matcher that is used to match the calibration and identification symbols while processing the rest of the document.

Among its approximately 5,300 symbols, the database contains about 1,600 mathematical symbols and 1,500 characters from different mathematical alphabets. The remaining symbols are mostly regular textual characters, accents, as well as additional scientific symbols, such as chemical or meteorological symbols.

As well as all the symbols from the standard *teTeX* distribution [10], we currently include the symbols from the following fonts and packages in the database: accents, amssymb, ar, bbding, bbm, bbold, bm, chemarr, chemarrow, dingbat, dsfont, esint, esvect, eurosym, euscript, fclfont, ifsym, latexsym, manfnt, mathabx, mathdots, mathrsfs, mbboard, nath, nicefrac, overrightarrow, phonetic, pifont, schwell, skak, skull, stmaryrd, suet, textcomp, tipa, trfsigns, trsym, txfonts, ulsy, undertilde, universa, upgreek, wasysym, wsuipa, yfonts, yhmath, zapfchan.

4   Assembling LaTeX Documents

At the moment we have essentially two algorithms available to produce documents with our OCR system. The first one recognises the glyphs in the input document and matches them against the glyphs in the databases. It then takes the closest matching glyph from the database, possibly applies some scaling to it, and places it at the position in the new document that corresponds to the position of the recognised glyph in the original file. While the results of this technique are visually nearly optimal and may have applications in the area of compression of images of scientific documents, the produced file is still not very useful for further processing, such as semantic text analysis or translation of mathematical expressions into the formal input syntax of other software systems. Nonetheless, it does provide an excellent basis for testing the efficacy of our glyph matching algorithm by providing an image which we can compare with the original image via an XOR based differencing function.

The resulting document provides us with information as to where and what (equivalence classes of) glyphs have to be placed and therefore provides constraints on which symbols are involved in the document. However, since the glyphs used can also be just parts of more complex symbols and can moreover be scaled, a significant amount of extra processing is required before the actual mathematical expression that is rendered can be reconstructed.

The aim of the second algorithm is to use the LaTeX commands associated by the database to the matched glyphs to assemble the final document. The algorithm roughly works in three steps:

1. The glyphs in the original document are identified.
2. an appropriate symbol is chosen from the database,
3. the LaTeX command for that symbol is put at the correct position in the output document.

The algorithm is identical to our first one up to the point where a matching glyph is chosen. Then there are essentially two cases to consider: (a) If the glyph matches with a symbol that consists of that one glyph alone we can simply pick it (the result may not be the correct symbol from a semantic point of view but the formatted output should be visually indistinguishable). (b) In the case that the best match for a recognised glyph is a glyph in the database that belongs to a symbol that is composed of multiple glyphs we cannot simply take that symbol since it might introduce glyphs into the result that have no counterpart in the original document. In this case we can consider two possible conflict resolution strategies:

1. We search all closely matching glyphs for one that is the only glyph of its associated symbol.
2. We search all closely matching glyphs for one whose sibling glyphs in its symbol are also matched in the appropriate relative position.

While approach 1 might not deliver necessarily the best matching glyph, it definitely will not introduce superfluous information into the document. But in some cases it will not be possible to find a symbol that matches acceptably well with the original glyph and approach 2 might be preferable (and in general, approach 2 is, of course, more correct from a semantic perspective), which forces a search over sets of glyphs of the particular area under consideration. In our current (first) implementation we have chosen to follow approach 1 by allowing for a small error threshold when matching glyphs and giving a preference to matching single glyph symbols over multi-glyph symbols within that threshold.

Once our algorithm has decided on an appropriate symbol, it retrieves the corresponding LaTeX command from the database and places it at the right position in the resulting document. Thereby it constructs a LaTeX picture environment whose measurements essentially corresponds to the bounding box given by the original document. The symbols are placed with single \put commands and, if necessary, put into *math* mode. In order to display symbols in the right size the algorithm uses the information provided by the database on the font size (from 5 to 20 points) of retrieved symbols. Depending on the desired point size of the final document, the command for a symbol is prefixed by the appropriate LaTeX command for changing font sizes. For instance, in the case when the algorithm

$$\lambda_u = \sum_{v \in V} \lambda_v \sum_{e \in E_{u \to v}} \left(r_{\kappa(e)} \right)^{s_1}, \qquad \sum_{u \in V} \lambda_u = 1, \qquad \lambda_u > 0.$$

Fig. 3. Original mathematical expression image

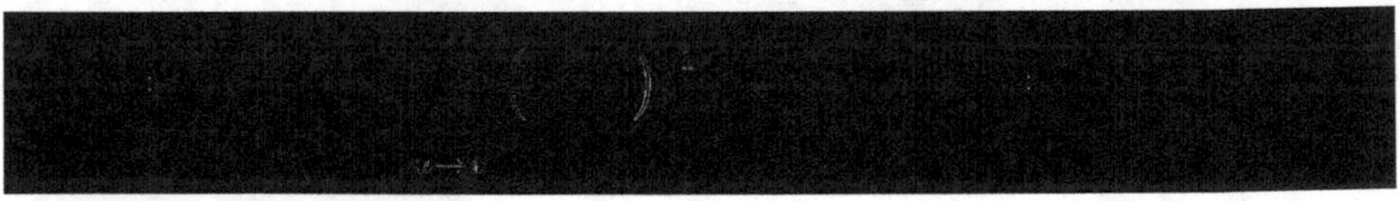

Fig. 4. Difference between Fig. 3 and Fig. 5 using XOR rendering

$$\lambda_u = \sum_{v \in V} \lambda_v \sum_{e \in E_{u \to v}} \left(r_{\kappa(e)} \right)^{s_1}, \qquad \sum_{u \in V} \lambda_u = 1, \qquad \lambda_u > 0.$$

Fig. 5. Generated mathematical expression image

retrieves a symbol of 11 point size and we assemble a 12 point document, the prefix will be \small. In the case when it retrieves a 14 point symbol, \large will be attached. This approach somewhat limits the number of available font sizes to those made available by LATEX's sizing commands. Moreover, it does not address the problem that symbols may need to be scaled horizontally and vertically differently in order to match the original glyphs. The second algorithm therefore does not give the same optimal results as the first algorithm which could insert the images with proper scaling. The right scaling for the LATEX commands will have to be addressed in future work.

We demonstrate the results of our algorithm with an example from a paper [11] we have experimented with that offers a large number of complex mathematical expressions. The particular expression we are interested in is given in Fig. 3 as it appears in the paper. The result of our OCR algorithm, as a comparison, is displayed in Fig. 5. Since the results are difficult to distinguish with the naked eye, we have combined both images using *exclusive-or rendering*, which is given in Fig. 4. Here, all pixels that show up in only one of the two images appear as white pixels. The difference in the rendering of the two expressions can be more easily explained when looking at the LATEX sources of the two expressions.

The source for the original expression is given in Fig. 6 whereas as the source for the output expression is displayed in Fig. 7. The latter is the input for the picture environment of dimension 3000×300 points in 12 point font size, where the unitlength is .12 points, which corresponds to 600 dpi.

If we now take a look at the symbols that are responsible for the differences in the two expressions we can see that they are caused by symbols that have not been recognised as the correct symbol or that are not properly scaled. In the latter category we have the two brackets, because large brackets in \large font size render slightly differently than in displaymath mode.

For the incorrectly recognised symbols the most obvious one is the small v on the righthand side of the arrow in the subscript of the second summation sign, which is recognised as an \upsilon. The arrow in this expression is also not recognised as an arrow in math mode but rather as one in text mode. The subscript 1 in the exponent s_1 is not recognised as a symbol from the mathrm font but as a symbol from a special text font with the \textoneoldstyle command. Furthermore, we can see that the equality sign, the only symbol here that is made up from more than one glyph, is not recognised as such but instead replaced by two vertical bar commands \HBar stacked on top of each other. Note that the

```
\begin{displaymath}
  \lambda_u = \sum_{v \in V} \lambda_v \sum_{e \in E_{u \to v}}
  \left(r_{\kappa(e)}\right)^{s_1}, \ \ \ \
  \sum_{u \in V} \lambda_u =1, \ \ \ \
  \lambda_u > 0.
\end{displaymath}
```

Fig. 6. LATEX source for expression in Fig. 3

```
\put(51,-145){\normalsize$\lambda$}
\put(105,-160){\scriptsize$\mathnormal{u}$}
\put(176,-152){\footnotesize\HBar}
\put(176,-172){\footnotesize\HBar}
\put(284,-144){\normalsize$\displaystyle\sum$}
\put(445,-145){\normalsize$\lambda$}
\put(499,-160){\scriptsize$\mathnormal{v}$}
\put(620,-144){\normalsize$\displaystyle\sum$}
\put(846,-150){\large$($}
\put(887,-145){\normalsize$\mathnormal{r}$}
\put(928,-160){\scriptsize$\kappa$}
\put(969,-160){\scriptsize$($}
\put(996,-160){\scriptsize$\mathnormal{e}$}
\put(1029,-160){\scriptsize$)$}
\put(1063,-150){\large$)$}
\put(1107,-87){\scriptsize$\mathnormal{s}$}
\put(1140,-104){\small\textoneoldstyle}
\put(1198,-204){\normalsize\textquoteright}
\put(1372,-144){\normalsize$\displaystyle\sum$}
\put(1533,-145){\normalsize$\lambda$}
\put(1587,-160){\scriptsize$\mathnormal{u}$}
\put(1657,-152){\footnotesize\HBar}
\put(1657,-172){\footnotesize\HBar}
\put(1766,-145){\normalsize$\mathrm{1}$}
\put(1815,-204){\normalsize\textquoteright}
\put(1988,-145){\normalsize$\lambda$}
\put(2042,-160){\scriptsize$\mathnormal{u}$}
\put(2118,-145){\normalsize\textgreater}
\put(2221,-145){\normalsize$\mathrm{0}$}
\put(2270,-197){\small\.{}}
\put(283,-262){\scriptsize$\mathnormal{v}$}
\put(319,-262){\scriptsize$\in$}
\put(366,-262){\scriptsize$\mathnormal{V}$}
\put(556,-262){\scriptsize$\mathnormal{e}$}
\put(589,-262){\scriptsize$\in$}
\put(636,-262){\scriptsize$\mathnormal{E}$}
\put(689,-278){\scriptsize$\mathnormal{u}$}
\put(725,-277){\scriptsize\textrightarrow}
\put(787,-278){\scriptsize$\upsilon$}
\put(1368,-262){\scriptsize$\mathnormal{u}$}
\put(1409,-262){\scriptsize$\in$}
\put(1456,-262){\scriptsize$\mathnormal{V}$}
```

Fig. 7. LATEX source for the expression in Fig. 5

\textoneoldstyle command belongs to the textcomp package and that the command \HBar is from the package ifsym package, which is not included in the standard LATEX distribution.

5   Discussion and Future Work

Besides working with documents that are already compiled from actual LaTeX source files we are currently experimenting with scanned images of documents. In particular, we have started experimenting with articles from the Transactions of the American Mathematical Society [9]. Within the repository of the JSTOR archive [2], images of all the back issues of this journal — starting 1900 — have been made available electronically. While the results of our approach for actual LaTeX documents are already nearly optimal, the reproduced files for scanned articles are still some way from a perfect translation.

Partly this is due to the fact that our implementation is new and time has not yet been available for tuning the system. In particular, we have not yet added the standard image preparation algorithms that one would expect in such a system (e.g., noise reduction filters, global and local deskewing algorithms etc.).

More seriously, our matcher is a metric based one and was originally developed on a much smaller database of glyphs. With such a small database the density of the occupied region of the metric space was low so that slightly distorted or noisy glyphs obtained from scanned images of non-LaTeX documents were still closer to an excellently matching glyph in the database (if there was an appropriate match at all) than to any other glyph. With the much higher density of the metric space region we now have (because the number of entries has vastly increased but the metric distance between the furthest separated objects and the effective dimensionality of the space has not), there is a much higher probability that a disturbance in the shape of a glyph will move it into the neighbourhood of a different glyph in the database. This is an indication of the importance of having a high quality database of glyphs when designing matchers for MOCR. Now that we have such a database, we intend to investigate refinements of our feature design and matcher implementation.

It is still our central hypothesis that, because of the range and coverage of symbols available from LaTeX, there will be few symbols in mathematical documents, from at least the last century, that will not be matched reasonably well by a corresponding LaTeX symbol. Furthermore, that a metric based symbol recogniser that works well on distorted or damaged LaTeX documents, will also work well on moderately high quality images from non-LaTeX mathematical documents such as can be found in the JSTOR repository [2]. This hypothesis requires proper testing.

As a first experimental implementation of an MOCR glyph database and an MOCR system, there are many missing features that need to be added and issues that need to be addressed. Some of these issues may require changes to the database, to the precompiled MOCR configuration information or just to the MOCR algorithm. As it is not yet clear which issue will require which type of change, we simply list the issues below:

- As first experiments with papers from the American Mathematical Society suggest, we need more fonts for ordinary alphanumeric symbols. Since our database is easily extensible, adding more fonts should not be a problem.

- We currently do not have a satisfactory strategy for handling symbols of variable size which do not expand in a simple scaling based manner, e.g., square root symbols. We are currently investigating the application of affine transformation invariants in dealing with the problem.
- We do not currently have a strategy for handling touching or broken glyphs in our MOCR system. Any solution that we arrive at may require extra information obtainable from the database.
- We need to add an image pre-processing front end for noise filtering, deskewing etc.
- Currently, our MOCR system identifies what it considers to be a best match for each target glyph and reports that. This is not satisfactory for a full featured system as there may be many excellent matches at the raw glyph comparison level and it may only be higher levels of the system that can disambiguate them. For this reason we intend to replace the current behaviour with returning a lazy list, filterable on annotation information such as point size and font family, in best match order. Hence higher levels of the system could choose easily between good matches to find one more suitable to the context.
- A longer term goal is to recognise complex mathematical objects from arrangements of single symbols in order to combine them to expressions that can be stated as the sort of commands that normally appear in LaTeX documents. There are a number of technologies, such as graph grammar rewriters, tree transformers etc., that may be required in such tasks.
- We do not currently deal with any aspect of diagrams, even basic ones such as line or curve identification.

More generally, symbol recognition is only one part of a full document analysis system for scientific texts. There are many pairs of symbols which are visually very similar (or even identical) but which are used with different intentions. These intentions can not be disambiguated by the symbol recogniser. At best, the recogniser can return a list of symbols which, from a purely visual point of view, are credible matches for the target symbol. Ideally this list can be ordered by visual similarity to the target symbol. A naïve document analysis system might simply choose the first element in the list as the matched symbol. A more sophisticated system would choose the best match based on contextual information, visual similarity as evidenced by the ordering in the match list, and resolution of the system of constraints that arise from the possible interdependent sets of choices of surrounding symbols. The design of the interface between the recogniser and other levels of the document analysis system will be critical in obtaining a high quality and efficient result. In particular, we believe that information transfer should not be unidirectional from the recogniser to the syntax and semantic analysis subsystems. Instead there should be a flow of data in both directions so that processing on both sides of the interface can be informed by (partial) results on the other. Our future work will include research on such an interface with particular emphasis on avoiding compromising acceptable processing performance.

6   Conclusion

We have developed a large database of glyphs used in mathematical texts, which we propose as a training set for OCR systems for scientific document analysis. Together with the database, we have developed an initial, if still basic, MOCR system that demonstrates the utility of the database. We have applied it to construct LaTeX processable source files from given images. It produced excellent results on papers formatted in LaTeX and we are currently conducting similar experiments with scanned original documents.

The database is easily extensible in terms of dealing with new characters and fonts as well as multiple occurrences of effectively the same characters. We have, however, not yet fully resolved the problems arising from multi-glyph symbols.

We deal with different point sizes of characters by translating them into appropriate LaTeX commands. While this approach does not fully solve the problem of scaling glyphs to the right size and is therefore not yet optimal, it already leads to acceptable results.

We consider the sensitivity of the recogniser to the precise shapes of the same symbol at different point sizes to be a success of the system: it is highly sensitive but, even when it chooses the wrong symbol, it chooses one which is visually very similar to the correct one. In general, it is our position that over-sensitivity can be tuned down and managed with appropriate syntactic and semantic feedback from other levels in a full document analysis system. However, we believe that under-sensitivity is much more difficult to compensate for.

Although the current database and associated software is still under active development, and in particular the MOCR system is still only a prototype, we already have promising results in our experiments and expect later versions of the system to be a valuable tool for the transcription of traditional mathematical documents into electronically managed knowledge.

Availability. The experimental version of the database is available upon request from the authors.

References

1. Robert H. Anderson.   *Syntax-Directed Recognition of Hand-Printed Two-dimensional Mathematics.* PhD thesis, Harvard University, Cambridge, MA, January 1968. Shorter version in M. Klerer and J. Reinfelds (eds), *Interactive Systems for Experimental Applied Mathematics*, pages 436–459, 1968, Academics Press.
2. The JSTOR scholarly journal archive. http://www.jstor.org/.
3. The JSTOR production process. http://www.jstor.org/about/process.html.
4. M.Suzuki, F.Tamari, R.Fukuda, S.Uchida, and T.Kanahori. Infty — an integrated ocr system for mathematical documents. In C.Vanoirbeek, C.Roisin, and E. Munson, editors, *Proceedings of ACM Symposium on Document Engineering*, pages 95–104, Grenoble, France, 2003.
5. Scott Parkin. The comprehensive latex symbol list. Technical report, CTAN, 29 September 2003. available at twww.ctan.org.

6. Alan Sexton and Volker Sorge. Database-driven mathematical character recognition. In Josep Llados and Liu Wenyin, editors, *Graphics Recognition, Algorithms and Applications*, LNCS, Hong Kong, August25-26 2005. Springer Verlag. to appear.

7. Alan P. Sexton, Alison Todman, and Kevin Woodward. Font recognition using shape-based quad-tree and kd-tree decomposition. In *3rd International Conference on Computer Vision, Pattern Recognition and Image Processing*, pages 212–215, Atlantic City, USA, Feb. 2000. Appears in Vol 2 of the collected proceedings of JCIS 2000, the Fifth Joint Conference on Information Sciences.

8. Masakazu Suzuki, Seiichi Uchida, and Akihiro Nomura. A ground-truthed mathematical character and symbol image database. Technical report, Faculty of Mathematics, Kyushu University, 6-10-1 Hakozaki, Higashi-ku, Fukuoka-shi, 812-8581 Japan, 2004. Available at http://www.inftyproject.org/AboutInftyCDB-1.pdf.

9. Transactions of the American Mathematical Society. Available as part of JSTOR at http://uk.jstor.org/journals/00029947.html.

10. The teTeX homepage. http://www.tug.org/teTeX/.

11. Peter Tiño and Barbara Hammer. Architectural Bias in Recurrent Neural Networks: Fractal Analysis. *Neural Computation*, 15(8):1931–1957, 2003. available at http://www.cs.bham.ac.uk/~pxt/PAPERS/rnn.frac.nc.fin.ps.gz.

Toward an Object-Oriented Structure
for Mathematical Text

Fairouz Kamareddine, Manuel Maarek, and J.B. Wells

ULTRA group, Heriot-Watt University
http://www.macs.hw.ac.uk/ultra/

Abstract. Computerizing mathematical texts to allow software access to some or all of the texts' semantic content is a long and tedious process that currently requires much expertise. We believe it is useful to support computerization that adds some structural and semantic information, but does not require jumping directly from the word-processing level (e.g., LATEX) to full formalization (e.g., Mizar, Coq, etc.). Although some existing mathematical languages are aimed at this middle ground (e.g., MathML, OpenMath, OMDoc), we believe they miss features needed to capture some important aspects of mathematical texts, especially the portion written with natural language. For this reason, we have been developing MathLang, a language for representing mathematical texts that has weak type checking and support for the special mathematical use of natural language. MathLang is currently aimed at only capturing the essential grammatical and binding structure of mathematical text without requiring full formalization.

The development of MathLang is directly driven by experience encoding real mathematical texts. Based on this experience, this paper presents the changes that yield our latest version of MathLang. We have restructured and simplified the core of the language, replaced our old notion of "context" by a new system of blocks and local scoping, and made other changes. Furthermore, we have enhanced our support for the mathematical use of *nouns* and *adjectives* with object-oriented features so that nouns now correspond to *classes*, and adjectives to *mixins*.

1 Introduction

From Euclid to Bourbaki, mathematicians have written their texts meticulously, in a precise, structured, and coherent form of natural language mixed with symbolic formula, which we call the Common Mathematical Language (CML). Is CML accurately reflected in current approaches to computerizing mathematics? If not, how can we make an improvement?

Approaches to computerizing mathematics. Computerizing mathematics is being done in various ways, each of which has advantages and disadvantages.

Mathematical word processing. The examples in figure 1 were included in this paper through the most basic kind of computerization. We typed the letters of the words of the text, and inserted LATEX commands like `\begin{definition}`

M. Kohlhase (Ed.): MKM 2005, LNAI 3863, pp. 217–233, 2006.

> **Definition 20.** *Of trilateral figures, an equilateral triangle is that which has its three sides equal, an isosceles triangle that which has two of its sides alone equal, and a scalene triangle that which has its three sides unequal.*
>
> Euclid [7–Book I]
>
> **Definition 1.** *A set with an associative law of composition, possessing an identity element and under which every elements is invertible, is called a* group. *[...] A group G is called finite if the underlying set of G is finite [...] A group [with operators] G is called* commutative *(or* Abelian*) if its group law is commutative.*
>
> N. Bourbaki [2–Chapter I, §4]

Fig. 1. Two examples of CML

and \end{definition} to guide the output. In this approach, a computer program can produce a visual representation of the CML, but a computer program will have great difficulty in automatically recognizing the semantic content of the LaTeX encoding even if the TeX macros are being carefully chosen as proposed by Kohlhase [13]. Even in the best case, LaTeX can not be expected to capture the semantic content of natural language text any better than OMDoc (see below).

Semantic markup languages. A more advanced solution is computerization of CML that records more semantic content. In the semantic markup languages MathML-Content (http://www.w3.org/Math/) and OpenMath (http://www. openmath.org/), symbolic formulas are encoded using a library of predefined symbols. OMDoc (http://www.mathweb.org/omdoc/) adds a theory level. There are many ways to write our examples from figure 1 in OpenMath/OMDoc using a mixture of structural and symbolic XML elements and chunks of natural language. A possible encoding of our examples in OpenMath/OMDoc is sketched[1] here:

```
<!-- First example -->
<theory name="Euclid-book-1">
  <symbol id="equilateral-triangle">
    <CMP>An equilateral triangle is [...]
<!-- Second example -->
<theory name="Group">
  <symbol id="*">
  <symbol id="E">
    <CMP>A set with  <OMOBJ>*</OMOBJ>, associative
         law of composition.
    <FMP>(a * b) * c = a * (b * c)
  <symbol id="e"> [...]
<theory name="FiniteGroup">
  <imports from="Group"> [...]
```

Natural language can only be stored in OMDoc in CMP elements as uninterpreted "blobs", while precise mathematical structure requires using symbolic encoding

---

[1] For readability and brevity, we show only the opening tag of each XML element for most elements; instead we use indentation to express nesting. We also use traditional mathematical output for OpenMath formulas instead of showing the XML tree.

(e.g., in FMP elements). Thus, for natural language mathematics, one must choose between retaining knowledge of the precise phrasing and presentation chosen by the mathematician, or capturing more of the structure via conversion to symbolic formula. Of course, one could do both like in our example above, keeping the uninterpreted natural language "blob" while adding a symbolic formula, but then the format does not support verifying they are mutually consistent. Generally, one does not expect formal checking of mathematics encoded in OMDoc.

Full formalization. Theorem Provers (TPs) such as Mizar (http://www.mizar. org/), Isabelle (http://www.cl.cam.ac.uk/Research/HVG/Isabelle/), and Coq (http://coq.inria.fr/) have made a tremendous contribution to computerizing mathematics, providing frameworks in which a full formalization can be written and verified automatically. However, they do not support important issues of mathematical text, such as control over presentation and phrasing and processing of the semantic structure. Furthermore, because full formalization is very expensive in human time, most mathematical texts are unlikely to be fully formalized, but might still benefit from some form of computerization.

Semi-formalization. Lighter TPs have been proposed, such as the work by Wiedijk [17] defining Formal Proof Sketches (FPS) as light Mizar proofs. An FPS article is a basically a Mizar article with holes. This approach reduces the expense of computerization via formalization (and also loses the certainty of full formalization), but does not appear to greatly improve control over presentation and phrasing and support for semantics-based manipulation.

Computerizing the mathematical vernacular. N. G. de Bruijn, founder of the Automath project [4], suggested capturing the essence of CML through his Mathematical Vernacular (MV) [5], a language with *substantives* (*nouns*), *adjectives*, and *flags*. Weak Type Theory (WTT) [12] adapted the ideas from MV in a type-theoretical fashion. To evaluate the practicality of MV and weak types for mathematical texts, we developed MathLang-WTT [11, 10].[2] (In related work, others have investigated translating from WTT into type theory [9, 8].) MathLang-WTT improved over WTT by internalizing *flags* and *blocks* and by implementing a type checker and various automated output views of MathLang documents which are faithful to CML. See [10] for a description (which is still applicable to MathLang) of these MathLang-WTT features.

Limitations of MathLang-WTT. Despite the features of MathLang-WTT, our plan to closely follow the expressiveness of CML in a computerized language was still not fully satisfied. Limitations of MathLang-WTT (and hence also of MV and WTT) appeared during the translation of *Euclid's Elements* [7] in describing mathematical entities such as *triangles* and *lines*. Consider the example from Euclid in figure 1. A triangle is intrinsically related to the three lines it is formed by, but encoding it in MathLang-WTT was unsatisfactory. One approach was for *each* triangle to define the three lines and the triangle separately and then to state their relation. This was awkward, and more importantly there was

---

[2] We call the old version MathLang-WTT to distinguish it from this paper's version.

clearly a missed opportunity to do some simple type checking, like complaining if there was an attempt to define some triangle as consisting of four surfaces (like a tetrahedron) instead of three lines. Another approach was to define a triangle-constructing function, but then the type system could not check that the result was a valid triangle and could be used where a triangle was required. Description of mathematical objects needed improvement.

In trying to solve these problems, we noticed that (a) N. G. de Bruijn's informal definition of substantives and adjectives could be better formalised in MathLang and (b) work in object-oriented programming carries useful clues.

Object-oriented concepts. Some programming language research has focused on allowing organizing programs in the way that seems most natural to the programmers. *Classes* are a way of packaging definitions so that it is easy to obtain not only instances (*objects*) but also multiple distinctly modified and extended variants (*subclasses*) via *inheritance*. *Mixins* [3] are abstract subclass generators that allow reusing modifications and extensions.

Classes and objects. In object-oriented programming, a class is usually defined by a set of *fields* and *methods*. An object is an encapsulated sub-program with an internal state that is an *instance* of a particular class. Classes define the common behavior of a group of objects. Fields are named values associated with each instance, while methods are named operations on the instances.

Inheritance. Class inheritance avoids repeating the definition of fields and methods shared by several classes. A new class can be defined by inheriting from an existing *parent* class, and the child's set of fields and methods will by default contain those of the parent.

Mixins. With simple class inheritance, to make two classes share a common set of new methods without duplicating the method definitions, the classes must inherit from an ancestor class containing the new methods. This may require radical rearrangement of an existing class hierarchy. To alleviate this problem, mixins are subclass definitions that are parameterized on their superclass, and thus act as functions from classes to classes. When a mixin is applied to a class, this makes a new subclass that adds or redefines fields and methods.

Contributions of this paper. The needs of encoding mathematical texts led to the design of the following new features for MathLang reported in this paper.

1. **Lighter abstract syntax and an accessible type system.** We simplified the syntax and type system of MathLang-WTT. The new syntax of MathLang contains only one kind of identifier in contrast to the variables/constants/binders of both WTT and MathLang-WTT (section 2 and 4).
2. **Generalised reasoning structure.** This paper refines MathLang-WTT's blocks to a simpler yet more general notion and replaces MathLang-WTT's flags and contexts by a more flexible and general *local scoping*. Our new *block* and *local scoping* constructs are cases of *steps*, which are MathLang's generic structuring concept (section 2).

3. **Turning nouns into classes.** We combine MV's substantives (inherited via the nouns of WTT and MathLang-WTT) and object-oriented classes to make MathLang's *nouns* (section 3.1). Nouns are conceptually similar to classes, *terms* are objects, and *sets* can be defined from nouns.

4. **Turning adjectives into mixins.** We combine MV's mathematical adjectives with object-oriented mixins to make MathLang's *adjectives*, which can be used in different ways with nouns, adjectives, sets, and terms (section 3.2).

2   A More Generic and Structured MathLang

This section shows how MathLang improved over MathLang-WTT by defining more generic and structured constructions.

One class of identifiers. In MathLang's predecessors WTT and MathLang-WTT, identifiers are separated into three disjoint sets: *variables*, *constants*, and *binders*. The rest of this paragraph briefly describes how identifiers work in the older MathLang-WTT: All three kinds of identifiers have a weak type, and this is all that variables have. Constants also have a definition and parameters (each parameter being a variable declaration). Each use of a constant is applied to arguments of the right weak type. Binders have parameters like constants, and one additional special parameter for the bound variable. Unlike variables and constants, binders can not be defined inside a document but can only be declared in the *preface*. Binders can not be given definitions; a statement using a binder can act as a definition but there is no way to indicate this.

In encoding texts, we found these restrictions of the different identifier kinds problematic, so MathLang instead now has just one kind of identifiers and distinguishes the uses via types. To fit binders in our new scheme and to allow declaring/defining new binders in documents, we replace the old single special parameter of each binder with a new kind of parameter with a *declaration type* usable with any identifier. For example, the binder $\forall$ might be declared as `forall (dec(a), stat) : stat`, making it an identifier with output type `stat` and two parameters: a declaration of an identifier of arbitrary type `a` and an expression of type `stat` (statement). An example using this identifier is the translation `forall (n:N, >=(n,0))` of the proposition $\forall n \in \mathbb{N}. n \geq 0$ (assuming $\mathbb{N}$, `>=` and 0 are already declared). Similarly, Russell's definite description binder ι (iota) could be declared with two parameters, a declaration and a statement. The first parameter is a variable that stands for the entire expression, and which should therefore have the same type: `iota (dec(a), stat) : a`. The expression $\iota n \in \mathbb{N}. (3 < n < 5)$ (meaning, "the unique $n \in \mathbb{N}$ s.t. $3 < n < 5$") would then be encoded as `iota (n:N, and(<(3,n),<(n,5)))`.

Simpler grouping and scoping. A fundamental idea of MathLang (inherited from MV) is capturing the grammatical and binding structure of a mathematical text. In MV and WTT, each *line* of a *book* has a context representing the set of assumptions about types of variables ("let x be a natural number") and truths ("suppose $x = y^2$ for some natural number y") used in the definition or statement made by the line. MV allows using flags as a secondary graphical 2-dimensional way of writing the current context in a book; an element repeated in the contexts of consecutive lines can be written as a flag whose *head* contains the repeated element and whose *flagstaff* goes through all the lines repeating the element. MV also has a secondary notion of blocks derived from flag nesting. (WTT could have used flags and blocks like MV, but this was never done.)

Unlike MV, MathLang-WTT directly supports flags and blocks rather than treating them as secondary syntax-sugaring notions derived from the contexts [11]. Upon careful examination of MathLang-WTT's flags and blocks, we found that they overlapped in function. MathLang-WTT's blocks allow grouping lines and sub-blocks and limiting to a block the scope of some of the constants defined in the block. MathLang-WTT's flag allow identifying a group of lines in which a context element is active.

In MathLang, we instead merged similar functionality. A *block*, written $\{step_1, \ldots, step_n\}$, is a sequence of statements. The *local scoping* construct $step_1 \triangleright step_2$ makes the declarations, definitions, and assertions inside $step_1$ assumptions used by $step_2$ and restricts declarations and definitions inside $step_1$ to be visible only in $step_2$. Both blocks and local scoping constructs are *steps*, as are declarations, definitions, and assertions. Steps can be of various sizes, such as the declaration of a variable, the definition of a function, a proof, or an entire book.

Example 1. Sequences of statements in a proof are represented by a block.

```
{ x.(y+1) = x.y';
  x.y' = x.y+x;
  x.y+x = x.y+x.1 }
```

Similarly each different sub-part of a proof as well as the overall proof is represented by a block. Sections and chapter are also blocks in MathLang as they decompose the text. For example, a proof by induction could be a block with two sub-blocks (note that the second sub-block carries a local scoping which holds the inductive hypothesis):

```
{ --A proof of P by induction--
  { --Proof of the base-- [...]; P(0) };
  { --Proof of the induction--
    { n:N; P(n) } |>  { [...]; P(n+1) } } }
```

An entire proof (e.g., a proof by contradiction) can be contextualised in a local scoping.

```
{ --Proof of the contradiction-- [...] }
     |> { --Statement proved by contradiction-- [...] }
```

Note that we write a block in braces { and } , the elements of the block are separated by a semi-colon ; . We write a local scoping with a step (which is the context), followed by the symbol |> (ASCII representation of $\triangleright$), followed by a step (in which the elements of the context will be available). In these examples we added some comments in between -- and cut some pieces of code ([...]). For readability, we make use of infix notation.

3   Abstraction with Nouns and Adjectives

3.1   Nouns as Classes

If we say that p is a demisemitriangle, one does not think of the set or the class of all demisemitriangles in the first place, but rather thinks of "demisemitriangle" as a type of p. It says what kind of things p is. [...] MV does not take sets as the primitive vehicles for describing elementhood but substantives (in the above example semidemitriangle is a substantive).

N. G. de Bruijn [5]

Nouns are abstractions that classify objects according to their common features. Nouns have an important place in some previous systems of representations of mathematics, such as WTT and MathLang-WTT, in which one of the weak types is *noun*. Nouns have used in translations of the first chapter of E. Landau's *Foundation of Analysis* [14] into WTT [16] and MathLang-WTT [11].

As already mentioned in section 1, we encountered limitations of the expressiveness of WTT-style nouns when we started translating Euclid's Elements [7]. Euclid starts his first chapter by defining basic geometric objects such as points, lines, figures, triangles, angles, etc. The definition of a line is as follows: *A line is breadthless length.* In MathLang-WTT, one way to write this is by defining *line* by forming a noun characterized by two statements: one that *line* "has length", the other that *line* is breadthless (does not "have breadth"). This uses a constant "has" which takes two nouns and returns a statement. This constant was unsatisfactory because it is hard to define its semantics precisely and because MathLang-WTT could not make any use of it for checking well-formedness. Because "has" deeply characterises the noun *line* and by consequence any concrete *line* — weak typed as "`term`" — created as a *line* instance, we felt it should be replaced by something that informs the language that *lines* have *length*, to allow approving of statements about the length of a line and disapproving of those about nonsense properties like its breadth, angle, weight, etc.

We found a solution in the concept of classes and objects in programming. A *line* is a class with one field *length*. Any instance of *line* is an object with a *length*. We characterise a line as breadthless in our translation with the absence of such a field. Table 1 gives more examples.

Consider the first definition in figure 1. Definition 20 of Euclid's example uses the noun *figure* (we see in section 3.2 how we encode the other nouns of

Table 1. Examples of noun definitions

Euclid's Elements	MathLang translation
A point is that which has no parts	`point := Noun`
A line is breadthless length	`line := Noun {length:term}`
A surface is that which has length and breadth only	`surface := Noun {length:term; breadth:term}`

this example). In the preceding definitions in [7], *figures* (*rectilinear figures*) are defined as *those contained by straight lines*. Therefore we define the noun `figure` with one field being the set of *straight lines* (we shorten it to *lines* in this example) and a statement precising that the figure is contained by this set of lines. The `Noun` constructor describes the noun with a step (in between braces { and }). The first unit of this step defines the field `sides`. Sides is a set of lines. The second unit of this step is a statement which uses an identifier `contained_by`. This identifier (declared earlier) takes a term and a set and returns a statement (`contained_by (term,set): stat`). The two parameters passed to this identifier are the future instance of the figure itself (encoded by the keyword `self`) and by the sides of the figure (field `sides` of `self`).

```
figure := Noun { sides : set(line);
                 contained_by(self,self.sides) }
```

Our second example in figure 1 is the definition of *group* by N. Bourbaki. We define **group** as a noun, The fields of this noun are identifiable in the text. The set E, the compositional law $*$ and the neutral element e. Two statements also define a group: the associativity of $*$ and the existence of an inverse of any element of E (we use an infix notation for =).

```
group := Noun { E:set; { a:E; b:E } |> *(a,b):E; e:E;
                forall (a:E, forall (b:E, forall (c:E,
                *(*(a,b),c) = *(a,*(b,c))))));
                forall (x:E, invertible(e,x)) }
```

3.2   Adjectives as Mixins

An adjective belongs to a substantive, and serves a double purpose: (i) to form a new substantive, and (ii) to form a new sentence.

N. G. de Bruijn [5]

According to *(i)*, an adjective is a function from noun to noun. An adjective, like *isosceles*, when applied to a noun like *triangle* creates a new noun *isosceles triangle*. In our system where nouns are classes, the adjectives will be mixins [6]. Intuitively, a mixin is a function from class to class. As in mixin calculi, an adjective can also be applied to an adjective to form a new adjective, to a term to form a new term, and to a set to form a new set (mapping the adjective across all members of the set). In MathLang, we call these constructions *refinements*. Following *(ii)*, we also incorporate the possibility that an existing term has the properties held by an adjective. For example one can describe a triangle ABC and demonstrate that this triangle is isosceles. The last line of this demonstration can be written in MathLang as the statement: `ABC << isosceles` (read ABC is isosceles). In our syntax we join this *adjective statement* to a *sub-noun statement*. The sub-noun statement $A \ll B$, given by N. G. de Bruijn in MV, states that *"every A is a B"*. For example, `triangle << trilateral figure`. We kept this notation in MathLang.

Let us see the use of these notions in our two examples. In the example taken from Euclid's *Elements*, several adjectives are defined. The noun *triangle* is defined as a refinement of the noun *figure* using the adjective *trilateral*. We define the adjective `trilateral` with the constructor `Adj`. The `Adj` constructor takes as a parameter the noun to be extended to form the new noun. In the case of our example, `trilateral` could only be applied to figures as it requires the field `sides`. The body of `Adj` is a step (similarly to the `Noun` constructor). In this step two specific objects are available. `self` which refers to the instances of the noun being defined (see section 3.1) and `super` which refers to the instance of the noun being refined (only needed when a component of the old noun is hidden by a component with the same name of the new noun). After the definition of `trilateral`, `triangle` is simply defined as a `trilateral figure`. We similarly define the adjectives *equilateral*, *isosceles* and *scalene* (We use an infix notation for the identifiers = (term,term):stat and != (term,term):stat and and (stat,stat):stat).

```
trilateral  := Adj (figure) { card(self.sides) = 3 };
triangle    := trilateral figure
equilateral := Adj (triangle) {
                   forall (side1:self.sides,
                     forall (side2:self.sides,
                       side1.length = side2.length)) }
isosceles   := Adj (triangle) {
                   exists (side1:self.sides,
                     exists (side2:self.sides,
                       side1 != side2
                       and side1.length = side2.length)) }
scalene  := Adj (triangle) {
                   forall (side1:self.sides,
                     forall (side2:self.sides,
                       side1.length != side2.length)) }
```

3.3 Multi Adjective Refinements

With adjectives we have an operation of simple inheritance between nouns. Let us see with this last example how multi adjective refinements work.

Our *group* example defines two adjectives. These adjectives for groups are *finite* and *Abelian*. *Finite* states that the set E of the group is finite. *Abelian* (or *commutative*) states that the operator of the group is commutative. In MathLang, we write the definitions of these adjectives as follow.

```
finite  := Adj (group) { finite_set (self.E) }
Abelian := Adj (group) {
    forall (x:self.E, forall (y:self.E, self.*(x,y) = self.*(y,x))) }
```

We could combine these two adjectives to obtain either `Abelian finite group` or `finite Abelian group`. In MathLang both expressions share the same type. Their meaning may differ as the statements introduced by the adjectives may overlap. It is for instance possible to define an `isosceles equilateral scalene triangle`. This expression is perfectly typable but of course would be considered as inconsistent even by pupils in primary schools. This reflects exactly the purpose of this first layer of MathLang which is to capture the structure of the text and its elements to allow, in a later stage, semantical analysis.

Table 2. Syntax of MathLang

ident, i	= denumerably infinite set of identifiers
label, l	= denumerably infinite set of labels
cvar, v	= denumerably infinite set of category variables

category, c	::=	**term**(exp) | **set**(exp) | **noun**(exp) | **adj**(exp, exp)	
		| **stat** | **dec**($category$) | *cvar*	
cident, ci	::=	*ident* | *exp.cident*	Identifiers anf fields
step, s	::=	*phrase*	Basic unit
		| **label** *label step*	Labelling
		| *step* ▷ *step*	Local scoping
		| {$\overrightarrow{step}$}	Block
phrase, p	::=	*exp*	
		| *cident*($\overrightarrow{ident}$) := *exp*	Definition
		| *ident*($\overline{exp}$) := *exp*	Definition by matching case
		| *ident* ≪ *cident*	Sub-noun and adjective statement
exp, e	::=	*cident*($\overline{exp}$)	Instance
		| *ident*($\overrightarrow{category}$) : *exp*	Elementhood declaration
		| *ident*($\overrightarrow{category}$) : *category*	Declaration
		| **Noun** {*step*}	Noun
		| **Adj**(exp) {*step*}	Adjective
		| *exp exp*	Refinement
		| **self** | **super**	Self and super
		| **ref** *label*	Referencing

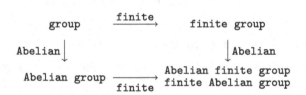

4   Language Description

Abstract syntax. The syntax of MathLang is given in Table 2. An arrow on top of a meta-variable represents a sequence of 0 or more meta-variables. For example $\overrightarrow{exp}$ is a sequence of *exp*. The elements of the sequence are separated with a comma "," in *ident, category* and *exp* and a semi-colon ";" in *step*.

Note the existence of a category constructor *noun* which describes a category expression and of a noun constructor *Noun* which describes a noun expression. In the following example, three identifiers with field a are defined: p is a noun, p' is a term instance of a noun, p'' is defined as a noun.

```
{ p:noun(Noun {a:term});
  p':Noun {a:term};
  p'' := Noun{a:term} }
```

We use the following notational conventions in this document:

1. When an identifier has no parameters we omit the (). E.g., we write *ident* in place of *ident*() and *ident* : *c* in place of *ident*() : *c*.
2. We do not leave double braces in noun and adjective expressions defined with a block step. E.g., we write Noun $\{s_1; \ldots; s_n\}$ and Adj (*e*) $\{s_1; \ldots; s_n\}$ instead of Noun $\{\{s_1; \ldots; s_n\}\}$ and Adj (*e*) $\{\{s_1; \ldots; s_n\}\}$.
3. We abbreviate category expressions to shorten the syntax of some term, noun and set categories. E.g., we write noun (resp. set and term) in place of noun(Noun $\{\{\}\}$) (resp. set(Noun $\{\{\}\}$) and term(Noun $\{\{\}\}$)).

Example 2. The following illustrate this syntactic sugaring:

1. We write x in place of x().
2. We write x:term in place of x():term.
3. We write Noun {x:term; >(x,0)} in place of Noun {{x:term; >(x,0)}}.
4. We write point:noun in place of point:noun(Noun{{}}).

Type system. We now present the typing rules of our language. Each typing rule has the form: *context* ⊢ *construction* **⦂** *type judgement* where:

– A *context* of typing is a *step* of the language (with two additional markers that hold the type of self and super and the labels). It represents the previous steps of reasoning in which the *expression* is to be typed.
– A *type judgement* is either an *atomic type* or a *type* where:
  atomic type = *Term*($\mathcal{T}$) ∪ *Set*($\mathcal{T}$) ∪ *Stat* ∪ *Noun*($\mathcal{T}$) ∪ *Adj*($\mathcal{T}, \mathcal{T}$) ∪ *Step* ∪ *cvar* ∪ *Dec*(*type*) ∪ *Def*(*type*)
  $\mathcal{T}$ is the set of mappings from *ident* to *type*.
  type is the set of mapping from sequences of *atomic type* to *atomic type*.
  T (resp. *at* and *t*) ranges over $\mathcal{T}$ (resp. *atomic type* and *type*).

Here are some functions used in the derivation rules of our type system.

$I : step \mapsto ident$	Gives the set of declared, defined and updated (sub-noun statement) identifiers in a step.
$dI : step \mapsto ident$	Gives the set of declared identifiers in a step.
$DI : step \mapsto ident$	Gives the set of defined identifiers in a step.
$L : step \mapsto label$	Gives the set of defined labels in a step.
dom(f)	Being the domain of the function f.
$T \uplus T'$	$T \cup \{ (i, T'(i)) \mid i \notin \text{dom}(T)\}$
$T \ominus T'$	$\{ (i, T(i)) \mid i \notin \text{dom}(T')\}$

And here is the subtyping relation between types and atomic types.

– *Term*(T) ≤ *Term*(T') *if* ∀*i* ∈ dom(T), $T(i) \le T'(i)$.
– *Set*(T) ≤ *Set*(T') *if* ∀*i* ∈ dom(T), $T(i) \le T'(i)$.
– *Stat* ≤ *Stat and Step* ≤ *Step*.
– *Dec*(t) ≤ *Dec*(t') *if* $t \le t'$.
– *Def*(t) ≤ *Def*(t') *if* $t \le t'$.
– *Noun*(T) ≤ *Noun*(T') *if* ∀*i* ∈ dom(T), $T(i) \le T'(i)$.

$$\frac{\vdash s \; \colon Step \qquad s \vdash e \; \colon Term(T) \qquad i \in \mathsf{dom}(T)}{s \vdash e.i \; \colon T(i)} \text{ IDENT-FIELD}$$

$$\frac{\vdash \{\vec{s};p\} \; \colon Step \qquad \{\vec{s}\} \vdash p \; \colon Dec(t) \qquad dI(p) = \{i\}}{\{\vec{s};p\} \vdash i \; \colon t} \text{ IDENT-DEC}$$

$$\frac{\vdash \{\vec{s};p\} \; \colon Step \qquad \{\vec{s}\} \vdash p \; \colon Def(t) \qquad DI(p) = \{i\}}{\{\vec{s};p\} \vdash i \; \colon t} \text{ IDENT-DEF}$$

$$\frac{\vdash \{\vec{s}; i_1 \ll ci_2\} \; \colon Step}{\{\vec{s}\} \vdash i_1 \; \colon Term(T_1) \qquad \{\vec{s}\} \vdash ci_2 \; \colon Adj(T_2, T_2') \qquad T_2 \le T_1}{\{\vec{s}; i_1 \ll ci_2\} \vdash i_1 \; \colon Noun(T_1 \uplus T_2)} \text{ IDENT-ADJ-TERM}$$

$$\frac{\vdash \{\vec{s}; s' \rhd s''\} \; \colon Step \qquad i \in I(s'') \qquad \{\vec{s}; s'; s''\} \vdash i \; \colon t}{\{\vec{s}; s' \rhd s''\} \vdash i \; \colon t} \text{ IDENT-LOCAL-SCOPING}$$

$$\frac{\vdash \{\vec{s}; s'\} \; \colon Step \qquad i \notin I(s') \qquad \{\vec{s}\} \vdash i \; \colon t}{\{\vec{s}; s'\} \vdash i \; \colon t} \text{ IDENT-SKIP-STEP}$$

These rules indicate how we retrieve the type of an identifier from the context. They decompose the *step* as context of typing to find the declaration IDENT-DEC, definition IDENT-DEF or the adjective statement IDENT-ADJ-TERM. In the case of a field of a term ($e.i$) the IDENT-FIELD rule applies first.

Fig. 2. Identifiers

- $Adj(T_1, T_2) \le Adj(T_1', T_2')$
  if $\forall i \in \mathsf{dom}(T_2), T_2(i) \le T_2'(i)$ *and* $\forall i \in \mathsf{dom}(T_1'), T_1'(i) \le T_1(i)$.
- $v \le v$.
- $(at_1, \ldots, at_n) \to at \le (at_1', \ldots, at_n') \to at'$ if $at \le at'$ *and* $\forall j \in [\![1 \ldots n]\!], at_j \le at'j$
  (after renaming of the category variables).
- $T \le T'$ *if* $\forall i \in \mathsf{dom}(T), T(i) \le T'(i)$.

Figures 2, 3, 4, 5, 6 and 7 compose MathLang type system. According to these typing rules the **group** identifier defined in section 3.1 has type $Noun(\{(\mathrm{E}, Set), (*, (Term, Term) \to Term), (\mathrm{e}, Term)\})$. Similarly the noun **triangle** and the adjective **isosceles** have respective types:

$Noun(\{(\texttt{sides}, Set(\{(\texttt{length}, Term)\}))\})$ and
$Adj(\{(\texttt{sides}, Set(\{(\texttt{length}, Term)\}))\}, \{(\texttt{sides}, Set(\{(\texttt{length}, Term)\}))\})$.

The type system prevents any misuse of identifiers' fields. For instance, let ABC be a declared triangle (ABC:triangle). This triangle is therefore a term with type $Term(\{(\texttt{sides}, Set(\{(\texttt{length}, Term)\}))\})$. According to our definition of triangle, the only defined field is **sides**, the set of lines composing a triangle. The expression ABC.sides refers to the sides of our triangle ABC. The set ABC.sides has type $Set(\{(\texttt{length}, Term)\})$.

The scopes of the identifiers depends on the location of the declaration or definition. Declarations could occur anywhere in an expression or could be an atomic step. We explain here the three possible cases: a declaration/definition in the flag

$$\frac{\vdash s : Step \qquad s \vdash e : Noun(T)}{s \vdash \texttt{term}(e)/\texttt{set}(e)/\texttt{noun}(e) : Term(T)/Set(T)/Noun(T)} \text{ CATEG-TERM/SET/NOUN}$$

$$\frac{\vdash s : Step \qquad s \vdash e : Noun(T) \qquad s \vdash e' : Noun(T') \qquad T \leq T'}{s \vdash \texttt{adj}(e, e') : Adj(T, T')} \text{ CATEG-ADJ}$$

$$\frac{\vdash s : Step}{s \vdash \texttt{stat} : Stat} \text{ CATEG-STAT} \qquad\qquad \frac{\vdash s : Step \qquad s \vdash c : at}{s \vdash \texttt{dec}(c) : Dec(() \rightarrow at)} \text{ CATEG-DEC}$$

$$\frac{\vdash s : Step}{s \vdash v : v} \text{ CATEG-VAR}$$

Category expressions are used in declarations. These category expressions set the category of the parameters and of the output of an identifier. Some category constructors (**term**, **noun**,**adj** and **set**) are parametrised by a noun expression.

Fig. 3. Categories (we use the symbol / to group the three similar rules)

part of a local scoping, a declaration/definition as atomic step in the body of a local scoping, a declaration as a parameter of an identifier. The first two are shared by definitions and declarations. The third one is declaration specific.

1. The first case is the presence of a declaration or a definition inside the flag of a local scoping. The introduced identifier is available in the step (and its sub-steps) covered by the flag-context. Here, an identifier x is declared in the context part of a local scoping. x is available in the part of this context that follows the declaration (3), and also in the body part (4) of the local scoping. But x is not available before being declared, in the preceding steps (1) as well as in the preceding part of the context (2) of the local scoping. x is also not available in the steps that follow the local scoping (5).

```
{ (1);
  { (2);
    x:term;
    (3) } |> { (4) };
  (5) }
```

2. The second case is a declaration or a definition as atomic step. The identifier is therefore available for all the following steps of the MathLang document. A declaration of a triangle is an atomic step of the sub-block of a block. The identifier **triangle** is not available before being declared (1) and (2) but is available in all what follows (3) an (4). The availability of **triangle** would have been identical if the declaration had been replaced by a definition.

```
{ (1);
  { (2);
    triangle:noun;
    (3) };
  (4) }
```

3. The last case is declarations as parameters of identifiers. If an identifier takes a declaration as a parameter, then the declared identifier is available

$$\frac{\vdash s \; \colon \; Step \qquad \forall j \in \llbracket 1 \dots n \rrbracket, \{s; e_1; \dots; e_{j-1}\} \vdash e_j \; \colon \; at'_j}{s \vdash ci(e_1, \dots, e_n) \; \colon \; at'} \text{INSTANCE}$$

where $s \vdash ci \; \colon \; (at_1, \dots, at_n) \to at$, $at' \notin cvar$, $(at_1, \dots, at_n) \to at \le (at'_1, \dots, at'_n) \to at'$ satisfiable

$$\frac{\vdash s \; \colon \; Step \qquad \{s; self : Term(T)\} \vdash s' \; \colon \; Step \qquad \forall i \in I(s'), \{s; self : Term(T); s'\} \vdash i \; \colon \; T(i)}{s \vdash \texttt{Noun} \; \{s'\} \; \colon \; Noun(T)} \text{NOUN}$$

$$\frac{\vdash s \; \colon \; Step \qquad s \vdash e \; \colon \; Noun(T) \quad T \le T' \quad \{s; super : Term(T); self : Term(T')\} \vdash s' \; \colon \; Step}{s \vdash \texttt{Adj} \; (e) \; \{s'\} \; \colon \; Adj(T, T')} \text{ADJ}$$

with $\forall i \in I(s'), \{s; super : Term(T); self : Term(T'); s'\} \vdash i \; \colon \; T'(i)$

$$\frac{\vdash s \; \colon \; Step \qquad s \vdash e_1 \; \colon \; Adj(T_1, T'_1) \qquad s \vdash e_2 \; \colon \; Noun(T_2)/Set(T_2)/Term(T_2) \qquad T_1 \le T_2}{s \vdash e_1 e_2 \; \colon \; Noun(T'_1 \uplus T_2)/Set(T'_1 \uplus T_2)/Term(T'_1 \uplus T_2)} \text{REFINEMENT}$$

$$\frac{\vdash s \; \colon \; Step \qquad s \vdash e_1 \; \colon \; Adj(T_1, T'_1) \qquad s \vdash e_2 \; \colon \; Adj(T_2, T'_2) \qquad T_1 \le T'_2 \qquad T'_1 \le T_2}{s \vdash e_1 e_2 \; \colon \; Adj(T_1 \uplus (T_2 \ominus T'_1), T'_1 \uplus T'_2)} \text{ADJ-REFINEMENT}$$

$$\frac{\vdash s \; \colon \; Step \qquad s \vdash i_1 \; \colon \; Noun(T_1) \qquad s \vdash ci_2 \; \colon \; Noun(T_2) \qquad T_2 \le T_1}{s \vdash i_1 \ll ci_2 \; \colon \; Stat} \text{SUB-NOUN}$$

$$\frac{\vdash s \; \colon \; Step \qquad s \vdash i_1 \; \colon \; Term(T_1) \qquad s \vdash ci_2 \; \colon \; Adj(T_2, T'_2) \qquad T_2 \le T_1}{s \vdash i_1 \ll ci_2 \; \colon \; Stat} \text{ADJ-TERM}$$

$$\frac{\vdash \{\overrightarrow{s}; self : at\} \; \colon \; Step}{\{\overrightarrow{s}; self : at\} \vdash \texttt{self} \; \colon \; () \to at} \text{SELF}$$

$$\frac{\vdash \{\overrightarrow{s}; super : at\} \; \colon \; Step}{\{\overrightarrow{s}; super : at\} \vdash \texttt{super} \; \colon \; () \to at} \text{SUPER} \qquad\qquad \frac{\vdash s \; \colon \; Step \qquad l \in L(s)}{s \vdash \texttt{ref} \; l \; \colon \; Step} \text{REF}$$

The typing of the parameter expressions should satisfy the type of the identifier for the *instantiation of the identifier* (INSTANCE rule). In the NOUN rule, **self** is added to the context for the typing of the step defining the noun. In ADJ, both **self** and **super** are added. A *refinement* creates a noun expression from an adjective and a noun. The set of components required to use the adjective should be a subset of the set of components of the noun.

Fig. 4. Expressions (we use the symbol / to group the similar refinement rules)

for the following parameters. Let us illustrate this with the encoding of an expression with the universal quantifier. We declare an identifier **binder** with a declaration as second parameter. We also declare an identifier **operator** with three parameters. In an expression using these two identifiers, a variable x is declared. This identifier x is not available before being declared (1) and (2). x is available in the parameters of the **binder** that follows the declaration of x (3). Finally x is neither available in the remaining part of the expression (4) nor in the steps that follow (5).

$$\frac{\vdash s \ \vdots \ Step \qquad s \vdash i_1 \ \vdots \ Noun(T_1) \qquad s \vdash ci_2 \ \vdots \ Noun(T_2)}{\dfrac{\text{dom}(T_2) \subseteq \text{dom}(T_1) \qquad \forall i \in \text{dom}(T_2), \ T_1(i) \le T_2(i)}{s \vdash i_1 \ll ci_2 \ \vdots \ Stat}} \text{ SUB-NOUN}$$

$$\frac{\vdash s \ \vdots \ Step \qquad s \vdash i_1 \ \vdots \ Term(T_1) \qquad s \vdash ci_2 \ \vdots \ Adj(T_2, T_2')}{\dfrac{\text{dom}(T_2) \subseteq \text{dom}(T_1) \qquad \forall i \in \text{dom}(T_2), \ T_1(i) \le T_2(i)}{s \vdash i_1 \ll ci_2 \ \vdots \ Stat}} \text{ ADJ-TERM}$$

Fig. 5. Phrases

$$\frac{\vdash s \ \vdots \ Step}{\dfrac{i \notin I(s) \qquad \forall j \in [\![1 \dots n]\!], \ s \vdash c_j \ \vdots \ at_j \qquad s \vdash e \ \vdots \ Noun(T)/Set(T)}{s \vdash i(c_1, \dots, c_n) : e \ \vdots \ Dec((at_1, \dots, at_n) \to Term(T))}} \text{ DEC-1}$$

$$\frac{\vdash s \ \vdots \ Step \qquad i \notin I(s) \qquad \forall j \in [\![1 \dots n]\!], s \vdash c_j \ \vdots \ at_j \qquad s \vdash c \ \vdots \ at}{s \vdash i(c_1, \dots c_n) : c \ \vdots \ Dec((at_1, \dots, at_n) \to at)} \text{ DEC-2}$$

$$\frac{\vdash s \ \vdots \ Step \qquad i \notin DI(s) \qquad \forall j, k \in [\![1 \dots n]\!], j \ne k \Rightarrow i_j \ne i_k}{\dfrac{\forall j \in [\![1 \dots n]\!], s \vdash i_j \ \vdots \ () \to at_j \qquad \cup_{j \in [\![1 \dots m]\!]} i_j = dI(s) \setminus \{i\}}{\dfrac{s \vdash e \ \vdots \ at \qquad \text{if } i \in dI(s) \text{ then } s \vdash i \ \vdots \ (at_1, \dots, at_n) \to at}{s \vdash i(i_1, \dots, i_n) := e \ \vdots \ Def((at_1, \dots, at_n) \to at)}}} \text{ DEF}$$

$$\frac{\vdash s \ \vdots \ Step \qquad \text{if } i \in I(s) \text{ then } s \vdash i \ \vdots \ (at_1, \dots, at_n) \to at}{\dfrac{\forall j \in [\![1 \dots n]\!], s \vdash e_j \ \vdots \ at_j \qquad s \vdash e \ \vdots \ at}{s \vdash i(e_1, \dots, e_n) := e \ \vdots \ Def((at_1, \dots, at_n) \to at)}} \text{ DEF-CASE}$$

Declarations and definitions introduce new identifiers. For a declaration, the category of the identifier could be explicitly expressed (DEC-2 rule) or an expression could be given that represents the elementhood of the identifier (DEC-1 rule). For a definition, the parameters could either identifiers (DEF rule) or expressions for definition by matching (DEF-CASE rule).

Fig. 6. Declarations and definitions

```
{ binder (term, dec(term), term): term;
  operator (term, term, term): term;
  [...] a((1), binder ((2), x:term , (3)), (4)) [...];
  (5) }
```

5 Conclusion and Future Work

To have MathLang being adopted by mathematicians is our aspiration. We are convinced that providing yet another concrete syntax will never make a mathematical language widely used. We are therefore focusing on interfacing MathLang with user-friendly tools. We are currently embedding MathLang concepts and type checking in the scientific editor TEXmacs with the development of a MathLang-plugin. This plugin is making full reuse of the mechanisms for rendering MathLang texts in their original CML forms. These mechanisms were

$$\frac{\vdash s_1 \; \colon Step \qquad s_1 \vdash s_2 \; \colon Step \qquad \{s_1; s_2\} \vdash \{\vec{s}\} \; \colon Step}{s_1 \vdash \{s_2; \vec{s}\} \; \colon Step} \text{ STEP-COMPOSITION}$$

$$\frac{\vdash s \; \colon Step \qquad s \vdash s' \; \colon Step \qquad \{s; s'\} \vdash s'' \; \colon Step}{s \vdash s' \triangleright s'' \; \colon Step} \text{ LOCAL-SCOPING}$$

$$\frac{\vdash s \; \colon Step \qquad s \vdash p \; \colon Stat/Dec(t)/Def(t)}{s \vdash p \; \colon Step} \text{ ATOMIC-STEP}$$

$$\frac{\vdash s \; \colon Step}{s \vdash self : t \; \colon Step} \text{ SELF-MARKER} \qquad\qquad \frac{\vdash s \; \colon Step}{s \vdash super : t \; \colon Step} \text{ SUPER-MARKER}$$

$$\frac{}{\vdash \{\} \; \colon Step} \text{ EMPTY-STEP} \qquad\qquad \frac{\vdash s \; \colon Step \qquad \{s; l : Label\} \vdash s' \; \colon Step}{s \vdash \texttt{label } l \; s' \; \colon Step} \text{ LABEL}$$

Only well typed statements, declarations or definitions could be *phrases* (ATOMIC-STEP rule). Phrases stand for atomic steps *Blocks*. Each element of a block should be a valid step in the context formed by the preceding elements of this particular element (BLOCK rule). As presented in section 2, a *local scoping* builds a step as a context for another step (LOCAL-SCOPING rule). See example 1 for examples of *steps*.

Fig. 7. Steps

presented in [10]. In parallel, we are implementing the MathLang's new features presented in this paper. These new features will be tested on already realised translations. New translations will benefit from the assistance of the editor and will gain in expressiveness with the new object oriented features of the language.

Concerning the language definition part, we believe that more flexible abstraction mechanisms could be added. For this purpose we will investigate the possibility to integrate in our system the notion of traits, a new member of the object-oriented programming.   We would also like to relate our low level encoding of groups to a previous work. In the computer algebra system Focal [15], *species* and *collections* are object oriented structures that have been used to create an algebraic hierarchy.   Finally we would be interested in comparing MathLang's nouns and adjectives with concepts and roles of Deductive Logics (DLs) [1] and in investigating existing research in mixin extension of DLs.

In this paper we proposed to capture the structure of mathematical with object-oriented features. We exposed the relevance of this approach with two examples and presented a type system for MathLang that incorporates these features. This work is a step of our larger aim to consider the encoding of the natural language parts when computerizing mathematical text.

References

1. Franz Baader, Diego Calvanese, Deborah McGuinness, Daniele Nardi, and Peter Patel-Schneider, editors. *The Description Logic Handbook: Theory, Implementation, and Applications.* Cambridge University Press, 2003.
2. Nicolas Bourbaki. *Elements of Mathematics - Algebra*, volume II. Hermann, 1974.

3. Gilad Bracha and Gary Lindstrom. Modularity meets inheritance. In *Proc. Int'l Conf. Computer Languages*, pages 282–290, 1992.
4. N.G. de Bruijn. The mathematical language Automath, its usage, and some of its extensions. *Lecture Notes in Mathematics*, 125:29–61, 1970.
5. N.G. de Bruijn. The mathematical vernacular, a language for mathematics with typed sets. In *Workshop on Programming Logic*, 1987.
6. M. Flatt, S. Krishnamurthi, and M. Felleisen. Classes and mixins. In *POPL '98*, pages 171–183, 1998.
7. Heath. *The 13 Books of Euclid's Elements*. Dover, 1956.
8. G. Jojgov and R. Nederpelt. A path to faithful formalizations of mathematics. In *MKM 2004*, volume 3119 of *LNCS*, pages 145–159, 2004.
9. G. Jojgov, R. Nederpelt, and M. Scheffer. Faithfully reflecting the structure of informal mathematical proofs into formal type theories. In *MKM Symposium 2003*, volume 93 of *ENTCS*, pages 102–117, 2004.
10. F. Kamareddine, M. Maarek, and J. B. Wells. Flexible encoding of mathematics on the computer. In *MKM 2004*, volume 3119 of *LNCS*, pages 160–174, 2004.
11. F. Kamareddine, M. Maarek, and J. B. Wells. MathLang: Experience-driven development of a new mathematical language. In *MKM Symposium 2003*, volume 93 of *ENTCS*, pages 138–160, 2004.
12. F. Kamareddine and R. Nederpelt. A refinement of de Bruijn's formal language of mathematics. *Journal of Logic, Language and Information*, 13(3):287–340, 2004.
13. Michael Kohlhase. Semantic markup for TEX/LATEX. Mathematical User-Interfaces Workshop, 2004.
14. Edmund Landau. *Foundations of Analysis*. Chelsea, 1951.
15. Virgile Prevosto, Damien Doligez, and Thérèse Hardin. Algebraic structure and dependent records. In *TPHOLs 2002*, volume 2410 of *LNCS*, pages 298–313, 2002.
16. Mark Scheffer. Formalizing Mathematics using Weak Type Theory. Master's thesis, Technische Universiteit Eindhoven, 2003.
17. F. Wiedijk. Formal proof sketches. In *TYPES 2003*, volume 3085 of *LNCS*, pages 378–393, 2004.

Explanation in Natural Language of $\bar{\lambda}\mu\tilde{\mu}$-Terms

Claudio Sacerdoti Coen*

Project PCRI, CNRS, École Polytechnique, INRIA, Université Paris-Sud
sacerdot@cs.unibo.it

Abstract. The $\bar{\lambda}\mu\tilde{\mu}$-calculus, introduced by Curien and Herbelin, is a calculus isomorphic to (a variant of) the classical sequent calculus LK of Gentzen. As a proof format it has very remarkable properties that we plan to study in future works. In this paper we embed it with a rendering semantics that provides explanations in pseudo-natural language of its proof terms, in the spirit of the work of Yann Coscoy [3] for the λ-calculus. The rendering semantics unveils the richness of the calculus that allows to preserve several proof structures that are identified when encoded in the λ-calculus.

1 Introduction

An important topic of Mathematical Knowledge Management (MKM) is the definition of standards for the representation of mathematical documents at different semantical levels (presentation, content, semantics using the terminology of [1]). The current situation for mathematical expressions is almost satisfactory: MathML Presentation is a W3C standard for the presentation level, and the lack of MathML rendering engines has been solved; OpenMath is a de facto standard for the content level, and several tools already integrate phrasebooks for communicating formulae in OpenMath according to a given content dictionary; the interactive theorem proving community is slowly starting to consider open formats for replacing the proprietary semantic encodings or just for communication with external tools. On the contrary, there is no mature format for proofs at the content level. The only candidate is the OMDoc standard, that integrates a module for proofs since its first version. However, the original format was not expressive enough for describing in a natural way the proofs of the HELM[1] library. Thus the proof module was redesigned almost from scratch in the MoWGLI[2] European Project, and the new proposal will be part of the forthcoming OMDoc 1.2 standard [7]. A rendering semantics (i.e. a default explanation of the proofs in a pseudo-natural language) is also provided by MoWGLI [2]. However, a serious third party evaluation of the new proposal has not been done and there exists no test suite of proofs that can be used to assess the flexibility of the format.

---

* Partially supported by 'MoWGLI: Math on the Web, Get it by Logic and Interfaces', EU IST-2001-33562.
[1] http://helm.cs.unibo.it
[2] http://mowgli.cs.unibo.it

M. Kohlhase (Ed.): MKM 2005, LNAI 3863, pp. 234–249, 2006.

To try to improve the situation the first step consists in fixing a few requirements that a proof format for the content level must satisfy. Here is our list:

1. **Flexibility.** It must be possible to encode both rigorous, human provided, proofs and proofs that are generated from their semantics level. The encoding should respect the *structure* of the proof, avoiding the identification of proofs that differ in their structure. What the structure of a proof is is already a non trivial question. For instance, proof nets or natural deduction identify more proofs than sequent calculus. For presentational purposes we are interested in identifying as few proofs as possible, up to their structure only. For instance, a top down proof should not be identified with its bottom up counterpart. However, a content encoding must identify proofs that have the same structure and that differ only up to rhetorical text.

2. **Annotations.** It must be possible to decorate the proof structure with rhetorical text. The rhetorical text is the presentational counterpart of the proof content. It is requested only for consumption by humans.

3. **Explanation in Natural Language.** The format must have a *rendering semantics* associated to it. That is, it must be possible to generate rhetorical text that describes the proof structure. The generated text is not required to be nice to read or close to the text that a mathematician would choose. Annotations are explicitly provided to deal with the situation where a nice presentational proof is required. The rendering semantics is useful, for instance, when the proof is automatically generated from a semantics proof — say, created using a proof assistant by mimicking a pen&paper proof — and the user needs to check whether the pen&paper proof that she wants to formalize and the proof generated by the proof assistant are actually the same proof.

4. **A Clear Semantics.** This is surely the most controversial point. On the one hand we are talking about a content level format, that should not be restricted to the proof steps that are correct in just one foundation and one logic; on the contrary it should capture the usual rigorous but informal style of the proofs of real world mathematicians. On the other hand it must describe a proof, and not a document with an arbitrary structure; it must allow for simple checks, as for references to hypotheses out of scope or for the well nesting of subproofs; it must allow for proof transformations, such as cut elimination. In other words, it must be as close as possible to a calculus without becoming a semantic encoding instead of a content level encoding.

The OMDoc 1.2 proof module strives to achieve the points 1–3. However, its semantics is someway defined a posteriori and it is not fully understood nor made explicit.

Via the Curry-Howard isomorphism, several λ-calculi can be seen as proof formats at the semantic level for the logics they are isomorphic to. As proof formats they can be equipped with a rendering semantics [3] and extending them with annotations is also a trivial exercise. However, they lack flexibility. A partial reason is that, being at the semantics level, they are bound to a precise logic. However, there are deeper reasons that are illustrated in Sect. 2 and that are not related to their focus on a particular logic. Thus they are not a good model to build a content level proof format on.

In a seminal paper in 2000 [5] Curien and Herbelin proposed the $\bar{\lambda}\mu\tilde{\mu}$-calculus that is isomorphic to (a variant of) the classical sequent calculus LK of Gentzen. I claim that this calculus is a perfect proof format at the semantics level and that it is inherently very flexible. To obtain a content level calculus from it it is just necessary to relax a bit its interpretation by decoupling it from its logic. Moreover, I also claim that it has several remarkable similarities with OMDoc 1.2 and in a future work I plan to make this relation explicit by providing a bisimulation of OMDoc into the $\bar{\lambda}\mu\tilde{\mu}$-calculus that respects the rendering semantics. As a preliminary step in that direction, in this paper I will provide a rendering semantics to the $\bar{\lambda}\mu\tilde{\mu}$-calculus that is extremely intuitive and unveils all the good features of the calculus as a proof format.

2   A $\bar{\lambda}\mu\tilde{\mu}$-Calculus Primer

The $\bar{\lambda}\mu\tilde{\mu}$-calculus [5] is an extremely elegant synthesis of the $\bar{\lambda}$-calculus of Herbelin [6] and the $\lambda\mu$-calculus of Parigot [9]. The $\bar{\lambda}$-calculus of Herbelin is a λ-calculus that is isomorphic to (a variant of) the *intuitionistic sequent calculus* LJ of Gentzen. The $\lambda\mu$-calculus of Parigot is a λ-calculus that is isomorphic to *classical, multi conclusions, natural deduction*. The $\bar{\lambda}\mu\tilde{\mu}$-calculus is isomorphic to (a variant of) the *classical sequent calculus* LK of Gentzen. However, the interest of the calculus is that it is not a simple merge of two existing calculi; on the contrary, it is greatly superior to both of them since it makes explicit for the first time at the syntactic level two fundamental dualities of the computation:

1. Terms vs Contexts
2. Call-by-name vs Call-by-value

We will explain the two dualities in detail. Before that, however, we notice that this result is, a posteriori, not very surprising. Indeed the classical sequent calculus is well known for its meta-theoretical properties, since it reveals the deep symmetries of the logical connectives that are hidden in natural deduction and since it can also be seen as a fine grained analysis of natural deduction, especially for cut elimination. Thus it is natural that a λ-calculus isomorphic to LK should be the best framework for the study of the symmetries of computation. What is not absolutely obvious, however, is that these two dualities are deeply connected with the flexibility of the proof format. Let's explain this.

Terms vs Contexts. A context is an expression with exactly one placeholder $\square$ for a "missing' term. The placeholder can be filled with a term to obtain a placeholder-free expression. The placeholder can be typed with the type of the expected term, and only terms of the expected type can fill the placeholder. A context can apply its placeholder to arguments ($\square\ \vec{t}$) or it can bind a name to it (**let** $x := \square$ **in** c) to refer to it later on, for instance to pass it to a function.

Dually, a term can be seen as an expression with exactly one placeholder $]_{-}[$ for a "missing context" that is "all around" the term. The placeholder can be filled with a context to obtain a placeholder-free expression. The placeholder can be typed with the type of the term, and only contexts that expects a term of the expected type can fill the placeholder. A term can wait for inputs from its context ($]\lambda\bar{x}.t[$) or — in languages with control operators like Scheme's CALL/CC — it can bind a name to it ($]\mu\alpha.c[$, μ is the binder and α the bound name) to refer to it later on.

Now consider an expression without placeholders and imagine it to be isomorphic to a proof of some thesis from some set of hypotheses. The expression can be broken to be seen as the composition of a term and a context whose placeholders are given "the same type" T (actually, a dual type; we will be more precise later). The term and the context can be thought respectively as "a proof of T from the hypothesis" and "a proof of the thesis from T". Thus in the term the type of the placeholder represents what must be proved as a first step in the proof, and the placeholder is the rest of the proof. In the context the type of the placeholder represents what was proved so far and the placeholder is the proof so far. The operators that are used to bind the placeholder in a term and in a context can be thought as ways of stating or manipulating the (local) conclusion(s) (for a term), or as ways of stating or manipulating the (local) hypotheses (for a context). This kind of manipulation is very frequent in pen&paper proofs, where an intermediate result can be claimed (binding a context), a label can be associated to intermediate results for further reference (binding a term), a proof of an intermediate result can be postponed (a context that binds its term is displayed before the term), or the current thesis can be reduced to another one by anticipating the rest of the proof (a term that binds a context is displayed after the context).

Call-by-name vs Call-by-value. What are the dynamics of call-by-value and call-by-name? The first strategy processes the arguments before processing the function; the second strategy processes the function until it needs to process the arguments. If you substitute "explains" or "prove" with "process" you will obtain the definition of the bottom-up and top-down proof styles. A bottom-up proof proves (process the argument) a result (the type of the argument) before using it later on (processing the function). A top-down proof prooves the thesis (process the function) until it has reduced the thesis to an easier one (the type of the argument) that is then proved (the argument is processed).

Usually, call-by-name and call-by-value are global strategies that are applied in the reduction of a functional program (a λ-expression). In the $\bar{\lambda}\mu\tilde{\mu}$-calculus, instead, there exists at the syntactic level both call-by-value related and call-by-name related redexes (and a third form of redexes whose strategy is not yet fixed and that can non-deterministically reduce towards one of the other two redexes, but this is not important in our discussion). Thus the $\bar{\lambda}\mu\tilde{\mu}$-calculus is flexible enough to distinguish between top-down and bottom-up proof steps, while this is not possible in the plain λ-calculus (unless we play tricks, as using β-expansion to "mark" bottom-up steps or we extend the calculus with a let . . . in construct that is native of the $\bar{\lambda}\mu\tilde{\mu}$-calculus).

Since we think that the intuition we just provided is someway deeper than the gory technical details we are shortly going to present, we prefer to reinforce it by explaining it again along a different axis. As we already said, the $\bar{\lambda}\mu\tilde{\mu}$-calculus is a beautiful synthesis of the $\bar{\lambda}$-calculus and the $\lambda\mu$-calculus, made completely symmetric by adding a $\tilde{\mu}$ operator (the **let** ... **in** in a more usual syntax). We give now the intuition about what is the contribution for flexibility (as a proof format) of each component.

$\bar{\lambda}$. The $\bar{\lambda}$-calculus establishes a Curry-Howard isomorphism with a sequent calculus. A sequent calculus identifies far fewer proofs than natural deduction, which is Curry-Howard isomorphic to the λ-calculus. In particular, top-down and bottom-up proofs are distinguished in a sequent calculus derivation (where it is recorded if the user eliminates a rule on the left hand side first or on the right hand side first). In natural deduction, instead, top-down vs bottom-up corresponds to the order of construction of the derivation (from the leafs to the root or from the root to the leaves), but both procedures at the end produce exactly the same tree (unless cuts are artificially introduced to mark the bottom-up steps). This is one reason why the sequent calculus provides a more fine-grained analysis of the process of construction of the derivation and, in our context, it gives more flexibility in proof representation.

$\tilde{\mu}$. The **let** $x : T := \Box$ **in** c (that we will soon write $\tilde{\mu}x : T.c$ to show the beautiful symmetries of the calculus) gives a label (x) to the last result proved ($\Box$) and it makes explicit its type T. The label is used later on to refer to the result. The type makes explicit what is the conclusion of the last proof step (the "last" proof step of $\Box$).

This construct is necessary for a proof format since it allows to reuse a subproof more than once, without replicating a proof, and since it is used to associate to a subproof its thesis. In the λ-calculus a redex can be used for sharing a proof, partially simulating the $\tilde{\mu}$. Moreover, since the semantics of $\tilde{\mu}$ is that of a bottom-up proof (since it gives a label to the previous proof step), redexes can be rendered as bottom-up proofs. Notice that in a typed calculus the binder in a redex also associates to the proof (the argument of the application in the redex) its thesis (its type). If we manage to avoid the redex trick, however, we have to guess the type that is no longer recorded by the binder. If the type system is decidable, the type can be automatically inferred. However, since we do not expect applications that adopt a proof format to integrate a type inference engine and since we want to impose no semantics (no choice of any type system) to our proof format, we need to pre-compute the type of the argument and explicitly store it in the proof format. Actually, we need to store the type of each subterm (we call this an inner-type in [1]).

The need for inner-types is evident when we recall that a λ-term is isomorphic to a derivation in natural language in the sense that you can obtain the λ-term from the derivation by erasing from the tree all the conclusions of the rule (i.e. what a user would keep in a pen&paper proof) since they can be inferred from the rules themselves (i.e. what a user throws away in a pen&paper proof) and the tree structure. Thus in our proof format we are obliged to reconstruct from the λ-term every inner-type, we need to keep the structure of the term, but we can throw away the term! (When the actual terms, i.e. the justifications of each proof step, are thrown away, we obtained a *proof sketch* in the terminology proposed by Wiedijk [10]).

Thus the $\bar{\lambda}\mu\tilde{\mu}$-calculus is superior to the λ-calculus since recording of the inner-types and the $\tilde{\mu}$ (or **let ... in**) is already part of the calculus, while it needs to be introduced in the λ-calculus.

Just to be precise, notice also that the $\tilde{\mu}$ construct can be simulated in the λ-calculus as a redex only from the point of view of the reduction. On the contrary, the typing rule for $\tilde{\mu}$ is not equivalent since in **let** $x : T := t$ **in** c we can type c under the assumption that x is equal to t, which is stronger than the assumption x has type T (for instance, when the type system admits dependent types).

μ. The control operator μ that binds the context of a term to reuse it later has a surprising role. It is introduced in the calculus to capture classical logic and, when the calculus is seen as a proof format, it is used to give a label and to state explicitly what is the thesis that is going to be proved next. The relation with classical logic is obvious: when multiple μ are in scope the expression has visibility of several possible conclusions at once, and it can dynamically choose to conclude any one of them. This clearly corresponds to a sequent with several conclusions.

However, a pen&paper proof, even a classical one, never uses multiple conclusions. Indeed, natural deduction with several conclusions (the logic the $\lambda\mu$-calculus of Parigot is Curry-Howard isomorphic to) is not natural at all, as the classical sequent calculus is not. Most mathematicians prefer to work in an intuitionistic natural deduction setting augmented with one or more equivalent classical axiom such as excluded middle or double negation elimination.

Thus we can easily argue that a proof format is not requested to support proofs with multiple conclusions, if not for completeness reasons. Thus we can argue that we will not need the ability of the μ constructor of associating a label to the thesis we want to prove next. Indeed in pen&paper proofs a thesis is never labelled. However a proof format does need a way to state what the user is going to prove next, since this construct is often used by mathematicians to clarify the proof or to postpone parts of it. Once again, this construct is already native in the $\bar{\lambda}\mu\tilde{\mu}$-calculus, and in the λ-calculus it can only be simulated with a redex. Notice, however, that too many different things must already be simulated with a redex in the λ-calculus. In other words, once again we realise that the λ-calculus is not expressive enough to be a reasonable proof format.

Hoping to have transmitted all of our intuitions to the reader, we are now ready to briefly dive into the details of the calculus. The syntax is described first. The reduction and typing rules can be found in the appendixes. For the metatheory and the proof of its remarkable properties the reader can consult the literature, starting from [5] where the calculus has been defined.

2.1 Syntax

The $\bar{\lambda}\mu\tilde{\mu}$-calculus has three syntactic categories: *terms* (that include term variables $x, y, z, \ldots$); *environments* — or contexts — (that include context or continuation variables $\alpha, \beta, \gamma, \ldots$); and *commands* obtained by replacing the placeholder of an environment with a term (or, dually, by replacing the placeholder of a term with an environment, as already explained).

For each syntactic category we give both the $\bar{\lambda}\mu\tilde{\mu}$-calculus and the usual syntax in λ-calculus notation.

		$\bar{\lambda}\mu\tilde{\mu}$-syntax	usual syntax
Term	$v ::= x$		x
		$\mid \lambda x : T.v$	$\lambda x : T.v$
		$\mid \mu\alpha : T.c$	
Environment	$E ::= \alpha$		
		$\mid v \circ E$	$E[(\square\ v)]$
		$\mid \tilde{\mu}x : T.c$	**let** $x : T := \square$ **in** c
Command	$c ::= \langle v \| E \rangle$		$E[v]$

The term variable x is bound by λ in v and by $\tilde{\mu}$ in c; the environment variable α is bound by μ in c. Notice the (syntactic for now) duality between μ and $\tilde{\mu}$. The only two constructors that have no syntactic dual are λ and $\circ$ (pronounced "cons"). In [5] the calculus is made perfectly symmetric by adding duals for λ and $\circ$. This extended version of the calculus is Curry-Howard isomorphic with classical subtractive sequent calculus [4]. We do not consider the subtractive case now, but we will comment on that in Sect. 4.

The "intuitionistic" fragment of the calculus, i.e. the fragment that is Curry-Howard isomorphic to the intuitionistic sequent calculus, is obtained by a simple syntactic restriction: only one environment variable is allowed (we denote it by $\star$ instead of using a Greek letter to make explicit that it is unique). Since only one variable is available, every μ constructor will override $\star$, so that only the latter continuation is in scope. This corresponds to the fact that the intuitionistic sequent calculus is obtained by restricting the sequents to have just one conclusion.

For the sake of completeness we give the reduction and typing rules of the calculus in App. A and B. They are taken without modification from [5]. The typing and reduction rules will not play any major role in the rest of the paper. However, we will exploit the possibility of inferring a type for each $\bar{\lambda}\mu\tilde{\mu}$-expression (by means of the typing rules) and of recording it directly in the term (by means of a μ or $\tilde{\mu}$-expansion rule, see App. A). In the λ-calculus it is also possible to infer the type of a subexpression, but the type cannot be recorded in the term without introducing explicit type assignment operators.

3   Structural Natural Language Rendering

We are now ready to provide (pseudo-)natural language rendering rules for the $\bar{\lambda}\mu\tilde{\mu}$-calculus. Before that we present similar rules for the λ-calculus, inspired by [3].

In both cases we attempt a *structural* translation, i.e. we try to associate to a term t its pseudo-natural language rendering $[\![t]\!]$ by structural recursion over t. We will also struggle to perform recursion over the direct subterms of t only and we will avoid processing the result of the recursive calls. Forcing the usual terminology, we will call *structural* a translation that respects all these properties.

Ideally, we would also require the translation to preserve the order of the subterms: if A and B are two sibling subterms in the proof and if A precedes B, than the rendering of A must precede that of B. This additional constraint — that surprisingly is satisfied for the $\bar{\lambda}\mu\tilde{\mu}$-calculus — makes extremely easy for a human being to "invert the transformation", building by hand the term from its rendering.

Our interest in a structural translation derives from our interest in the properties of the calculus as a proof format. For sure with complex, non-structural translations we can improve the generated text, aiming at more natural sentences. However, we claim that a good proof format must have a simple rendering semantics: if generating natural language for the proofs encoded in the proof format requires major proof transformations we consider this a serious fault of the proof format. Moreover, especially when we are interested in generating explanations of formal proofs proved with an interactive or automatic theorem prover, we do require the rendering semantics to be simple and structural to avoid loosing confidence on the correctness of the proof we are examining.

For technological reasons, every proof format should be equipped with an XML concrete syntax, imposing XSLT as the standard language for describing transformations on the document. Notice that the expressive power of XSLT (1.0) is, in practice, extremely close to our second definition of structural transformation. Indeed XSLT does not allow to process the result of a recursive call (a Result Tree Fragment) and only simple recursive functions can be described in a concise way[3].

3.1 λ-Calculus

$$\llbracket x \rrbracket \quad := \text{ consider } x \qquad\qquad \llbracket (\ldots (t\ t_1)\ \ldots\ t_n) \rrbracket := \llbracket t_1 \rrbracket$$

$$\text{we proved } T_1\ (H_1)$$
$$\ldots$$

$$\llbracket \lambda x : T.t \rrbracket := \text{ suppose } T\ (x) \qquad\qquad \llbracket t_n \rrbracket$$
$$\llbracket t \rrbracket \qquad\qquad\qquad\qquad \text{we proved } T_n\ (H_n)$$
$$\llbracket t \rrbracket$$
$$\text{we proved } T\ (H)$$
$$\text{by } H, H_1, \ldots, H_n$$

---

[3] XSLT is a Turing complete purely functional language. However, Turing completeness derives from the fact that a Result Tree Fragment (a tree) can be converted to a string for further processing and that every data type (e.g. the state of a Turing machine) can be encoded in a string and manipulated with general recursion. In practice, however, working with strings is quite cumbersome in an ad-hoc language designed to transform trees.

At first we observe that the transformation is not really structural since for the case of application we process the inner term t_1 before the outer term t_n. Notice that in the $\bar{\lambda}\mu\tilde{\mu}$-calculus the application $(\ldots(t\ t_1)\ldots t_n)$ is turned inside out, becoming $\langle t || t_1 \circ (\ldots \circ (t_{n-1} \circ t_n)\ldots)\rangle$ and making the transformation structural!

We can now repeat several of the observations we already made when discussing the intuitions about the $\bar{\lambda}\mu\tilde{\mu}$-calculus. In every rule one or more inner-types T are lacking and type inference is required to reconstruct them beforehand. Since there is just one construct, application, to derive new facts, bottom-up and top-down proofs are identified. Thus we need to represent all the proofs in the same way. The rule we provided renders every proof step in a bottom-up way, processing the t_i before using them to conclude T'. In this case, not only inner-types T_i are missing for the t_i, but also fresh labels H_i. A structural rule to render applications in a mixed bottom-up/top-down way can be easily provided:

$$[\![(t\ t_1)]\!] := [\![t]\!]$$
$$\text{we proved } T\ (H)$$
$$\text{by } H \text{ we reduce the thesis to } T_1$$
$$[\![t_1]\!]$$

where T and T_1 are the inner types of t and t_1.

However, the latter rule does not solve the lack of flexibility that derives from having to choose a uniform style of rendering every applications (bottom-up vs top-down). Notice also that the latter rule introduces a mixed proof style since t is rendered as a bottom-up step. This cannot be avoided unless an ad-hoc rule is provided for redexes.

The solution that provides more flexibility by forcing a particular interpretation of redexes can be obtained adding the rule

$$[\![(\lambda x : T.t\ t_1)]\!] := [\![t_1]\!]$$
$$\text{we proved } T\ (x)$$
$$[\![t]\!]$$

and by replacing the rule for application with

$$[\![(x\ t_1)]\!] := \text{by } x \text{ we reduce the thesis to } T_1$$
$$[\![t_1]\!]$$

Notice that in this way we impose a normal form on the λ-terms: every application $(t\ t_1)$ where t is not a variable must be β-expanded to $(\lambda x : T.(x\ t_1)\ t)$, that is semantically equivalent (according to our rendering semantics) to the mixed top-down bottom-up rule for application given before. We will not spend more time on improvements for the rendering semantics of the λ-calculus, since in the $\bar{\lambda}\mu\tilde{\mu}$-calculus these problems simply disappear.

We conclude by showing as a small example the λ-term that corresponds to a proof of $A \Rightarrow (A \Rightarrow B) \Rightarrow (B \Rightarrow C) \Rightarrow C$ and its structural natural language rendering. We use superscripts to record in the λ-term the inner-type and label of a sub-term.

$\lambda H : A.\lambda AB : A \to B.$
  $\lambda BC : B \to C.$
    $(BC\ (AB^{AB':A\Rightarrow B}\ H^{H':A})^{K:B})^{.C}$

suppose A (H), suppose $A \Rightarrow B$ (AB)
suppose $B \Rightarrow C$ (BC)
consider H; we proved A (H')
consider AB; we proved $A \Rightarrow B$ (AB')
by AB', H' we proved B (K)
consider BC; we proved $B \Rightarrow C$ (BC')
by BC', K

Notice again that, due to lack of structurality, it is difficult to transform the λ-term to its textual counterpart looking at the λ-term only. Building the λ-term from the text — a plausible operation if we consider the calculus a proof format — is even more complex. Indeed, the natural language really corresponds to the equivalent (up to β-expansions) λ-term $\lambda H : A.\lambda AB : A \to B.\lambda BC : B \to C.(\lambda H' : A.(\lambda AB' : A \to B.(\lambda K : B.(\lambda BC' : B \to C.(BC'\ K)\ BC)\ (AB'\ H'))\ AB)\ H)$ that is simpler for a human to render in natural language, but still quite annoying since the eyes must wonder back and forth between the λ-abstractions and their arguments in redexes. Only introducing **let ... in** and replacing redexes with them it becomes possible to read the term in natural language (and to produce the term by hand from the natural language!) without any major effort.

3.2 $\bar{\lambda}\mu\tilde{\mu}$-Calculus

We provide now a similar but completely structural rendering semantics for the $\bar{\lambda}\mu\tilde{\mu}$-calculus. According to the intuitions we provided, we should associate one or more sentences to a term, a textual context (i.e. a text with a placeholder) to an environment and we should render a command by filling the placeholder of its environment with the text obtained by its term. However, we anticipate that our semantics is so well behaved that the placeholder (that we will leave implicit) is always at the beginning of the text. Thus rendering a command simply amounts to concatenating the two generated texts.

$$[\![\langle v || E \rangle]\!] := [\![v]\!]\ [\![E]\!]$$

$[\![x]\!]$ $:= \text{by } x$ $[\![\alpha]\!]$ $:= \boxed{\leftarrow}\ \text{done}$

$[\![\lambda x : T.t]\!] := \text{suppose } T\ (x)$ $[\![t \circ E]\!]$ $:= \text{and } [\![t]\!]$
        $[\![t]\!]$ $[\![E]\!]$

$[\![\mu\alpha : T.c]\!] := \text{we need to prove } T$ $[\![\tilde{\mu}x : T.c]\!] := \text{we proved } T\ (x)$
        $\boxed{\hookrightarrow}\ [\![c]\!]$ $[\![c]\!]$

The symbols $\boxed{\hookrightarrow}$ and $\boxed{\leftarrow}$ stand for the increase/decrease of the indentation.

We provide as an example two $\bar{\lambda}\mu\tilde{\mu}$-terms that correspond to two different proofs of $A \Rightarrow (A \Rightarrow B) \Rightarrow (B \Rightarrow C) \Rightarrow C$.

Fully bottom-up proof:

$\mu\star : A \to (A \to B) \to (B \to C) \to C$ we need to prove $A \Rightarrow (A \Rightarrow B) \Rightarrow (B \Rightarrow C) \Rightarrow C$

$\langle \lambda H : A . \lambda AB : A \to B.$ suppose A (H); suppose $A \Rightarrow B$ (AB)

$\lambda BC : B \to C.$ suppose $B \Rightarrow C$ (BC)

$\mu\star : C.$ we need to prove C

   $\langle AB \| H \circ \tilde{\mu} K : B.$ by AB and by H we proved B (K)

   $\langle BC \| K \circ$ by BC and by K

   $\star \rangle\rangle$ done

$\| \star \rangle$ done

Fully top-down proof:

$\mu\star : A \to (A \to B) \to (B \to C) \to C$ we need to prove $A \Rightarrow (A \Rightarrow B) \Rightarrow (B \Rightarrow C) \Rightarrow C$

$\langle \lambda H : A . \lambda AB : A \to B.$ suppose A (H); suppose $A \Rightarrow B$ (AB)

$\lambda BC : B \to C.$ suppose $B \Rightarrow C$ (BC)

$\mu\star : C.$ we need to prove C

   $\langle BC \|$ by BC

   $\mu\star : B.$ and we need to prove B

     $\langle AB \|$ by AB

     $\mu\star : A.$ and we need to prove A

       $\langle H \|$ by H

       $\| \star \rangle$ done

     $\| \star \rangle$ done

   $\| \star \rangle$ done

$\| \star \rangle$ done

As made obvious by the two examples, all the rendering rules are not only structural, but they also preserve the order of the subterms. Thus it is very easy to read a $\bar{\lambda}\mu\tilde{\mu}$-term from left to right mentally producing the corresponding natural language. Dually, it is very easy to translate a proof sketch or a pen&paper proof to a $\bar{\lambda}\mu\tilde{\mu}$-term, a fundamental property for a proof format.

Notice also that indentation directives are "already present" in the term: indentation must be incremented when a μ is found and it must be decremented when a $\star$ is met. Moreover, an indented sub-proof can easily be hidden to the user by an interactive interface, showing only its thesis. The HELM library[4] adopts this strategy to increase usability by giving to the user a partial form of control over the level of details. The user can simply click on a hidden proof to unfold it, requesting more details.

In Sect. 3.1 we did not consider indentation directives. However, indentation rules cannot be avoided in the transformation to make explicit the scope of the hypotheses. Indeed the user can be deceived by the too simplified rendering semantics proposed for the λ-calculus. This does not happen for the $\bar{\lambda}\mu\tilde{\mu}$-calculus.

While the proof of the first example seems very readable, that of the second example is not. However, if you replace "and we need to prove" with the more appealing (and equally semantically faithful) sentence "we reduce the thesis to" you will get a totally reasonable text: "... we need to prove C; by BC we reduce the thesis to B; by AB we reduce the thesis to A; by H done ...". This and

---

[4] http://helm.cs.unibo.it

other similar improvements can be implemented by trading off a little bit the property of the transformation being structural.

4   Generalization and Improvements

The structural rendering semantics provided in the previous section confirms our intuitions about the fact that the $\bar{\lambda}\mu\tilde{\mu}$-calculus is a natural candidate for being a good proof format. Indeed it satisfies properties 1 (flexibility), 3 (explanation in natural language) and 4 (a clear semantics) given in the introduction. Property 2 (annotations) can be easily obtained by associating rhetorical text to each constructor of an expression. Since expressions are rendered in a structural way from left to right, associating placeholders in the rhetorical text to subexpressions is often as simple as matching the i-th placeholder with the i-th direct subexpression (instead of permutating the direct subexpressions or picking a subexpression that is deeper in the term).

As Yann Coscoy did for the λ-calculus [3] and as it has been done in a more incisive way in the MoWGLI project, we can trade the naturality of the generated text with the complexity of the rendering semantics. Since we are already starting from a much more structural and simple semantics and since the language is much richer, we can hope to obtain better results.

As we did for the $\bar{\lambda}\mu\tilde{\mu}$-calculus, it is also possible to get rid of expressions that have a weird explanation in natural language by imposing a normal form on the terms. An example is the renaming $\tilde{\mu}$ redex: $\langle H || \tilde{\mu}K : T.c \rangle$ ("by H we proved $T(K)$; $[\![c]\!]$") can be reduced to $c\{K/H\}$. More generally, redexes correspond to cuts and cuts are detours in the proof. Cut elimination (i.e. reduction) can be applied to get rid of the unwanted detours. Due to lack of space we omit the analysis of the weird redexes associated to the semantics we provide and of the associated normal form that solves the problem. We only remark that to reach the normal form it is sufficient to either reduce the redexes or η-like-expand one subexpression of the redex according to the reduction rules of the calculus.

Of course, a calculus that is Curry-Howard isomorphic to the implicative fragment of the propositional calculus is not very interesting. The $\bar{\lambda}\mu\tilde{\mu}$-calculus can be easily extended to be in correspondence with stronger logics. In particular, we modified Fellowship[5], an experimental sequent calculus based proof assistant for first order logic developed by Florent Kirchner, to produce extended $\bar{\lambda}\mu\tilde{\mu}$-calculus proof terms. We have also already defined and implemented the structural rendering semantics for the extended calculus and we plan to enhance the generated text in the near future. The extension to the constructors that are related to the other connectives (negation, conjunction, disjunction and first order universal and existential quantification) have reserved no surprises and have not broken any good property of the calculus.

According to our initial claim, we can exploit the $\bar{\lambda}\mu\tilde{\mu}$-calculus in two different ways. Either as a format for proof terms in a proof assistant or as a general proof format. In the first case we should ask whether the non-standard sequent

---

[5] http://www.lix.polytechnique.fr/Labo/Florent.Kirchner/fellowship

calculus the calculus is isomorphic too is reasonable to develop proofs in. Our short experience with Fellowship shows that as a sequent calculus it is indeed very interesting and pleasant to work with, especially for automation purposes. Indeed the distinguished formula acts as the linear hypothesis/continuation that must be eliminated next, reducing the search space. As a result Fellowship exposes only three tactics, axiom, cut and elim. The latter does not need as an argument the formula that must be eliminated, since it always act on the current distinguished (or focused) formula. More on this subject can be found in the literature about the calculus.

With respect to natural deduction, we remark that sequent calculus is always clumsier to work with interactively. We can easily adapt a natural deduction based system to produce $\bar{\lambda}\mu\tilde{\mu}$-calculus proof terms, since the sequent calculus is more fine grained than natural deduction. However, according to our initial remarks, we need to do it very carefully to obtain proof terms that record all the details of the process of construction of the proof, without identifying, for instance, top down and bottom up proofs. Concretely doing it by adapting an existent proof assistant based on natural deduction is other future work we plan to start.

We already remarked that classical proofs are usually presented in an intuitionistic logic extended with classical axioms, and not by handling multiple conclusions at once. Thus, to render classical proofs in Fellowship, we implemented a simple translation from classical $\bar{\lambda}\mu\tilde{\mu}$-expressions (i.e. expressions were there are occurrences of continuation variables that are not bound from the innermost enclosing μ binder) to intuitionistic $\bar{\lambda}\mu\tilde{\mu}$-expressions. Of course, to do so we need to introduce in the calculus a family of distinguished constants $\mathbf{EM}_T$ that inhabits the excluded middle for each type T. The translation can be easily implemented by structural recursion over the $\bar{\lambda}\mu\tilde{\mu}$-expressions:

$$\mathcal{F}_\sigma(\mu\alpha : T.c) := \mu\alpha : T.\mathcal{F}_\sigma(c) \text{ if } \alpha \text{ is used intuitionistically}$$
$$\mathcal{F}_\sigma(\mu\alpha : T.c) := \mu\alpha : T.\langle\mathbf{EM}_T||\lambda H : T.H \circ \lambda H : \neg T.\mathcal{F}_{(\alpha,T,H)::\sigma}(c) \text{ otherwise}$$
$$\mathcal{F}_\sigma(\alpha) \qquad := \tilde{\mu}x : \neg T.\langle H||x \circ \xi\rangle \text{ if } (\alpha, T, H) \in \sigma$$
$$\mathcal{F}_\sigma(\alpha) \qquad := \alpha \text{ otherwise}$$

All the other cases call $\mathcal{F}_\sigma$ recursively over each subexpression. The distinguished constant $\mathbf{EM}_T$ has type $(T \rightarrow T') \rightarrow (\neg T \rightarrow T') \rightarrow T'$ for each type T', and ξ ("ex falso sequitur quodlibet") is a distinguished continuation of type $\perp$ (i.e. a continuation that expects a term of type $\perp$ to conclude the proof). When the calculus is extended with disjunction, $\mathbf{EM}_T$ can be typed as $T \vee \neg T$ and $\mathcal{F}_\sigma$ must be slightly modified to use the $\bar{\lambda}\mu\tilde{\mu}$-calculus constructor that corresponds to case analysis (elimination of $\vee$ on the left hand side of a sequent).

Notice that the translation is purely syntactical and it does not depend on the typing judgement or on the reduction rules.

The translation is particularly effective, allowing to unveil the mathematical intuition that underlies a proof developed in a multi-conclusion sequent calculus (and that is usually extremely complex to grasp looking at the derivation only). However, the translation often introduces lots of redexes that complicate the proof. Automatic elimination of weird redexes is probably mandatory as a post-processing step to obtain natural proofs.

5   Conclusions and Perspectives

We have found yet another remarkable property of the $\bar{\lambda}\mu\tilde{\mu}$-calculus: it admits a very simple and structural rendering semantics, i.e. a translation from expressions (that are Curry-Howard isomorphic to proofs) to pseudo-natural language text. The calculus is so rich that it is able to differentiate between bottom-up and top-down proof steps, and it permits to label each intermediate result, also stating its thesis. Translating by hand a pen&paper proof sketch into a $\bar{\lambda}\mu\tilde{\mu}$-expression preserving the natural language (up to the rhetorical text) is also quite simple. Annotations can be added later on to the term to retrieve the original language.

Our impression is that, as a proof format, the $\bar{\lambda}\mu\tilde{\mu}$-calculus is as flexible as OMDoc. We plan to make this statement more precise in a forthcoming paper by providing a mutual translation between OMDoc and the $\bar{\lambda}\mu\tilde{\mu}$-expressions that respects the rendering semantics provided to both calculi in this work and in the MoWGLI project (for OMDoc).

We have also extended Fellowship, a proof assistant prototype developed by Florent Kirchner for first order logic, to record proofs as $\bar{\lambda}\mu\tilde{\mu}$-expressions, and we have equipped it with natural language rendering of the proofs. The rendering semantics implemented, being almost the one described in this paper, already produces readable explanations, but we plan to improve them in the near future by introducing new normal forms for $\bar{\lambda}\mu\tilde{\mu}$-expressions that avoid unnatural proof constructions. We have also implemented a translator of $\bar{\lambda}\mu\tilde{\mu}$-expressions to Coq proof terms and we are implementing a similar translator to Mizar and Isabelle/ISAR scripts. Automatic generation of OMDoc documents is also planned. In all these cases the aim is not only that of showing that a translation is possible, but also understanding the relative expressivity of these languages as proof formats by trying to preserve the rendering semantics of the $\bar{\lambda}\mu\tilde{\mu}$-expressions. Fellowship can be downloaded from

http://www.lix.polytechnique.fr/Labo/Florent.Kirchner/fellowship

References

1. A. Asperti, F. Guidi, L. Padovani, C. Sacerdoti Coen and I. Schena. "Mathematical Knowledge Management in HELM". In Annals of Mathematics and Artificial Intelligence, 38(1): 27–46, May 2003.
2. A. Asperti, C. Sacerdoti Coen. "Stylesheets to intermediate representation" (prototypes D2.c-D2.d) and I. Loeb, "Presentation stylesheets" (prototypes D2.e-D2.f), technical reports of MoWGLI (project IST-2001-33562).
3. Y. Coscoy. *Explication textuelles de preuves pour le calcul des constructions inductives*. PhD. thesis, Université de Nice-Sophia-Antipolis, 2000.
4. T. Crolard. "Subtractive logic". In Theoretical computer science, 254:1–2(2001), 151–185.
5. P. Curien, H. Herbelin. "The duality of computation". In Proceedings of the Fifth ACM SIGPLAN International Conference on Functional Programming (ICFP'00), ACM, SIGPLAN Notices 35(9), ISBN:1-58113-2-2-6, 233–243, 2000.

6. H. Herbelin. *Séquents qu'on calcule: de l'interprétation du calcul des séquents comme calcul de lambda-terms et comme calcul de stratégies gagnantes.* PhD. thesis, 1995.
7. M. Kohlhase. *OMDoc: An Open Markup Format for Mathematical Documents (Version 1.2).*
8. S. Lengrand. "Call-by-value, call-by-name, and strong normalization for the classical sequent calculus". In B. Gramlich and S. Lucas editors, Electronic Notes in Theoretical Computer Science, 86(4), Elsevier, 2003.
9. M. Parigot. "λμ-calculus: An algorithmic interpretation of classical natural deduction". In Proc. of the International Conference on Logic Programming and Automated Reasoning (LPAR), LNCS 624.
10. F. Wiedijk. "Formal Proof Sketches". In S. Berardi, M. Coppo and F. Damiani eds., Types for Proofs and Programs: Third International Workshop, TYPES 2003, LNCS 3085, 378–393, 2004.

A $\bar{\lambda}\mu\tilde{\mu}$-Calculus Reduction Rules

We present the reduction rules both in $\bar{\lambda}\mu\tilde{\mu}$-calculus syntax and in the usual λ-calculus syntax, omitting the contextual rules to propagate reduction everywhere in an expression. As usual, the reduction rules correspond to cut elimination.

$\bar{\lambda}\mu\tilde{\mu}$-syntax

$\langle \mu\alpha : T.c || E \rangle \quad \triangleright \ c\{E/\alpha\}$

$\langle v || \tilde{\mu}x : T.c \rangle \quad \triangleright \ c\{v/x\}$

$\langle \lambda x : T.v_1 || v_2 \circ E \rangle \triangleright \langle v_2 || \tilde{\mu}x : T.v_1 \circ E \rangle$

usual syntax

$E[\mu\alpha : T.c] \quad \triangleright \ c\{E/\alpha\}$

let $x : T := v$ **in** $c \triangleright c\{v/x\}$

$E[(\lambda x : T.v_1 \ v_2)] \ \triangleright \ E[\textbf{let}\ x{:}T{:=}v_2\ \textbf{in}\ v_1]$

We report just a few standard observations on the calculus that can be found and are explained in [5]. First of all notice that the μ and $\tilde{\mu}$ reduction rules are perfectly dual, whereas the rule for λ is asymmetric. Its dual rule is present in the subtractive system. Secondly, notice that the μ and $\tilde{\mu}$ rules form a critical pair. Giving priority to the μ rule imposes a call-by-value strategy to the calculus; the dual priority leads to call-by-name. Finally, observe that any redex is a command, but that there are commands that are not redexes. There exists a variant of the calculus where every command is a redex [8]. We have not investigated yet the property of these as proof formats.

The rules we have just presented are similar (and related) to β-reduction rules in the λ-calculus. The $\bar{\lambda}\mu\tilde{\mu}$-calculus can also be extended with rules that correspond to η-expansion. These rules are important for us since we can use them to put expressions in a normal form before rendering them in pseudo-natural language.

$$\mu\text{-expansion:} \quad v \Rightarrow \mu\alpha : T.\langle v || \alpha \rangle$$
$$\tilde{\mu}\text{-expansion:} \quad E \Rightarrow \tilde{\mu}x : T.\langle x || E \rangle$$

In the previous two rules T is the type of v (respectively of E). Type inference is required in the general case to compute T. However, for each term v (or environment E) we can always precompute its type once and for all, recording it explicitly in the expression by means of a μ-expansion (a $\tilde{\mu}$-expansion). This property is exploited when the calculus is used as a proof format.

B $\bar{\lambda}\mu\tilde{\mu}$-Calculus Typing Rules

A typing judgement is associated to each syntactic category of the calculus:

$$\Gamma \vdash v : T | \Delta, \quad \Gamma | E : T \vdash \Delta, \quad c : (\Gamma \vdash \Delta)$$

In all three kind of judgements the context Γ is a list of assumptions (i.e. a list of typed term variables $x_i : T_i$) and Δ is a list of continuations (i.e. a list of typed context variables $\alpha_i : T_i$). Notice that types associated to terms are differentiated from types associated to environments (i.e. the type expected for the term that will fill the placeholder). The former are written on the left hand side of the turnstile, whereas the latter are written on the right hand side.

A command is typed with the sequent $\Gamma \vdash \Delta$ that associates a type to every free variable in the command. Terms and environments are typed with sequents that associate types to every free variable and that are "enriched" with a distinguished formula, on the right hand side for terms and on the left hand side for environments. The distinguished formula is the type of the term or, dually, the type of the placeholder.

Table 1. Typing rules

$$(\text{CUT}) \quad \frac{\Gamma \vdash v : T | \Delta \qquad \Gamma | E : T \vdash \Delta}{\langle v || E \rangle : (\Gamma \vdash \Delta)}$$

$$(\text{AX-R}) \quad \frac{}{\Gamma ; x : T \vdash x : T | \Delta} \qquad\qquad \frac{}{\Gamma | \alpha : T \vdash \alpha : T ; \Delta} \quad (\text{AX-L})$$

$$(\text{IMPL-R}) \quad \frac{\Gamma ; x : T \vdash v : T' | \Delta}{\Gamma \vdash \lambda x : T.v : T \rightarrow T' | \Delta} \qquad \frac{\Gamma \vdash v : T | \Delta \qquad \Gamma | E : T' \vdash \Delta}{\Gamma | v \circ E : T \rightarrow T' \vdash \Delta} \quad (\text{IMPL-L})$$

$$(\tilde{\mu}) \quad \frac{c : (\Gamma \vdash \alpha : T ; \Delta)}{\Gamma \vdash \mu \alpha : T.c : T | \Delta} \qquad \frac{c : (\Gamma ; x : T \vdash \Delta)}{\Gamma | \tilde{\mu} x : T.c : T \vdash \Delta} \quad (\mu)$$

The Curry-Howard correspondence with classical sequent calculus should be evident from the typing rules given in Table 1 (where the distinguished formula can be considered at first just as a normal formula).

Observe that the symmetries of the calculus are perfectly respected at the typing level. For instance a term is given type $A \rightarrow B$ (on the right hand side of the sequent) when it waits for an input of type A to provide an output of type B. Dually an environment is given type $A \rightarrow B$ (on the left hand side of the sequent) when it provides an input of type A and waits for an output of type B.

Engineering Mathematical Knowledge

Achim Mahnke[1] and Jan Scheffczyk[2]

[1] Universität Bremen
amahnke@tzi.de
[2] Universität der Bundeswehr München
jan.scheffczyk@gmx.net

Abstract. Due to their rapidly increasing amount, maintaining mathematical documents more and more becomes an engineering task. In this paper, we combine the projects MMiSS[1] and CDET.[2] That way, we achieve major benefits for mathematical knowledge management: (1) Semantic annotations relate mathematical constructs. This reaches beyond mathematics and thus fosters integration of mathematical content into a broader context. (2) Fine-grained version control enables change management and configuration management. (3) Semi-formal consistency management identifies violations of user-defined consistency requirements and proposes how they can be best resolved.

1 Introduction

The corpus of electronically available mathematical knowledge increases rapidly. Usually, mathematical objects are embedded in and related to different kinds of documents like articles, books, or lecture material, the domain of which can be different from mathematics, e.g., engineering or computer science. Therefore, maintaining high-quality mathematical knowledge becomes a non-trivial engineering task for teams of authors that must be supported.

In this paper, we combine MMiSS, a general-purpose approach for maintaining structured documents that are semantically annotated, and CDET, a general-purpose approach for semi-formal consistency management. That way, we achieve major benefits for mathematical knowledge management (MKM), which pay particularly for teams of authors:

1. Ontology-driven semantic annotations relate mathematical constructs, which reaches beyond mathematics and thus fosters automatic analysis and integration of mathematical content into a broader context.
2. Fine-grained version control enables sophisticated change management and configuration management.

---

[1] MMiSS: Multimedia in Safe and Secure Systems – see www.mmiss.de
[2] CDET: The Consistent Document Engineering Toolkit – see www.unibw.de/inf2/ CDET

M. Kohlhase (Ed.): MKM 2005, LNAI 3863, pp. 250–266, 2006.

3. Consistency management identifies violations of user-defined requirements and proposes how these violations can be best resolved, i.e., repaired.

MMiSS supports semantic markup for structured LaTeX documents and modeling of semantic dependencies through the integration of formal domain knowledge representation methods (ontologies) into the process of authoring mathematical documents. Using ontologies to structure semantic markup makes our approach highly adaptable to the application at hand and avoids ambiguities that can arise from overloading mathematical symbols or leaving such symbols implicit. MMiSS provides a repository, which keeps track of document revisions and which facilitates the explicit reuse and sharing of document parts.

During the development of semantically related documents, quality control w.r.t. user-defined requirements has proven crucial in a range of applications; see, e.g., [1, 2]. This kind of *consistency management* is particularly useful for teams of authors.[3] Inconsistencies (i.e., violations of user-defined requirements) are natural during the editing process and must be tolerated (at least temporarily). CDET (1) provides automated consistency checking of user-defined consistency requirements, (2) tolerates inconsistencies, and (3) suggests prioritized repairs [3]. CDET can check consistency at various granularity levels of formal, informal, and semi-structured content and integrates fully with authors' established practices. The MMiSS repository supports easy process integration, efficient consistency checking, and temporal consistency rules (which restrict the development of documents over time). Although CDET can work on MMiSS documents only, employing ontologies results in highly configurable and flexible consistency management.

This paper proceeds as follows: First, we discuss related work. In Sect. 3 we give a short survey of MMiSS and illustrate the benefits for MKM. In Sect. 4 we present typical consistency requirements. In Sect. 5 we illustrate consistency management by CDET. In Sect. 6 we evaluate our approach and outline future research directions.

2   Related Work

MKM systems often concentrate on the representation of mathematical objects themselves, which facilitates the exchange of core mathematical content (formulas, theories etc.), the use of mathematical tools like computer algebra systems, and the management of change for these objects (e.g., [4, 5, 6, 7, 8]). Approaches for representing and formalizing the dependencies between formal parts of mathematical documents and their informal counterparts are needed but rarely found (the HELM approach described in [4] does this to a limited extent).

---

[3] Our notion of "consistency" means to fulfill user-defined requirements; as opposed to the mathematical term, which means that a theory has a model. Consistency checking roughly corresponds to model checking.

With respect to structuring and interrelation of document objects, MMiSS is comparable to OMDoc [9] – in fact, MMiSS documents can be translated to OMDoc and vice versa. The main difference lies in the ontological layer, which allows authors of MMiSS documents to create new structural elements or semantic relations by inheriting the predefined ones. Our import and export facility for OMDoc supports exchanging documents that adhere to the OM-Doc standard. The translation includes the mapping of element types (like defi nition, lemma etc.) and most of OMDoc's basic relations between these elements.[4]

In contrast to systems managing the consistency and change of mathematical objects themselves (e.g., doing theorem proving) [8, 10], we provide consistency management on structured documents. Therefore, our approach cannot be used for theorem proving. Instead, we can use theorem provers in order to check semantic relations on embedded mathematical objects. For example, we can require that every theorem is proven; then CDET can use a theorem prover to detect whether a potential proof really holds.

Ontologies have been used to formalize the semantics of mathematical objects in a number of approaches, but they either concentrate on the pure mathematical contents [11] or they exploit ontological representations for certain functionalities only, e.g., Web services or search facilities [12, 13]. Semantic annotations for LaTeX documents rely on naming only [14, 15]. In order to foster integration of mathematical documents into a broader context and to support automated reasoning over mathematical documents by different tools, semantic annotations themselves should be defined and structured by means of ontologies.

The CPoint [16] system addresses the problem of creating semantical annotated mathematical documents and is very similar to MMiSS in its goals. The main difference to MMiSS is that CPoint is an invasive editor for PowerPoint, whereas MMiSS is built for LaTeX authoring. CPoint exports OMDoc documents, which can be viewed as instance documents according to the (fixed) ontology OMDoc provides for mathematical documents. MMiSS authors are creating knowledge on the instance (or fact) level, too, but the underlying ontology can be extended if necessary. During the MMiSS project, a tool for visualizing ontologies has been developed, which is similar to the CPointGraphs module.

Despite of its importance, consistency management w.r.t. user-defined requirements appears rarely addressed in MKM systems, which severely hinders collaborative maintenance of mathematical documents. Regarding formal consistency checking, we find a lot of related work in the field of software engineering, e.g., the consistency management tool xlinkit [2] (for further details see [3]). In contrast to xlinkit, which checks distributed documents, CDET employs a repository for better process integration and temporal consistency checks.

---

[4] Formulas are not converted to MathML or OpenMath, but MMiSS contains a package with a restricted set of macros for mathematical symbols and operators, called MathLight, which could be translated to MathML.

3   Writing Mathematical Documents Using MMiSS

MMiSS is a general-purpose editing environment, mostly used for preparing lectures [17]. Here, we use MMiSS for maintaining mathematical documents.

MMiSS provides a LaTeX package with environments and commands for creating cleanly structured documents. Their high-level structure is given by sections, which can be nested to an arbitrary depth. For formal content (like mathematics or programs), a set of environments has been defined; e.g., a structural entity called "theory" groups constituent formal units like assertions (theorems, lemmas, corollaries etc.). Fig. 1 shows a portion of a theory document written with MMiSSLaTeX that includes several environments containing mathematical content: definitions for the concepts of "Signature" and "Algebra" (lines 6–8 and 10–13), a conjecture (lines 15–18), and its proof (lines 20–22). Using these environments, the author states, which parts of his document bear mathematical content.

3.1   Ontologies

Ontologies provide the means for establishing a semantic structure. An ontology is a formal explicit description of concepts in a domain of discourse [18]. The

```
                          ── MMiSSLaTeX ──
1   \Class{Algebra}{algebra}{}
2   \Class{Signature}{signature}{}
3   \Class{TermAlgebra}{term algebra}{Algebra}
4
5   \begin{Section}[Label=AlgSpec, Title={Signatures, Terms and ...}]
6     \begin{Definition}[Label=DefSignature, Title=Signature]
7      A \Def{Signature} \SigmaSig{} is given by ...
8     \end{Definition}
9
10    \begin{Definition}[Label=DefAlgebra, Title={$\Sigma$-Algebra}]
11    An algebra $A= (S_A, \Omega_A)$ for a \Reference{Signature} ...
12    \Relate{requires}{DefAlgebra}{DefSignature}
13    \end{Definition}
14
15    \begin{Conjecture}[Label=TermAlgebra, Title={The Term Algebra}]
16    \Def{TermAlgebra}
17    \Ref[Terms]{TermsOverSets} ... form an~\reference{SigmaAlgebra}.
18    \end{Conjecture}
19
20    \begin{Proof}[Label=TermAlgebraProof]
21    \Relate{proves}{TermAlgebraProof}{TermAlgebra} For each sort ...
22    \end{Proof}
23  \end{Section}
```

Fig. 1. Abridged example of the MMiSSLaTeX document theory.tex

MMiSSLATEX package for ontologies provides a set of easy-to-use macros for the declaration of ontologies. Within the prelude of each MMiSS document, the author builds up an ontology, which covers the domain of this document. The ontology is used by the author to place semantic annotations and thereby build semantic relations between document elements.

The first two lines of our example (Fig. 1) show the declaration of concepts (as ontology classes) that should be defined within the document. The concept Signature is defined by the command \Def{Signature} in line 7. MMiSS can, therefore, derive that the surrounding definition environment forms the definition for this concept. References to the concept of a signature thus link to this element (e.g., line 11 shows such a reference).

MMiSS provides relation types, which capture deeper meaning than such references, e.g., the *proves* relation. The conjecture labeled TermAlgebra (line 15) is proven by the proof TermAlgebraProof (line 20), thus changing the state of the conjecture into a theorem. We express the proves relation by the MMiSSLATEX statement \Relate{proves}{TermAlgebraProof} {TermAlgebra} (in line 21), which creates a proves-link between these two objects. The proves relation is predefined in the so-called MMiSS systems ontology and is a subrelation of *reliesOn*. The reliesOn relation reflects that two structural entities semantically rely on each other. Relations are important for consistency management: If there are changes to entities which others rely on, the consistency is suspicious to be violated; deletion of such an entity breaks consistency. In our example, if the proves relation is broken, the theorem turns into a conjecture again.

3.2   Benefits for Maintaining Mathematical Documents

One must not underestimate the effort of developing an ontology. Particularly for teams of authors, the benefits of using ontologies clearly outweigh these costs.

Without ontologies we often encounter difficulties when we reuse or share material: The exact meaning of a concept is unclear, or different terms are used for the same semantic concept, or the same term is used for different semantic concepts. For example, consider the term "algebra", which is frequently used in different contexts. Whereas a human user can often discriminate from the context, a tool must have unambiguous information: we would certainly expect a hyperlink to lead to the correct target definition. In MMiSSLATEX ontologies, the author associates a default phrase to each declared class, object, or relation. If the same phrase is attached to different semantic concepts, the user is presented with the appropriate term; the author states which concept he means by referencing the corresponding class or object in his domain ontology.

Using ontologies to represent authors' knowledge is domain independent, so it is easy to associate mathematical facts with aspects of other domains, e.g., computer science, engineering science, or pedagogics. Particularly, relationships to pedagogical knowledge can be useful in the preparation of eLearning material – recent trends in the field of eLearning put strong emphasis on rich pedagogical metadata for Learning Objects. MMiSS provides the relation type *illustrates*, which can be used to link examples or illustrations to the concepts they explain.

These links can be explored to support pedagogically motivated consistency rules – see R 4 in Sect. 4.

Since domain ontologies developed along with the documents can be translated to Semantic Web standards like OWL, mathematical MMiSS documents can easily be converted into resources for the Semantic Web.[5] Existing tools like semantics-aware search engines or visualization tools can be utilized.

Besides of using ontologies, maintaining mathematical documents benefits from unique features of the MMiSS repository. Fine-grained (XML-based) version control on the object level supports concurrent changes of documents by providing merge functions. The repository can store variants of objects, so that different presentations (variants like natural language or slides and scripts etc.) can be generated out of the same document. MMiSS provides a sophisticated editing environment, which respnects the rich structure of mathematical documents and semantic relations between mathematical (and non-mathematical) contents. Therefore, MMiSS provides a good basis for fine-grained consistency requirements, which in turn are fundamental prerequisites to automated consistency management.

4 Example Consistency Requirements

In practice, the most important step towards automated consistency management is to define a notion of "consistency," i.e., to develop consistency requirements based on the structure and semantics of the documents. These consistency requirements are informal and thus tool independent. They will be formalized once the tool for actually managing consistency has been chosen.

The advantages of our approach can be shown best in a scenario where a team of authors is creating and maintaining highly interrelated documents stored in the MMiSS repository. Let the repository contain a significant amount of mathematical documents of various kinds (books, articles, lecture material, scripts etc.), which is very likely to exist in a community of practice. In this paper, we use the particular scenario of two mathematicians, one of them (lets say A) has written a paper in which he develops a theory about Signatures and Terms (theory.tex in Fig. 1). A colleague (B) is preparing lecture material explaining parts of this work (lecture.tex in Fig. 2). Both documents are structured using the MMiSS facilities and are stored in the repository along with other material (papers, scripts etc.) referencing A's theory.

In this scenario, several consistency requirements arise. Usually, the requirements below *will be broken* during the development of MMiSS documents, even if only one author is editing the documents. In fact, for teams of authors consistency management is particularly helpful. Here, the additional formal effort produces multiple payoffs. Most of the requirements are considered "weak" – we explicitly permit violations. Some requirements are, however, so important that

---

[5] In MMiSS ontologies relations can be specified with properties like 'is a strict order relation' – not all of them can be expressed in OWL.

```
                            ─── MMiSSLaTeX ───
 1   \begin{Section}[Label=sec1, Title={Algebras}]
 2     \begin{Example}[Label=ExampleAlgebra, Title=An Algebra Example]
 3       \Def{ExampleAlgebra} An Example (in CASL): the natural numbers ...
 4       \Relate{illustrates}{ExampleAlgebra}{DefAlgebra}
 5     \end{Example}
 6
 7     \begin{Definition}[Label=DefSigmaAlgebra, Title=SigmaAlgebra]
 8       A \Def{SigmaAlgebra} ...
 9       \Relate{requires}{DefSigmaAlgebra}{TermAlgebra}
10     \end{Definition}
11   \end{Section}
12
13   \begin{Section}[Label=sec2, Title={Signatures}]
14     \begin{Example}[Label=ExampleSignature, Title=A Signature Example]
15       \Def{ExampleSignature} ...
16       \Relate{illustrates}{ExampleSignature}{DefSignature}
17     \end{Example}
18   \end{Section}
```

Fig. 2. Abridged example of the MMiSSLaTeX document lecture.tex

authors must not check in documents violating these requirements – such requirements are "strong". As a means of quality control, some weak requirements become strong for released documents.

The local domain ontology is a kind of signature stating the concepts which the author "promises" to define and to explain in this document. Therefore, it is natural to require:

R 1. (weak – strong for released documents)   All objects promised in the local ontology of a document are defined in the document body.

"Deeper-level" consistency requirements are imposed by the interaction between references to ontology components within the document and the structural document entities themselves. For example, consider the proves link between the proof TermAlgebraProof and the conjecture TermAlgebra in theory.tex. The proves relation is predefined in the systems ontology and is a subrelation of reliesOn. Thus, we require:

R 2. (weak)   A proven assertion should not change its formulae significantly.

By "significant" we mean, e.g., that the wording of a theorem may change without affecting its proof; also, the formula of an assertion may be weakened (i.e., the formula of the old version implies the formula of the new version). We should not expect too much from this requirement, just clear pointers to possible inconsistencies. Still, it is mathematicians who re-validate the proofs.

In MMiSS the author indicates the status of an assertion by using the appropriate element type, e.g., a conjecture (see Fig. 1). By proving this conjecture somewhere in his document, author A promotes it to a theorem. Authors, referencing this theorem trust in its status when using it for developing their theory.

If – at a later stage – the theorem is demoted to a conjecture again, because the proof turned out to be wrong, the mathematical objects relying on this theorem are suspicious to be broken. Thus, it is reasonable to require:

R 3. (weak)   The status of a relied-on assertion should not be demoted.

In lectures, it is good practice to connect an example explicitly to the illustrated concept (see lines 4, 16 in Fig. 2). Often, these concepts correspond to symbols defined in the context of a certain theory. Theories use other theories, thus inducing a dependency relationship on the corresponding symbols (concepts) which should be denoted by the originating author by means of the ontology (see line 12 in Fig. 1).[6]

In his lecture, author B can explore this knowledge by requiring that the examples he is developing are ordered according to the dependencies of the explained concepts, i.e., if concept x (somehow) reliesOn concept y, and he is going to explain both, than the example for y should be given first. If the course is developed by a group of lecturers (an often found situation), this requirement is likely to be broken.

R 4. (weak – strong for released documents)   Examples should be ordered according to the dependencies of the addressed concepts.

5   Pragmatic Consistency Management

For managing user-defined consistency of (mathematical) MMiSS documents we employ CDET, a general-purpose consistency management approach. In Sect. 5.1 we introduce our consistency management architecture, combining MMiSS and CDET. In Sect. 5.2 we formalize the consistency requirements from the previous section, in order to enable consistency management (shown in Sect. 5.3). CDET considerably benefits from MMiSS, i.e., fine-grained version control and semantic annotations that are structured by an ontology. Technical details about CDET are beyond the scope of this paper, see [3] and the CDETWeb site.

5.1   The CDET Consistency Management Approach

Fig. 3 illustrates the CDET consistency management approach. In their day-to-day work, mathematicians check out documents, edit them, and check them in again via one of the methods offered by MMiSS. Consistency requirements are formalized by consistency rules in a variant of first-order predicate logic. Strong rules must be adhered to, weak rules may be violated. Rules can be customized to specific projects, e.g., lectures or courses. Since users can formalize their own rules, consistency management is tunable to the application at hand. In order to ensure well-formedness, rules are type-checked against the functions

---

[6] The reliesOn relation seems appropriate for these symbol dependencies, but further analysis may show that a subrelation of reliesOn suits better. In our example we use the subrelation *requires*.

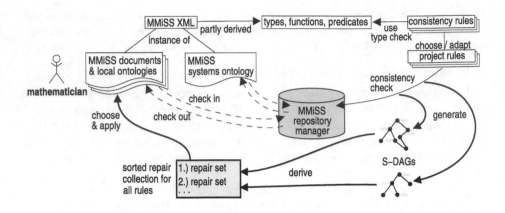

Fig. 3. Consistency management by CDET – simplified overview

and predicates they use. Functions and predicates are implemented in the functional programming language Haskell [19]. That way, we can handle document properties inexpressible by first-order rules like traversing recursive document structures, which we encounter in MMiSS XML document structures. Moreover, Haskell's Foreign Function Interface supports to integrate mathematical tools like theorem provers for sophisticated content evaluation.

At a check-in, CDET checks the MMiSS repository for consistency and generates appropriate repairs, from which mathematicians can choose. Repairs are generated in two steps, in order to avoid exponential computational complexity of repair enumeration: (1) during consistency checking CDET generates for each rule an S-DAG (Suggestion-carrying Directed Acyclic Graph); (2) on demand CDET derives one repair collection from all S-DAGs. The MMiSS repository is locked during step (1) only, which is performed incrementally. An S-DAG visualizes inconsistencies and plausible repair actions for *one* rule. S-DAGs do not include all possible repairs but only a few repairs that require the least changes to the repository. S-DAG reduction is a major benefit in practice. The repair collection contains alternative repair sets, each of which includes repairs for *all* rules. Since the repair collection can grow large (hundreds of alternatives), it is sorted w.r.t. a user-defined preference metric.

5.2   Formalizing Consistency Rules

In order to check consistency across document versions, CDET employs a variant of temporal predicate logic, which explicitly quantifies over repository states.[7] That way, we can formalize rules that restrict the development of documents over time. Annotations guide repair generation by supplying domain knowledge.

---

[7] A repository state represents the accumulated check-ins up to a given point in time.

> **R 1.** At all repository states t, we require for all MMiSS documents *doc* that each
> object *obj* promised in the local ontology is defined; i.e., in *doc* there exists a
> definition *def* equal to *obj*.
>
> $\forall t^{\mathsf{KEEP}} \in \mathbf{repStates} \bullet \forall doc^{\mathsf{KEEP}} \in \mathbf{repMMiSSDocs}(t) \bullet \forall obj \in \mathbf{objects}(\mathcal{O}_{doc}) \bullet$
>
> $\exists def \in \mathbf{allDefs}(doc) \bullet \mathbf{cont}(def) = \mathbf{name}(obj) \left\{ \begin{array}{l} \{\text{False: } def.\mathbf{cont} \rightsquigarrow \mathbf{name}(obj)\}, \\ \{\text{False: } obj.\mathbf{name} \rightsquigarrow \mathbf{cont}(def)\} \end{array} \right\}$

R 1 first quantifies over all repository states t, provided by **repStates**. Second, we parse all MMiSS documents with the help of **repMMiSSDocs**(t). We determine all objects declared in the local domain ontology $\mathcal{O}_{doc}$ of a document *doc* by **objects**($\mathcal{O}_{doc}$). For all these objects we require a suitable definition (\Def{...} in MMiSSLaTeX), where **cont**(*def*) returns the content of the definition *def*. We get all definitions in a MMiSS document *doc* by **allDefs**(*doc*), which also identifies their locations, comparable to XPath axes. Notice that the definition *def* may occur *anywhere* in the document *doc*. Inconsistencies can be resolved by either changing the definition content towards the object name or vice versa. This is expressed by the *hint collection* {{False: *def*.**cont** $\rightsquigarrow$ **name**(*obj*)}, {False: *obj*.**name** $\rightsquigarrow$ **cont**(*def*)}}.

A predicate can be annotated by a hint collection. Each hint set within a hint collection is an alternative. A user-supplied hint indicates how the truth value of a predicate can be flipped. That way, hints provide domain knowledge for repair generation. A hint b: $v.\mathbf{f} \rightsquigarrow e$ proposes to change the field $\mathbf{f}$ of the variable v to the term e if the predicate evaluates to the truth value b. Notice that *any* well-typed term e can be used here. Therefore, hints provide great flexibility to react to inconsistencies; e.g., the current repository state of the violation can be evaluated, which supports repair strategies that change over time. Due to static typechecking of hints, repairs that change document content respect the document structure. In addition, we can annotate quantified variables by ^{KEEP} in order to avoid repairs for these variables.

> **R 2.** At all repository states, we require for all theory documents *thD* with previous
> version thD_p the following: If the previous version $assert_p$ of an assertion *assert* has
> been proven then all formulae of *assert* are implied by the formulae of $assert_p$.
>
> $\forall t^{\mathsf{KEEP}} \in \mathbf{repStates} \bullet \forall thD^{\mathsf{KEEP}} \in \mathbf{repTheoryDocs}(t) \bullet$
>
> $\forall thD_p^{\mathsf{KEEP}} \in \mathbf{repTheoryDocs}(\mathbf{prevState}(t)) \bullet \mathbf{dId}(thD) = \mathbf{dId}(thD_p) \Rightarrow$
>
> $\forall assert^{\mathsf{KEEP}} \in \mathbf{allAssertions}(thD) \bullet \forall assert_p^{\mathsf{KEEP}} \in \mathbf{allAssertions}(thD_p) \bullet$
>
> $\left(\begin{array}{l} \mathbf{label}(assert) = \mathbf{label}(assert_p) \quad \wedge \\ \exists pr_p^{\mathsf{KEEP}} \in \mathbf{allProofs}(thD_p) \bullet \exists r_{pr_p}^{\mathsf{KEEP}} \in \mathbf{allRelates}(pr_p) \bullet \\ \exists rel_{P_p}^{\mathsf{KEEP}} \in \mathbf{subRels}(\mathcal{O}_{\mathbf{prevState}(t)} \cup \mathcal{O}_{thD_p}, \mathbf{proves}) \bullet \\ \mathbf{relation}(r_{pr_p}) = \mathbf{name}(rel_{P_p}) \quad \wedge \quad \mathbf{target}(r_{pr_p}) = \mathbf{label}(assert) \end{array} \right) \Rightarrow$
>
> $\forall f^{\mathsf{DEL}} \in \mathbf{formulae}(assert) \bullet \mathbf{formulae}(assert_p) \mapsto f$

R 2 relates consecutive versions of theories. We get the current version *thD* by **repTheoryDocs**(t) and the previous version thD_p by **repTheoryDocs**(**prevState**(t)). For each assertion (theorem, conjecture, proposition etc.) *assert* in *thD* we retrieve its previous version $assert_p$ in thD_p. R 2 only affects assertions that have been proven previously, i.e., there exists a proof pr_p with a proves relation r_{pr_p} targeting the assertion $assert_p$. We determine all proves relations by first joining the systems ontology at the previous state $\mathcal{O}_{\text{prevState}(t)}$ and the local domain ontology $\mathcal{O}_{thD_p}$. Then $\texttt{subRels}(\mathcal{O}_{\text{prevState}(t)} \cup \mathcal{O}_{thD_p}, \texttt{proves})$ returns the reflexive transitive closure of all subrelations of the proves relation. If for the previous version $assert_p$ there exists a proof then each formula f of $assert$ should be logically implied by all formulae of $assert_p$, in order to ensure validity of the proof. Inconsistencies can be resolved by deleting formulae f that are not implied, which we express by annotating the variable f by ^{DEL}. We denote logical implication by the predicate $\mapsto$, which currently is implemented as a simple "element of" relation. For specific applications this implementation can be replaced by a more sophisticated variant since predicates are implemented in Haskell. Other tools, e.g., theorem provers, can be employed via Haskell's Foreign Function Interface.

R 3. At all repository states, we require for all theory documents thD with previous version thD_p the following: If an assertion $assert$ is relied on by any MMiSS document and the previous version $assert_p$ was proven then the new version $assert$ should be proven, too.

$\forall t^{\text{KEEP}} \in \texttt{repStates} \bullet \forall thD^{\text{KEEP}} \in \texttt{repTheoryDocs}(t) \bullet$

$\forall thD_p^{\text{KEEP}} \in \texttt{repTheoryDocs}(\text{prevState}(t)) \bullet \texttt{dId}(thD) = \texttt{dId}(thD_p) \Rightarrow$

$\quad \forall assert^{\text{KEEP}} \in \texttt{allAssertions}(thD) \bullet \forall assert_p^{\text{KEEP}} \in \texttt{allAssertions}(thD_p) \bullet$

$$
\begin{pmatrix}
\texttt{label}(assert) = \texttt{label}(assert_p) \quad \wedge \\
\begin{pmatrix}
\exists doc^{\text{KEEP}} \in \texttt{repMMiSSDocs}(t) \bullet \exists r^{\text{KEEP}} \in \texttt{allRelatesDoc}(doc) \bullet \\
\exists rel_R^{\text{KEEP}} \in \texttt{subRels}(\mathcal{O}_t \cup \mathcal{O}_{doc}, \texttt{reliesOn}) \bullet \\
\texttt{relation}(r) = \texttt{name}(rel_R) \quad \wedge \quad \texttt{target}(r) = \texttt{label}(assert)
\end{pmatrix} \wedge \\
\begin{pmatrix}
\exists pr_p^{\text{KEEP}} \in \texttt{allProofs}(thD_p) \bullet \exists r_{pr_p}^{\text{KEEP}} \in \texttt{allRelates}(pr_p) \bullet \\
\exists rel_{P_p}^{\text{KEEP}} \in \texttt{subRels}(\mathcal{O}_{\text{prevState}(t)} \cup \mathcal{O}_{thD_p}, \texttt{proves}) \bullet \\
\texttt{relation}(r_{pr_p}) = \texttt{name}(rel_{P_p}) \quad \wedge \quad \texttt{target}(r_{pr_p}) = \texttt{label}(assert_p)
\end{pmatrix}
\end{pmatrix} \Rightarrow
$$

$$
\begin{pmatrix}
\exists pr \in \texttt{allProofs}(thD) \bullet \exists r_{pr} \in \texttt{allRelates}(pr) \bullet \\
\exists rel_P^{\text{KEEP}} \in \texttt{subRels}(\mathcal{O}_t \cup \mathcal{O}_{thD}, \texttt{proves}) \bullet \\
\texttt{relation}(r_{pr}) = \texttt{name}(rel_P) \ \{\{\text{False:}\ r_{pr}.\text{relation} \rightsquigarrow \texttt{name}(rel_P)\}\} \quad \wedge \\
\texttt{target}(r_{pr}) = \texttt{label}(assert) \ \{\{\text{False:}\ r_{pr}.\text{target} \rightsquigarrow \texttt{label}(assert)\}\}
\end{pmatrix}
$$

R 3 relates multiple documents at consecutive repository states. For each assertion (theorem, conjecture, proposition etc.) $assert$ in a theory document thD we retrieve its previous version $assert_p$ in thD_p. R 3 only affects assertions that are relied on by *any* formal construct in *any* MMiSS document doc. We require any reliesOn relation r within doc, which particularly includes the requires relation. Therefore, r must be equal to a subrelation of the reliesOn relation in the joined ontology $\mathcal{O}_t \cup \mathcal{O}_{doc}$. Also, R 3 only affects assertions that have been proven previously — again, we require *any* proves relation rel_{P_p} within the proof pr_p. If for the previous version $assert_p$ there exists a proof and the new version $assert$ is relied on then the new version $assert$ should be proven, too. The assertion $assert$ may be proven by a different proof or by a different proof strategy (expressed by a different proves subrelation). We propose to adapt the relation type and target of the current proof to resolve inconsistencies.

R 4. At all repository states, we require for all examples ex_1 and ex_2 in a lecture leD: If ex_1 illustrates a formal construct def of a theory document thD and def relies on a formal construct that is illustrated by ex_2 then ex_2 should appear before ex_1.

$\forall t^{\text{KEEP}} \in \texttt{repStates} \bullet \forall leD^{\text{KEEP}} \in \texttt{repLectDocs}(t) \bullet$
$\forall ex_1^{\text{KEEP}} \in \texttt{allExamples}(leD) \bullet \forall ex_2^{\text{KEEP}} \in \texttt{allExamples}(leD) \bullet$

$\left(\begin{array}{l} \exists r_{ex_1}^{\text{DEL}} \in \texttt{allRelates}(ex_1) \bullet \exists rel_{I_1}^{\text{KEEP}} \in \texttt{subRels}(\mathcal{O}_t \cup \mathcal{O}_{leD}, \texttt{illustrates}) \bullet \\ \exists thD^{\text{KEEP}} \in \texttt{repTheoryDocs}(t) \bullet \exists def^{\text{KEEP}} \in \texttt{allFormalConstr}(thD) \bullet \\ \texttt{relation}(r_{ex_1}) = \texttt{name}(rel_{I_1}) \quad \wedge \quad \texttt{target}(r_{ex_1}) = \texttt{label}(def) \quad \wedge \\ \exists r_{def}^{\text{KEEP}} \in \texttt{allRelates}(def) \bullet \exists rel_R^{\text{KEEP}} \in \texttt{subRels}(\mathcal{O}_t \cup \mathcal{O}_{thD}, \texttt{reliesOn}) \bullet \\ \texttt{relation}(r_{def}) = \texttt{name}(rel_R) \quad \wedge \\ \exists r_{ex_2}^{\text{DEL}} \in \texttt{allRelates}(ex_2) \bullet \exists rel_{I_2}^{\text{KEEP}} \in \texttt{subRels}(\mathcal{O}_t \cup \mathcal{O}_{leD}, \texttt{illustrates}) \bullet \\ \texttt{relation}(r_{ex_2}) = \texttt{name}(rel_{I_2}) \quad \wedge \quad \texttt{target}(r_{ex_2}) = \texttt{target}(r_{def}) \end{array} \right) \Rightarrow$

$\texttt{num}(ex_2) \leq \texttt{num}(ex_1) \quad \{\{\text{False: } ex_2.\texttt{loc} \rightsquigarrow \texttt{loc}(ex_1)\}, \{\text{False: } ex_1.\texttt{loc} \rightsquigarrow \texttt{loc}(ex_2)\}\}$

In R 4 we parse all lectures by `repLectDocs(t)`. The example ex_2 should appear before the example ex_1 if ex_1 illustrates a definition that relies on a formal construct illustrated by ex_2. Examples are consecutively numbered by the `num` attribute. If they appear in wrong order we propose to switch their locations.

Formalizing consistency rules is due to experts in the field of formal logic. Employing the MMiSS ontologies alleviates the effort of formalization; e.g., subrelations do not need to be hard-coded inside the rules. The systems ontology and the local domain ontology can be seen as parameters to a consistency rule.

5.3 Managing Inconsistencies

Inconsistencies in our example and repairs are visualized by S-DAGs. A repair proposes an action that resolves an inconsistency, e.g., by changing document content.

The structure of an S-DAG resembles that of a consistency rule in miniscope form (i.e., negations and quantifiers are pushed inwards, implications are replaced by disjunctions). Nodes represent logical connectives or atomic formulae; edges target the subformulae of a connective. Universal nodes ⱽ and existential nodes ∃ represent universal and existential quantification, respectively. Outgoing edges carry value bindings to the quantified variable and alternative repair actions. A value represents a repository state, a document, or document content, blamed for inconsistencies. An action proposes to either add a value to (Add), or change a value within (Chg), or delete a value from the domain of the quantifier node (Del). Conjunction nodes ⋀ stand for conjunctions; disjunction nodes ⋁ stand for disjunctions. Negation nodes are omitted for brevity. A predicate leaf contains an atomic formula that causes inconsistencies and its truth value.

S-DAGs only include repairs that require little modifications to the repository. Generating all possible repairs has proven infeasible. Also, S-DAGs include individual repairs only; i.e., repairs are not combined. CDET derives a repair collection for all inconsistencies of all rules in a second step (see below). We separate S-DAG generation from the computationally expensive derivation of the repair collection, in order to lock the repository during S-DAG generation only (which is done incrementally).

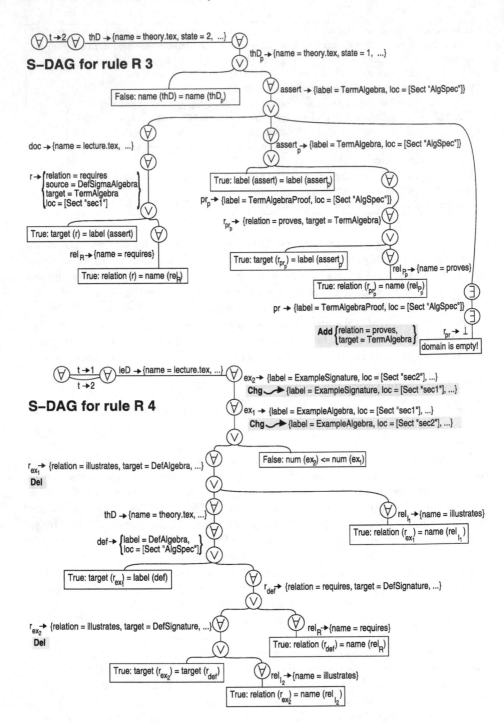

Fig. 4. Generated S-DAGs for the rules R 3 and R 4 (other rules are fulfilled)

Fig. 4 shows the S-DAGs generated by CDET for our example rules. The S-DAG for R 3 represents an inconsistency in the documents theory.tex and lecture.tex at state 2. That is because the assertion TermAlgebra is required by the lecture (left part), the assertion TermAlgebra has been proven at state 1 (middle part) and at state 2 there is no proof for TermAlgebra (lower right part). By the lower right leaf "domain is empty" the S-DAG informs that the proof TermAlgebraProof is lacking a proves relation. Consequently, CDET proposes to Add a proves relation targeting TermAlgebra.

R 4 is violated at state 1 and state 2. Within lecture.tex the example ExampleAlgebra appears before ExampleSignature. ExampleAlgebra illustrates the definition DefAlgebra and ExampleSignature illustrates the definition DefSignature both of which we find within theory.tex. This is an inconsistency because DefAlgebra requires DefSignature. Therefore, the examples should be reversed (upper right part of the S-DAG) or the illustrates relations should be deleted. CDET proposes to delete the illustrates relations due to our explicit request.

From all S-DAGs, CDET derives one repair collection on demand. The repair collection contains alternative repair sets; within each set, all repairs together resolve all inconsistencies for all rules. CDET guarantees that (1) each repair set is a real alternative that is not expressed by other repair sets and (2) the repairs within each set do not contradict each other.

A repair consists of four components: (1) affected rules and variables (a repair can affect multiple rules), (2) the domain term of the variables (as given by the rules), (3) variable bindings necessary to calculate the domain of the repair, and (4) the proposed repair action.[8] Naturally, repairs might violate consistency, which we can determine only after a new consistency check.

For our example, CDET derives a repair collection containing three alternative repair sets. Below you find the top-ranked repair collection at the repository state 2.

$$
\begin{aligned}
&\text{Rep } \{\text{R } 3(r_{pr})\} \quad \text{allRelates}(pr) \\
&\quad \left\{ \begin{array}{l} pr \mapsto \{\text{label} = \text{TermAlgebraProof}, \text{loc} = [\text{Sect AlgSpec}], \ldots\}, \\ thD \mapsto \{\text{name} = \text{theory.tex}, \ldots\}, t \mapsto 2 \end{array} \right\} \\
&\quad \text{Add } \{\text{relation} = \text{proves}, \text{target} = \text{TermAlgebra}\} \\
&\text{Rep } \{\text{R } 4(ex_2)\} \quad \text{allExamples}(leD) \quad \{leD \mapsto \{\text{name} = \text{lecture.tex}, \ldots\}, t \mapsto 2\} \\
&\quad \text{Chg } \{\text{label} = \text{ExampleSignature}, \text{loc} = [\text{Sect sec2}]\}.\text{loc} \leadsto [\text{Sect sec1}] \\
&\text{Rep } \{\text{R } 4(ex_1)\} \quad \text{allExamples}(leD) \quad \{leD \mapsto \{\text{name} = \text{lecture.tex}, \ldots\}, t \mapsto 2\} \\
&\quad \text{Chg } \{\text{label} = \text{ExampleAlgebra}, \text{loc} = [\text{Sect sec1}]\}.\text{loc} \leadsto [\text{Sect sec2}]
\end{aligned}
$$

CDET proposes to add a proves relation to the proof TermAlgebraProof in the document theory.tex, in order to resolve an inconsistency for R 3. In the document lecture.tex the locations of the examples ExampleSignature and ExampleAlgebra should be switched, in order to resolve the inconsistencies of R 4. Of course, there are many more repairs possible, e.g., to delete the illustrates relation of either example. This is, however, considered "too expensive". Users can express their preferences by further annotating hints by repair costs and defining a metric for sorting the repair collection (see [3] for details). Here, we have

---

[8] Repairs also contain a rating, which we omit here for brevity.

preferred changing content to adding content, which is preferable to deleting content or even whole documents.

6   Evaluation, Conclusion, Outlook

In this paper, we propose an approach to engineering mathematical knowledge by combining MMiSS and CDET. MMiSS provides a sophisticated environment for maintaining mathematical documents fostering reuse, sharing, and ontology-structured semantic interrelation also to non-mathematical content. Besides other benefits, ontologies alleviate consistency management w.r.t. user-defined requirements by CDET. Formal consistency rules define consistency, leaving no room for misinterpretations. CDET shows how inconsistencies can be best resolved by S-DAGs and the repair collection.

The major efforts for our approach are the definition of an ontology and distilling and formalizing consistency rules.

To our experience, the commands for building domain ontologies are easy to learn. The biggest effort is to structure the domain knowledge and to express it by ontological means. But so far, our experiences suggest that – for the task of preparing eLearning content – this step leads to clearer course structures and lecture material of higher quality. We suppose that this also applies to the preparation of mathematical documents. Although we have feedback of more than 20 authors of MMiSS courses,[9] the task of formally evaluating our approach with a larger set of authors is on our agenda.

The cost of user-defined consistency management is influenced mostly by determining the actual consistency requirements informally and defining document structures. Formalizing consistency rules requires technical effort of experts in the domain of formal logic.

Since consistency rules can be reused for many projects, the advantages of our combined approach outweigh its costs, particularly for teams of authors. Document engineering benefits from clear document structures. Formal rules and automatic consistency checks give precious insights and lead to fruitful discussions about "consistency". CDET benefits from semantic annotations and their structuring by an ontology. For example, by employing ontologies to determine subrelations, consistency rules become quite flexible. They can be reused without adaptations if the ontology changes, e.g., new subrelations are introduced. Version control by MMiSS supports efficient consistency checking and the formalization of temporal consistency rules, which relate different document versions.

Our idea of engineering mathematical documents can be transferred to any mathematical editing environment. For example, we find a large amount of quality requirements in the current OMDoc specification [9] that cannot be checked using native XML methods. For example, such requirements involve multiple documents, demand a specific structure based on specific attribute values, or are just best practices that should hold but may be violated from time

---

[9] During the MMiSS project more than 20 lectures have been prepared.

to time. We conjecture that many cases can be handled by CDET. Also, OMDoc could benefit from structuring semantic relations as we do in MMiSS. We look forward to transferring our experiences to OMDoc.

A field to be addressed is the integration of different ontologies: Upper ontologies, emerging mathematical ontologies, and the local ontologies created within the documents. MMiSS provides an import mechanism with hiding, revealing, and renaming of ontology components. Although this solves many problems on the syntactical level, integration on the logical level is a research area itself.

Moreover, we plan to enhance our approach as follows: (1) We reduce its formal overhead (e.g., formalization of hints and definition of preference metrics). (2) We improve usability of CDET's output (e.g., converting repairs to natural language or employing interactive graph viewing tools such as uDraw(Graph)[10] for handling S-DAGs). (3) We integrate mathematical tools such as theorem provers. (4) By further experiments, we want to determine a set of basic consistency rules for MKM that can be refined and adapted to specific MKM projects.

References

1. Nuseibeh, B., Easterbrook, S., Russo, A.: Leveraging inconsistency in software development. Computer **33** (2000) 24–29
2. Nentwich, C., Emmerich, W., Finkelstein, A.: Consistency management with repair actions. In: Proc. of the 25th Int. Conf. on Software Engineering 2003, Portland, OR, ACM Press (2003)
3. Scheffczyk, J., Borghoff, U.M., Rödig, P., Schmitz, L.: Managing inconsistent repositories via prioritized repairs. In: Proc. of the 2004 ACM Symp. on Document Engineering, Milwaukee, WI, ACM Press (2004) 137–146
4. Asperti, A., Padovani, L.: Mathematical knowledge management in HELM. [20]
5. Buchberger, B.: Mathematical knowledge management in Theorema. [20]
6. Rudnicki, P., Trybulec, A.: Mathematical knowledge management in Mizar. [20]
7. Cruz-Filipe, L., Geuvers, H., Wiedijk, F.: C-CoRN, the constructive Coq repository at Nijmegen. [21] 88–103
8. Autexier, S., Hutter, D., Mossakowski, T., Schairer, A.: The development graph manager MAYA (system description). In: Algebraic Methodology and Software Technology. Volume 2422 of LNCS., Springer-Verlag (2002) 495–502
9. Kohlhase, M.: OMDoc: Towards an internet standard for mathematical knowledge. In: Proc. of Artificial Intelligence and Symbolic Computation, AISC'2000. LNAI, Springer-Verlag (2001) See also http://www.mathweb.org/omdoc.
10. Hutter, D.: Management of change in structured verification. In: Proc. of the 15th IEEE Int. Conf. on Automated Software Engineering (ASE'00), Grenoble, France, IEEE Computer Society (2000) 23–34
11. Fürst, F., Leclère, M., Trichet, F.: Contribution of the ontology engineering to mathematical knowledge management. [20]
12. Marchiori, M.: The mathematical semantic web. [22] 216 – 223
13. Caprotti, O., Dewar, M., Turi, D.: Mathematical service matching using description logic and OWL. [21] 73–87

---

[10] See www.informatik.uni-bremen.de/uDrawGraph/en/home.html

14. Kohlhase, M., Anghelache, R.: Towards collaborative content management and version control for structured mathematical knowledge. [22] 147 – 161

15. Kohlhase, M.: Semantic markup for TEX / LATEX. In: Proc. of the Mathematical User-Interfaces Workshop at MKM'04, Bialowieza, Poland (2004)

16. Kohlhase, A., Kohlhase, M.: CPoint: dissolving the author's dilemma. [21] 175–189

17. Krieg-Brückner, B., et al.: Multimedia instruction in safe and secure systems. In: Recent Trends in Algebraic Development Techniques. Volume 2755 of LNCS., Springer-Verlag (2003) 82–117

18. Uschold, M., Grüninger, M.: Ontologies: Principles, methods and applications. Knowledge Engineering Review **11** (1996) 93–155

19. Peyton Jones, S.: Haskell 98 Language and Libraries: The Revised Report. Cambridge Univ. Press (2003)

20. Buchberger, B., Caprotti, O., eds.: Proc. of 1st Int. Wkshp. on Mathematical Knowledge Management, Linz, Austria, Johannes Kepler University (2001)

21. Asperti, A., et al., eds.: Proc. of 3rd Int. Conf. on Mathematical Knowledge Management. Volume 3119 of LNCS., Bialowieza, Poland, Springer-Verlag (2004)

22. Asperti, A., et al., eds.: Proc. of 2nd Int. Conf. on Mathematical Knowledge Management. Volume 2594 of LNCS., Bertinoro, Italy, Springer-Verlag (2003)

Computational Origami of a Morley's Triangle

Tetsuo Ida, Hidekazu Takahashi, and Mircea Marin*

Department of Computer Science,
University of Tsukuba,
Tsukuba 305-8573, Japan
{ida, hidekazu, mmarin}@score.cs.tsukuba.ac.jp

Abstract. We present a computational origami construction of Morley's triangles and automated proof of correctness of the generalized Morley's theorem in a streamlined process of solving-computing-proving. The whole process is realized by a computational origami system being developed by us. During the computational origami construction, geometric constraints in symbolic and numeric representation are generated and accumulated. Those constraints are then transformed into algebraic relations, which in turn are used to prove the correctness of the construction. The automated proof required non-trivial amount of computer resources, and shows the necessity of networked services of mathematical software. This example is considered to be a case study for innovative mathematical knowledge management.

1 Introduction

Computational origami[1] is a scientific discipline to study origami systematically using computers. It includes, among others, the mathematical study of paper folds, e.g. modeling of origami by algebraic and symbolic means, simulation of origami, and proving the correctness of geometric properties of constructed origami.

In the framework of computational origami we studied the construction of a Morley's triangle and automated proofs of Morley's theorem. Morley's theorem states that the three points of intersection of the adjacent interior trisectors of the angles of any triangle form an equilateral triangle. Morley's theorem can be generalized by taking into account the intersections of the exterior trisectors as well. For a given angle α ($0 < \alpha < \pi$), we have one pair of interior trisectors (primary trisectors) producing the pair of angles $(\alpha/3, 2\alpha/3)$, and two pairs of exterior trisectors producing the pairs of angles $((\pi + 2\alpha)/3, (2\pi + \alpha)/3)$ and $((2\pi + 2\alpha)/3, (4\pi + \alpha)/3)$. Therefore, we have 3^3 triangles formed by the intersections of the adjacent trisectors. The generalized Morley's theorem states that

---

* This research is supported by the JSPS Grants-in-Aid for Scientific Research No. 17300004 and No. 17700025.
[1] Origami is a Japanese word meaning a piece of folding paper or methodology of folding paper.

M. Kohlhase (Ed.): MKM 2005, LNAI 3863, pp. 267–282, 2006.
© Springer-Verlag Berlin Heidelberg 2006

out of the 27 triangles constructible by the intersections of the adjacent trisectors, 18 triangles are equilateral. Proofs of (generalized) Morley's theorem were published by several researchers since Morley gave his construction in 1898. Bogomolny provides in his web page various information about Morley's theorem including its historical accounts and proofs [2].

In this paper, we present a computational origami construction of Morley's triangles and prove automatically the correctness of the generalized Morley's theorem in a streamlined process of solving-computing-proving. The process is realized by the computational origami system (in this paper we denote our system by *Computational Origami*) [16] which is being developed by us. The automated proof of the generalized Morley's theorem was first published by Wu [18], and a concise explanation of his proof is given in [17]. What are new in our study are the computational origami construction and streamlined automated proof. Seemingly different kinds of knowledge in mathematical sciences, *i.e.* origami and automated theorem proving, are integrated in a common framework and, moreover, processed coherently. In the case of automated proof, mathematical knowledge about origami represented symbolically as a set of equalities is systematically transformed into a system of polynomials. The generated system of polynomials is sent to the theorem prover *Theorema* [4], which then establishes the theorem by using the Gröbner bases method.

The rest of the paper is organized as follows. After briefly explaining the principles of origami construction in Sect. 2, in Sect. 3 we show a stepwise origami construction of Morley's triangle. In Sect. 4 we give our automated proof of the generalized Morley's theorem. In Sect. 5 we discuss a method of connecting *Computational Origami* to *Theorema*. In Sect. 6 we summarize our results and indicate directions for future research.

2   Principles of Origami Construction

An origami is to be folded along a specified line on the origami called *fold line*. The line segment of a fold line on the origami is called a *crease*, since the consecutive operations of fold and unfold along the same fold line create crease(s) of the fold line on the origami. In Subsection 2.1 we recall the six basic origami axioms proposed by Huzita [13, 12] to fold origami. Each of his axioms prescribes a rule for constructing a fold line, which can be determined by either points, lines or a combination of them. It is known that Huzita's origami axiom set is more powerful than the ruler-and-compass method in Euclidean geometry [9]. Namely, origami can not only construct geometric objects constructible by the ruler-and-compass method, but also geometric objects that are impossible by the ruler-and-compass method. One of them is a trisector of an angle [5]. From this fact it is clear that Morley's triangles can not be constructed by the ruler-and-compass method.

Now let us see how to construct a geometric object with origami. First, we have some notational convention in this paper. We denote points by single capital letters $A, B, C, \ldots$, possibly subscripted, the line passing through points X and Y by XY, and the segment between points X and Y by $\overline{XY}$.

We define an origami[2] $\Box ABCD$ together with the set Π of constructed points $\{A, B, C, D\}$ and the set Γ of constructed lines $\{AB, BC, CD, DA\}$. We then start the origami construction from the initial origami. We make a fold on the origami by applying one of the axioms given below, possibly followed by unfolding. A fold of the origami gives rise to a set of points of coincidence of the fold line and the lines in Γ, resulting in new Π and Γ.

2.1   Huzita's Axioms

In [13], Huzita proposes the following axioms for origami geometry:

(O1) Given two points in Π, we can make a fold along the fold line that passes through them.

(O2) Given two points in Π, we can make a fold to bring one of the points onto the other.

(O3) Given two lines in Γ, we can make a fold to superpose the two lines.

(O4) Given a point P in Π and a line m in Γ, we can make a fold along the fold line that is perpendicular to m and passes through P.

(O5) Given two points P and Q in Π and a line m in Γ, either we can construct the fold line that passes through Q and we can make a fold along this fold line to superpose P and m, or we can decide that the construction of a fold line is impossible.

(O6) Given two points P and Q in Π and two lines m and n in Γ, either we can construct a fold line and make a fold along this fold line to superpose P and m, and Q and n, simultaneously, or we can decide that the construction of a fold line is impossible.

The axiom set gives the principles of origami based on an operational view of paper folds, hence the axioms would be better termed as *basic constructions*. Hatori proposed an additional axiom [11]:

(O7) Given a point P in Π and two lines m and n in Γ, we can construct the fold line that is perpendicular to n and we can make a fold along this fold line to superpose P and m.

He further showed that (O6) is sufficient to make all the folds by (O1)-(O5) and (O7). Indeed, (O1)-(O5) and (O7) are the degenerate cases of (O6). This does not imply, however, that (O6) is enough in practice. (O6) will produce at most three fold lines, and we would need to specify an additional parameter to select the desired fold line. Mathematical models of the set of constructed points by the folds by the applications of the axiom set are studied by Alperin [1].

---

[2] Note that we abuse the word *origami* to mean the methodology, the piece of paper, and the geometric object that is being constructed by means of paper folds, as we do in Japanese.

2.2 Implementation of Origami Axioms

The constructions described by axioms (O1)-(O7) are realized by function OFold of *Computational Origami*, which is implemented in *Mathematica*. Function OFold needs, as arguments, several constructed lines and points to compute the fold line(s) and to determine the face to be moved. Note that the types of the arguments and the argument keywords can discriminate the operations to be performed unambiguously. The figures that we will show in the next section are generated by the calls of OFold.

(O1) OFold[X, Along$\rightarrow PQ$]
(O2) OFold[P, Q]
(O3) OFold[RS, UV]
(O4) OFold[X, AlongPerpendicular $\rightarrow \{P, RS\}$]
(O5) OFold[P, RS, Through $\rightarrow Q$]
(O6) OFold[P, RS, Q, UV]
(O7) OFold[P, RS, AlongPerpendicular $\rightarrow UV$]

OFold[X, Along $\rightarrow PQ$] in (O1) makes a fold along the line PQ. The words Along, AlongPerpendicular and Through to the left of $\rightarrow$ are keywords. All the faces containing the value of X (i.e., the coordinate of point X) are to be moved. In all the cases we have hidden optional parameters which tell the system which faces of the origami should be moved (with keyword Move) and which directions (mountain or valley). For instance, in (O2), Move $\rightarrow P$ is implicit.

3   Origami Construction of a Morley's Triangle

Our origami construction of a Morley's triangle is depicted in Figures 1 – 6.

We construct a Morley's triangle inside the triangle $\triangle ABE$ (step 5) in the following way:

1. trisection of $\angle EAB$ (step 6 ... step 14)
2. trisection of $\angle EBA$ (step 15 ... step 21)
3. trisection of $\angle AEB$ (step 22 ... step 32)

Then, the Morley's triangle inside $\triangle ABE$ can be observed easily (step 33). Each step of the origami construction is performed by the call of function OFold and function Unfold. For the construction of the trisector we use Abe's method [8, 10].

Abe's Method

We first assume that the initial origami is a square paper $\square ABCD$ (Step 1). Let E be an arbitrary point on segment $\overline{CD}$. Later we will extend this formulation, which will allow us to put point E to be on the line that extends the segment $\overline{CD}$. Note that even this extension is not general enough to claim the correctness of the construction and the proof of the generalized Morley's theorem

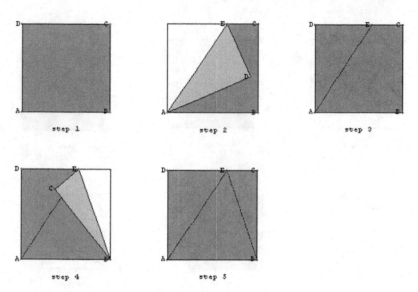

Fig. 1. Origami construction of a Morley's triangle: Steps 1–5

thereafter. We will further allow the initial origami to be an arbitrary rectangle. Then our claim will be justified. However, we will, for the moment, stick to the above configuration for the sake of the clarity of explanation and of the ease of computation in the automated proof.

The steps 1-5 are preliminary steps for constructing the triangle $\triangle ABE$. In these steps we apply (O1) at steps 2 and 4, and we unfold the origami at steps 3 and 5.

At step 6, we apply (O2). Note that in the figures we show only the constructed points that are necessary for later constructions.

Step 8 is the crucial step of Abe's method. We apply (O6). We make a fold to superpose points D and A to line AE and line GF, respectively. Finding a fold line in (O6) amounts to solving a cubic equation that describes the geometric constraints among the involved points. Hence we have the (at most) three fold lines as shown in step 8. At this step we need to interact with the system to specify which fold line we want to use. In our example we choose the one indicated as 3 in the figure. It is easily seen that the other fold lines also trisect the angle: line 1 for angle $(\pi - \angle EAB)$ and line 2 for angle $(2\pi - \angle EAB)$.

At steps 9 and 10, we make a fold along the fold line 3 and unfold. Let the points at which points G and A are placed in step 9 be I and H, respectively. This means that I and H are in Π at step 10. Now we can observe that $\angle EAI$ and $\angle HAB$ trisect $\angle EAB$. This can be proved automatically using the same technique that we are going to expound in Sect. 4. See paper [14] for details.

The trisector of $\angle EBA$ is similarly constructed. Finding the trisector of $\angle AEB$ is more involved, but the construction can be easily read off from the figures. The final origami is shown in step 33. We see the triangle $\triangle WLU$, which

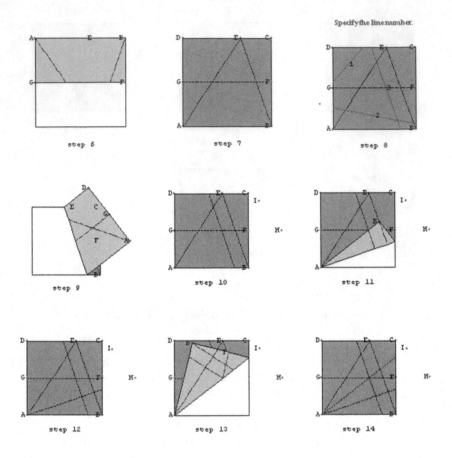

Fig. 2. Origami construction of a Morley's triangle: Steps 6–14

turns out (and is proved) to be equilateral for any E on segment $\overline{CD}$ excluding C and D. Note that Abe's method is not applicable to the cases where E is on point C or on D.

4   Proof of Morley's Theorem

4.1   Algebraic Formulation of Morley's Theorem

After we perform the construction steps described in the previous section, we switch the mode of computation from *construction* to *proof* and obtain the figure shown in Figure 7. We generate it by the proof-support tool, in order to facilitate geometric reasoning with constructed points. Note that the Morley's triangle in Figure 7 is $\triangle VLT$ instead of $\triangle WLU$. Points T and V overlap with points U and W, respectively, and hence they are not visible in Figure 6. We use points T and V since they will lead to less number of variables in the proof.

Specify the line number.

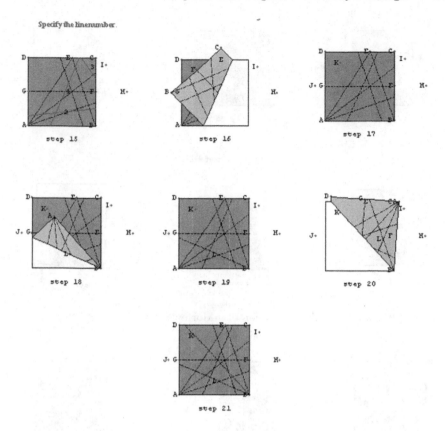

step 15   step 16   step 17

step 18   step 19   step 20

step 21

Fig. 3. Origami construction of a Morley's triangle: Steps 15–21

To avoid unnecessary clutter, we display only the creases and points which are relevant for stating the assumptions and conclusion of the generalized Morley's theorem. We already proved in [14] that Abe's construction makes the segments $\overline{A_1A_8}$ and $\overline{A_1V_{28}}$ trisect $\angle E_1A_1B_1$, $\overline{B_1B_{14}}$ and $\overline{B_1T_{26}}$ trisect $\angle A_1B_1E_1$, and $\overline{E_1E_{24}}$ and $\overline{E_1V_{28}}$ trisect $\angle B_1E_1A_1$. Therefore, Morley's triangle is $\Delta V_{28}L_{16}T_{26}$. We will prove that $\Delta V_{28}L_{16}T_{26}$ is equilateral by showing that

$$(|\overline{V_{28}L_{16}}| = |\overline{L_{16}T_{26}}|) \wedge (|\overline{L_{16}T_{26}}| = |\overline{T_{26}V_{28}}|)$$

where $|\overline{XY}|$ denotes the distance between points X and Y. To achieve this, we represent both the assumptions and the conclusion as systems of polynomial equalities which capture the algebraic properties of the origami construction. The conclusion can be specified by:

$$(|\overline{V_{28}L_{16}}|^2 = |\overline{L_{16}T_{26}}|^2) \wedge (|\overline{L_{16}T_{26}}|^2 = |\overline{T_{26}V_{28}}|^2) \tag{1}$$

which lends itself to polynomial equalities over $\mathbb{Q}$.

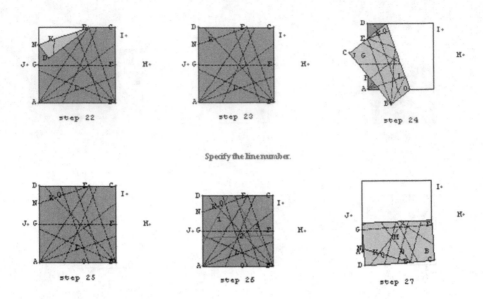

Specify the line number.

Fig. 4. Origami construction of a Morley's triangle: Steps 22–27

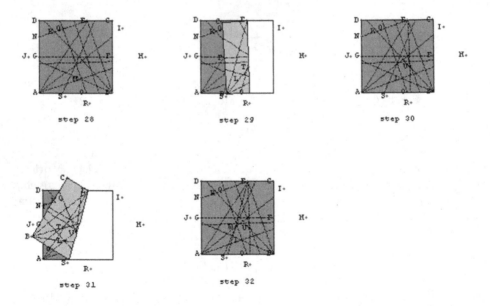

Fig. 5. Origami construction of a Morley's triangle: Steps 28–32

However, the specification of the hypotheses requires careful analysis. With *Computational Origami* we can retrieve automatically the geometric constraints of the objects of interest, and transform them into algebraic form. This is possible because *Computational Origami* keeps the algebraic properties of the origami

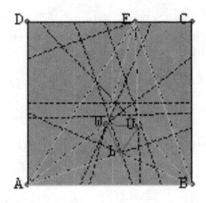

step 33

Fig. 6. Morley's triangle: final step

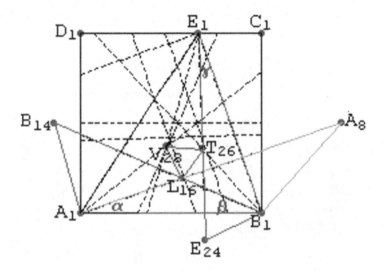

Fig. 7. Morley's triangle for support of proof

during the whole origami construction process. The collected constraints are then transformed to polynomial equalities.

It turns out that thus-obtained set of the polynomial equalities, however, is too general, and hence it is too weak as the premise of the proof. Namely, it allows the generation of 27 triangles, 9 out of which are not equilateral. One may attribute this to the fact that our system does not record the selection of the primary trisectors in steps 8, 15 and 26. However, even if we did so, we would not be able to prove with the algebraic method based on Gröbner bases that one Morley's triangle inside the given triangle is equilateral.

An illustration of the possible 27 triangles is given in Appendix A. The figures in Appendix A indicate that 18 triangles out of 27 are equilateral. We can single out these 18 triangles, by imposing the condition

$$\angle A_8 A_1 B_1 \pm \angle B_{14} B_1 A_1 \pm \angle E_{24} E_1 B_1 = \pm \pi/3 \ (\text{mod } 2\pi) \tag{2}$$

which is equivalent to

$$\text{Cos}[\angle A_8 A_1 B_1 \pm \angle B_{14} B_1 A_1 \pm \angle E_{24} E_1 B_1]^2 - \frac{1}{4} = 0 \tag{3}$$

Let us denote

$$p_1 := \text{Cos}[\angle A_8 A_1 B_1], p_2 := \text{Cos}[\angle B_{14} B_1 A_1], p_3 := \text{Cos}[\angle E_{24} E_1 B_1],$$
$$q_1 := \text{Sin}[\angle A_8 A_1 B_1], q_2 := \text{Sin}[\angle B_{14} B_1 A_1], q_3 := \text{Sin}[\angle E_{24} E_1 B_1],$$
$$d_1 := |\overline{A_1 A_8}|, d_2 := |\overline{A_1 B_1}|, d_3 := |\overline{B_1 B_{14}}|,$$
$$d_4 := |\overline{B_1 A_1}|, d_5 = |\overline{E_1 E_{24}}|, d_6 := |\overline{E_1 B_1}|.$$

Then, by straightforward trigonometric manipulations, we obtain the set of algebraic constraints

$$
\begin{aligned}
\mathcal{C}_{\mathcal{X}} = \{ & (-\frac{1}{4} + (p_1 p_2 p_3 - p_3 q_1 q_2 - p_2 q_1 q_3 - p_1 q_2 q_3)^2) \\
& (-\frac{1}{4} + (p_1 p_2 p_3 - p_3 q_1 q_2 + p_2 q_1 q_3 + p_1 q_2 q_3)^2) \\
& (-\frac{1}{4} + (p_1 p_2 p_3 + p_3 q_1 q_2 - p_2 q_1 q_3 + p_1 q_2 q_3)^2) \\
& (-\frac{1}{4} + (p_1 p_2 p_3 + p_3 q_1 q_2 + p_2 q_1 q_3 - p_1 q_2 q_3)^2) = 0 \} \cup \\
& \{ p_1^2 + q_1^2 - 1 = 0, \ p_2^2 + q_2^2 - 1 = 0, \ p_3^2 + q_3^2 - 1 = 0 \} \cup \\
& \{ d_1^2 - |\overline{A_1 A_8}|^2 = 0, \ d_2^2 - |\overline{A_1 B_1}|^2 = 0, \ d_3^2 - |\overline{B_1 B_{14}}|^2 = 0, \\
& d_4^2 - |\overline{B_1 A_1}|^2 = 0, \ d_5^2 - |\overline{E_1 E_{24}}|^2 = 0, \ d_6^2 - |\overline{E_1 B_1}|^2 = 0 \} \cup \\
& \{ |\overline{A_8 B_1}|^2 - (d_1^2 + d_2^2 - 2 d_1 d_2 p_1) = 0, \\
& |\overline{B_{14} A_1}|^2 - (d_3^2 + d_4^2 - 2 d_3 d_4 p_2) = 0, \\
& |\overline{E_{24} B_1}|^2 - (d_5^2 + d_6^2 - 2 d_5 d_6 p_3) = 0 \}
\end{aligned}
$$

In this way, we reduce the generalized Morley's theorem to the following statement:

If the origami satisfies the algebraic constraints $\mathcal{C} \cup \mathcal{C}_{\mathcal{X}}$, where $\mathcal{C}$ is the set of constraints generated automatically, then it satisfies the constraints (1).

Let $\mathcal{D}$ be an arbitrary set of polynomial equalities, $\mathcal{P}_{\mathcal{D}}$ be the set of the polynomials $\{ h \mid h = 0 \in \mathcal{D} \}$, and $\text{Ideal}(\mathcal{S})$ be the ideal generated by all the elements in polynomial set $\mathcal{S}$. According to the ideal-variety correspondence, the proof of the above statement amounts to proving that the polynomials for $|\overline{V_{28} L_{16}}|^2 - |\overline{L_{16} T_{26}}|^2$ and $|\overline{L_{16} T_{26}}|^2 - |\overline{T_{26} V_{28}}|^2$ are in the radical of $\text{Ideal}(\mathcal{P}_{\mathcal{C} \cup \mathcal{X}})$.

Thus the problem is equivalent to the following ideal membership problem (cf. [6–Chapter 4, §2]):

$$1 \in \text{Ideal}(\mathcal{S}) \tag{4}$$

where

$$\mathcal{S} = \mathcal{P}_{\mathcal{C} \cup \mathcal{C}_\mathcal{X}} \cup \{(k_1(|\overline{V_{28}L_{16}}|^2 - |\overline{L_{16}T_{26}}|^2) - 1)(k_2(|\overline{L_{16}T_{26}}|^2 - |\overline{T_{26}V_{28}}|^2) - 1)\}$$

and k_1 and k_2 are new variables introduced with Rabinowitch trick.

The ideal membership problem can be solved constructively by computing the Gröbner basis of $\mathcal{S}$. Namely, according to the theory of Gröbner bases [3], the statement (4) is true iff the reduced Gröbner basis of $\mathcal{S}$ is $\{1\}$.

4.2   Proof by *Computational Origami*

What follows in this subsection is the illustration of how we can realize this proof plan with *Computational Origami*.

First, we gather all geometric constraints that were accumulated during the 33 steps of the origami construction.

```
allprops = GatherProperty[ ]
```

Next, we select only the relevant properties of the objects for proving the generalized Morley's theorem:

```
props = SelectProperty[allprops, {{"L", 16}, {"T", 26}, {"V", 28}},
                AddBase → {{"E", 1}}]
```

Recall that points L_{16}, T_{26} and V_{28} form the Morley's triangle and point E_1 is an arbitrarily given point. The search for relevant constraints to be used in the proof starts from these points.

In order to translate these constraints into algebraic form, we fix the coordinate system to be Cartesian with point A being the origin, by calling the function CoordinateMapping. The call

```
cmap = CoordinateMapping[props,
          InitialShape → SquareP[{Point[0, 0], 1}, {"A", "B", "C", "D"}],
          InitialPoints → {{{"E", 1}, Point[u, 1]}}]
```

returns a mapping table, which is stored in cmap and will be used to translate props into algebraic form. Variable u denotes the x-coordinate of point E_1.

The call of function ToAlgebraic returns the set of polynomials in premise$_0$, *i.e.* the set of all the polynomials of the left-hand side of the equalities in set $\mathcal{C}$ explained before:

```
premise₀ = ToAlgebraic[props, cmap]
```

Then, we compute premise:

$\text{premise} = \text{premise}_0 \cup \{$
$\quad (-\frac{1}{4} + (p_1p_2p_3 - p_3q_1q_2 - p_2q_1q_3 - p_1q_2q_3)^2)$
$\quad (-\frac{1}{4} + (p_1p_2p_3 - p_3q_1q_2 + p_2q_1q_3 + p_1q_2q_3)^2)$
$\quad (-\frac{1}{4} + (p_1p_2p_3 + p_3q_1q_2 - p_2q_1q_3 + p_1q_2q_3)^2)$
$\quad (-\frac{1}{4} + (p_1p_2p_3 + p_3q_1q_2 + p_2q_1q_3 - p_1q_2q_3)^2),$
$\quad p_1^2 + q_1^2 - 1, p_2^2 + q_2^2 - 1, p_3^2 + q_3^2 - 1,$
$\quad d_1^2 - \text{Distance}[A_1, A_8]^2, d_2^2 - \text{Distance}[A_1, B_1]^2,$
$\quad d_3^2 - \text{Distance}[B_1, B_{14}]^2, d_4^2 - \text{Distance}[B_1, A_1]^2,$
$\quad d_5^2 - \text{Distance}[E_1, E_{24}]^2, d_6^2 - \text{Distance}[E_1, B_1]^2,$
$\quad \text{Distance}[A_8, B_1]^2 - d_1^2 - d_2^2 + 2d_1d_2p_1,$
$\quad \text{Distance}[B_{14}, A_1]^2 - d_3^2 - d_4^2 + 2d_3d_4p_2,$
$\quad \text{Distance}[E_{24}, B_1]^2 - d_5^2 - d_6^2 + 2d_5d_6p_3\}$

Finally, we compute the set of polynomials

$S = \text{premise} \cup \{ (k_1(\text{Distance}[V_{28}, L_{16}]^2 - \text{Distance}[L_{16}, T_{26}]^2) - 1)$
$\quad (k_2(\text{Distance}[L_{16}, T_{26}]^2 - \text{Distance}[T_{26}, V_{28}]^2) - 1)\}$

and check whether the (reduced) Gröbner basis of S is $\{1\}$. For this purpose, we call the *Mathematica* function GroebnerBasis:

$$\text{GroebnerBasis}[\text{S}, \text{vars}, \qquad\qquad\qquad\qquad\qquad (5)$$
$$\text{CoefficientDomain} \rightarrow \text{RationalFunctions},$$
$$\text{MonomialOrder} \rightarrow \text{DegreeReverseLexicographic}]$$

In this call, vars is a subset of the variables in S. The Gröbner basis computation will be carried out in the domain of polynomials whose variables are in vars and whose coefficients are in $\mathbb{Q}(u_1, \ldots, u_n)$ of rational functions, where $u_1, \ldots, u_n$ are the variables of S − vars. The variables $u_1, \ldots, u_n$ are independent variables and the variables in vars are dependent variables. In this case, variable u which is the x-coordinate of E_1 is only the independent variable.

The polynomial set S has 102 polynomials and the number the variables is 88. The computation of the above Gröbner basis is highly sensitive to the monomial ordering. With degree-lexicographic ordering[3], the call (5) returned $\{1\}$ in 1653.57 seconds on a Linux server with AMD Athlon 64 processor, 2.4 MHz CPU, and 4 GB RAM.

4.3   Proof of the General Case

The construction of a Morley's triangle and the proof can be easily generalized. We need to discuss the generalization in two separate issues. As for the construction, an initial triangle $\triangle ABE$ similar to arbitrary shapes of a triangle can be constructed by changing relative positions of the two edges of the triangle inside the initial square origami. We have to distinguish the two cases:

1. $A(0,0), B(u_1, 0), E(u_2, 1)$
2. $A(0,0), B(1,0), E(u_1, u_2)$

---

[3] We changed all the subscripted variables v_i for arbitrary v explained in the paper by vi for running the program.

Either by taking the initial origami to be a rectangular of u_1 by 1 or taking the initial origami to be a square and taking the first step construction by applying (O5) with parameters D, AD and E, we can construct an arbitrary initial triangle $\triangle ABE$ in both cases. To enable the first type of construction, *Computational Origami* has the functionality to specify a rectangle as the initial origami. All the *Mathematica* program code for this general case is published in [15].

As for the proof, the first case is sufficient to establish the proof since the generated algebraic specification covers the general case. The computation of the Gröbner basis is carried out in the domain of polynomials whose coefficients are in $\mathbb{Q}(u_1, u_2)$ of rational functions. With degree-lexicographic ordering, the computation of the Gröbner basis took 58136.1seconds (over 16 hours) on the same server.

5 Communication with *Theorema*

We envisage that in the near future mathematical software will be running independently in a networked environment, offering services to a community of common interest, say symbolic computation community, in the grid. We can receive the service from the grid, instead of loading packages in the same machine or establishing manually connections to remote servers. Our work presented in this paper would need such an infrastructure that provides access to mathematical software. *Computational Origami* may comfortably run on a laptop computer, but automated proof of the generalized Morley's theorem is beyond the power of ordinary laptop computers.

To make a further step to realize such a vision, we have developed a simple interface to *Theorema* that is running on a remote server. To establish communication with the *Theorema* server, we first establish a link with *Theorema* through some available port, say 8000, at the machine whose name is `thmserver.score.cs.tsukuba.ac.jp` (in this illustration). An appropriate arrangement on the remote `thmserver` side is necessary. This arrangement is not shown in this paper.

```
thma = LinkConnect["8000@thmserver.score.cs.tsukuba.ac.jp",
                   LinkProtocol → "TCPIP"];
```

We define

$$\texttt{concl} = \{\, \texttt{Distance}[V_{28}, L_{16}]^2 - \texttt{Distance}[L_{16}, T_{26}]^2,$$
$$\texttt{Distance}[L_{16}, T_{26}]^2 - \texttt{Distance}[T_{26}, V_{28}]^2 \,\}$$

We then send the data stored in the variables `vars`, `premise` and `concl` via the link `thma`, and wait for the proof to come from the *Theorema* server:

```
SendTheoremaFormula[thma, vars, premise, concl, "Morley"];
```

When the proof text arrives, we save it in the file `MorleyTriangleProof.nb`.

NotebookSave[LinkRead[thma], "MorleyTriangleProof.nb"];

Finally, we close the link:

LinkClose[thma];

The generated proof text in file MorleyTriangleProof.nb is a human readable sequence of statements and formulas. The proof is structured into two independent proof problems, each of which is reduced to computing reduced Gröbner basis. Since most of the formulas in the two proofs generated automatically by the prover are quite lengthy, we do not include the proof text in this paper.

6   Conclusion

We have shown the origami construction of a Morley's triangle and the automated proof of Morley's theorem. *Computational Origami* not only performs the simulation of origami construction, but also proves the correctness of the construction by communicating with *Theorema*, which has the implementation of the Gröbner basis algorithm [3].

While studying Morley's theorem, we see the following challenges. The first is the heavy requirement of computer resources, i.e., CPU time and memory. The computation is not only time-consuming, but also very sensitive to the monomial ordering that we have to specify for the Gröbner bases computation. The problem remains unsolved.

To view the problem indifferent perspective, we are convinced of the necessity of a computing environment such as symbolic computing grid on which we can easily access mathematical knowledge services such as Gröbner bases computation and perform experimental computation without paying much attention to internal mathematical representation. For the present study, both *Computational Origami* and *Theorema* are implemented in *Mathematica*. Hence, both systems can share the common mathematical knowledge base. We are working to generalize the communication outlined in the previous section by shifting to a service oriented framework with advanced mechanisms to discover mathematical services, control their life cycle, and guarantee a certain quality of service[7].

References

1. R. C. Alperin. A Mathematical Theory of Origami Constructions and Numbers. *New York Journal of Mathematics*, 6:119–133, 2000.
2. A. Bogomolny. Morley's Miracle. http://www.cut-the-knot.org/triangle/Morley, ©1996-2005.
3. B. Buchberger. Ein algorithmisches Kriterium für die Lösbarkeit eines algebraischen Gleichungssystems. *Aequationes mathematicae*, 4(3):374–383, 1970.
4. B. Buchberger, C. Dupre, T. Jebelean, F. Kriftner, K. Nakagawa, D. Văsaru, and W. Windsteiger. The Theorema Project: A Progress Report. In M. Kerber and M. Kohlhase, editors, *Symbolic Computation and Automated Reasoning : The Calculemus-2000 Symposium*, pages 98–113, St. Andrews, Scotland, August 6-7, 2000.

5. T. L. Chen. Proof of the impossibility of trisecting an angle with euclidean tools. *Mathematics Magazine*, 39:239–241, 1966.

6. D. Cox, J. Little, and D. O'Shea. *Ideals, Varieties, and Algorithms: An Introduction to Computational Algebraic Geometry and Commutative Algebra*. Springer Verlag, second edition edition, 1997.

7. D. Ţepeneu and T. Ida. MathGridLink-Connecting Mathematica to the Grid. In *Proceedings of the 6th International Mathematica Symposium (IMS 2004)*, 2004.

8. K. Fushimi. *Science of Origami*. October 1980. A supplement to Saiensu (in Japanese).

9. R Geretschläger. *Geometric Constructions in Origami*. Morikita Publishing Co., 2002. In Japanese, translation by Hidetoshi Fukagawa.

10. K. Haga. *Origamics Part I: Fold a Square Piece of Paper and Make Geometrical Figures (in Japanese)*. Nihon Hyoronsha, 1999.

11. K. Hatori. K's origami - fractional library - origami construction. http:// origami.ousaan.com/library/conste.html, 2005.

12. T. Hull. Origami and geometric constructions. http://www.merrimack.edu/ ~thull/omfiles/geoconst.html, 2005.

13. H. Huzita. Axiomatic Development of Origami Geometry. In *Proceedings of the First International Meeting of Origami Science and Technology*, pages 143–158, 1989.

14. T. Ida and B. Buchberger. Proving and Solving in Computational Origami. In *Analele Universitatii din Timisoara*, volume XLI, Fasc. Special of *Mathematica-Informatica*, pages 247–263, 2003.

15. T. Ida, H. Takahashi, D. Ţepeneu, and M. Marin. Morley's Theorem Revisited through Computational Origami. In *Proceedings of the 7th International Mathematica Symposium (IMS 2005)*, 2005.

16. H. Takahashi, D. Ţepeneu, and T. Ida. A System for Computational Origami progress report. In *Proceedings of the 6th International Mathematica Symposium (IMS 2004)*, 2004.

17. D. Wang. *Elimination Practice: Software Tools and Applications*. Imperial College Press London, 2004.

18. W.T. Wu. Basic principles of mechanical theorem proving in elementary geometry. *Journal of Automated Reasoning*, 2:221–252, 1986.

A   Triangles Generated by All Trisectors

The triangles drawn in the bold lines are given triangles, and those in the thinner lines (red in color) are the triangles formed by the intersections of neighboring trisectors. Although the sizes of the given triangles are different, they are similar. Eighteen of the constructed triangles are Morley's triangles.

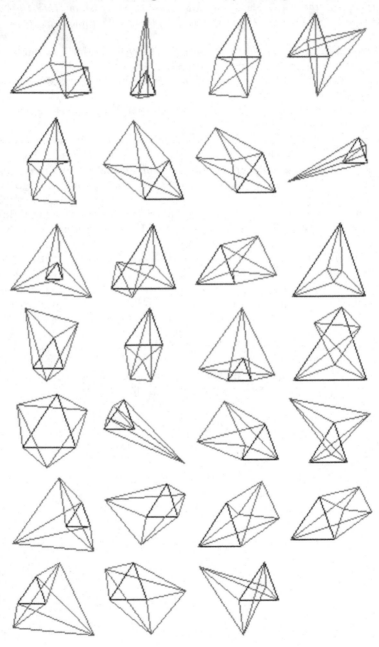

Designing Diagrammatic Catalogues of Types of Basic Interval Equation: A Case Study*

Zenon Kulpa

Institute of Fundamental Technological Research, Polish Academy of Sciences,
ul. Świętokrzyska 21, 00-049 Warsaw, Poland
zkulpa@ippt.gov.pl

Abstract. The use of diagrammatic representations as *catalogues of cases* is analyzed using an example of the catalogue of types of the basic interval equation $a \cdot x = b$. The procedure of finding and describing the types is outlined and a number of different diagrammatic and tabular catalogues are presented and their drawbacks and merits discussed. Suggestions for other solutions, like different forms of the catalogue and interactive catalogue are included. Some preliminary guidelines for designing such catalogues are formulated as well.

1 Introduction

As advocated in the previous *MKM* Conference paper [6], diagrams can be used for efficient representation of complex mathematical knowledge. They offer readable general comprehension of some part of knowledge "at a glance," allowing also for representation of precise structural relationships. One of the several kinds of uses of mathematical diagrams proposed in that paper is a *catalogue of cases*. The purpose of such a catalogue is to group a number of similar objects, types of objects, or reasoning cases, with the main emphasis on comparing the listed objects and delineate differences and similarities between them. The current state of research on mathematical diagrams does not provide any ready for use guidelines for the design of such catalogues. Thus, this paper is structured as a case study—a detailed exposition of problems and experiences with some particular set of mathematical data and various approaches to catalogue them. On the basis of these experiences, an attempt is made to formulate some preliminary guidelines for the design of such catalogues.

In general graphic design practice [15, 16] a notion of *multiples* is used, meaning structures built from similar repeating components. Multiples allow representation of a number of similar objects, facilitating their comparison end enhancing the dimensionality of otherwise flat diagramming medium [16]. Some kinds of catalogues of cases can be designed as such multiples. Other forms are also possible, like region maps or graphs (networks).

---

* The paper was supported by the grant No. 5 T07F 002 25 (for years 2003-2006) from the KBN (State Committee for Scientific Research).

In this paper, various forms of the catalogue of structural types of the interval equation $a \cdot x = b$ are presented and discussed. The detailed analysis of solutions of this simple interval equation was first conducted in [3] and its basic types were listed there diagrammatically. Another, simplified form of this basic catalogue was included as an example in [6] as well. However, there is a much greater number of intermediate and degenerate types of this equation. That makes compiling, handling and use of the complete catalogue of types rather troublesome without more attention to proper design, as discussed in this paper. Several basic forms of the complete catalogue were listed for reference in the report [7], but the design issues were not discussed either there or in other works that included various versions of the catalogues [3, 5, 8]. This paper is devoted to the design issues of the catalogues. Further extensions and improvements of the catalogues (like an interactive version) are also proposed, and some general design guidelines are formulated. Several issues are only sketched, as work on them is still under way.

To make the paper self-contained, basic material on intervals, interval linear equations and interval space diagrams is also included, together with the diagrammatic procedure of solving the basic equation and finding its structural types. These details are needed to fully understand the structure and contents of the catalogues and relative merits and usability of their various versions. However, some of the details can be skipped by the reader not interested in the exact meaning of data items included in the catalogues.

The importance of the basic equation $a \cdot x = b$ itself and its solutions comes from the fact that, as was shown in [4, 5, 8], various characterizations of solution sets of the general many-dimensional system of interval equations $A \cdot x = b$ are provided by solution sets of that simple one-dimensional equation with coefficients a and b obtained as functions of appropriate coefficients of the general system of equations.

2   Real Intervals and Interval Equations

Interval analysis, a new approach to reliable numerical computing allowing for proper tackling of inexact data and rounding errors, is based on the notion of a *real interval*. Generally, a (proper) *real interval*, say u, is defined as a pair of real numbers $u = [\underline{u}, \overline{u}]$, so that its endpoints (*beginning* $\underline{u}$ and *end* $\overline{u}$) obey the inequality $\underline{u} \leq \overline{u}$. For most purposes, real intervals can be identified with the closed set of numbers $u = \{x \mid \underline{u} \leq x \leq \overline{u}\} \subset \mathbb{R}$. For real intervals, two other parameters are in use, namely *midpoint* $\check{u}$ and *radius* $\hat{u}$, so that $\check{u} = (\underline{u} + \overline{u})/2$ and $\hat{u} = (\overline{u} - \underline{u})/2$. That leads to the *centred* notation for intervals (see e.g. [9]), namely $\check{u} \pm \hat{u} = [\check{u} - \hat{u}, \check{u} + \hat{u}]$. An interval is called *thick* if $\underline{u} < \overline{u}$ (or $\hat{u} > 0$); *thin* (or *point*) interval if $\underline{u} = \overline{u}$ (or $\hat{u} = 0$). Point intervals can be for most purposes identified with the corresponding real number, i.e., $[x, x] = x \in \mathbb{R}$. An interval for which $\underline{u} = -\overline{u}$ (or $\check{u} = 0$) is called zero-symmetric or just symmetric. Intervals not containing zero can be positive or negative, according to the relations $u > 0 \Leftrightarrow \underline{u} > 0$ and $u < 0 \Leftrightarrow \overline{u} < 0$.

Occasionally, a concept of an *exterval* will be also used here. A *real exterval e* is a sort of interval-like object that contains infinity,i.e., the set $]\bar{e}, \underline{e}[= (-\infty, \bar{e}] \cup [\underline{e}, +\infty) \subseteq \mathbb{R}$ (with $\bar{e} < \underline{e}$). A *one-sided exterval* is the set $]\bar{e}, \infty[= (-\infty, \bar{e}]$ or $]-\infty, \underline{e}[= [\underline{e}, +\infty)$. When $\bar{e} \geq \underline{e}$, then $]\bar{e}, \underline{e}[= \mathbb{R}$.

An important parameter of an interval is the function rex (for *relative extent*), first introduced in [2] and defined as $\text{rex}\, u = \hat{u} / \check{u}$. Its variant named κ (kappa) in [9] is sometimes more convenient: $\kappa\, u = \hat{u} / |\check{u}| = |\text{rex}\, u|$ (for proper intervals). For u containing 0 we have $\kappa\, u \geq 1$ while $0 \leq \kappa\, u < 1$ for u without 0. It is assumed to equal infinity for symmetric intervals (including 0).

When coefficients of the matrices A and b in the system of linear equations $A \cdot x = b$ are allowed to be intervals, the formula is usually called a system of interval *linear equations* [10, 12]. Precisely speaking, however, it is not *linear* (as the space of intervals is not a linear space), and usually is not treated as a system of *equations* either. The name "equation" is justified in the situation when one considers the *interval solution* (called also *algebraic solution*, or *formal solution* [14]) to the system. This solution is defined as an interval x_I which fulfills the equation $A \cdot x_I = b$ in the sense of interval arithmetic. In most cases, other definitions of a solution are considered, usually as sets of *real* vectors (not necessarily intervals), defined as follows (see e.g. [14]):

United Solution Set: $\quad\quad \Xi(A, b) = \{x \in \mathbb{R}^n \mid A \cdot x \cap b \neq \emptyset\} =$
$$= \Xi_{\exists\exists}(A, b) = \{x \in \mathbb{R}^n \mid (\exists \tilde{A} \in A)(\exists \tilde{b} \in b) \tilde{A} \cdot x = \tilde{b}\},$$

Control Solution Set: $\quad \Xi_{\supseteq}(A, b) = \{x \in \mathbb{R}^n \mid A \cdot x \supseteq b\} =$
$$= \Xi_{\exists\forall}(A, b) = \{x \in \mathbb{R}^n \mid (\forall \tilde{b} \in b)(\exists \tilde{A} \in A) \tilde{A} \cdot x = \tilde{b}\},$$

Tolerance Solution Set: $\quad \Xi_{\subseteq}(A, b) = \{x \in \mathbb{R}^n \mid A \cdot x \subseteq b\} =$
$$= \Xi_{\forall\exists}(A, b) = \{x \in \mathbb{R}^n \mid (\forall \tilde{A} \in A)(\exists \tilde{b} \in b) \tilde{A} \cdot x = \tilde{b}\}.$$

None of the above is actually a solution to the original equation. They are sets of real solutions to a system of interval *relational expressions*, with different relations put in the place of the equal sign, namely :

$A \cdot x \; \mathcal{I} \; b$ for the set $\Xi(A, b)$,
$A \cdot x \supseteq b$ for the set $\Xi_{\supseteq}(A, b)$,
$A \cdot x \subseteq b$ for the set $\Xi_{\subseteq}(A, b)$, respectively.

The relation symbol "$\mathcal{I}$," meaning $S \; \mathcal{I} \; T \iff S \cap T \neq \emptyset$ was introduced here for convenience. With this meaning of the interval relational expressions, the equation $A \cdot x = b$ will define the solution set $\Xi_=$ equal to $\Xi_{\supseteq} \cap \Xi_{\subseteq}$, different than the interval solution. From the definitions it follows also that $\Xi_{\subseteq} \subseteq \Xi$ and $\Xi_{\supseteq} \subseteq \Xi$.

In the one-dimensional case, the matrix A shrinks to a single interval a, as does the vector b. The relational expression becomes thus one of $a \cdot x \Diamond b$, where $\Diamond \in \{\mathcal{I}, \supseteq, \subseteq, =\}$. Diagrammatic analysis of solution sets for this case proves to be indispensable for diagrammatic analysis of the general multidimensional case. That analysis is based on the one-dimensional *radial* and *parallel* cuts through the solution space. As demonstrated in [4, 5, 8], the arrangement of solution sets along these cuts is provided by solutions of some one-dimensional equation whose

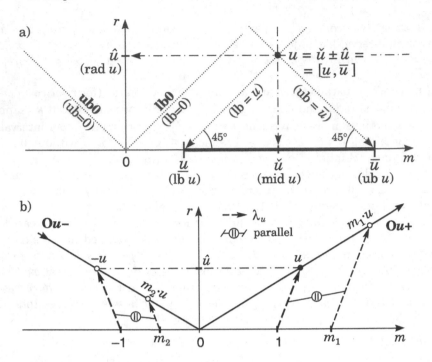

Fig. 1. The MR-diagram representation of the space of real intervals (a); interval axis, negation of intervals, and multiplication by real numbers (b)

coefficients are determined by the general equation coefficients and the direction of the cut.

3   Interval Space Diagram

The basis for the diagrammatic approach to interval analysis is the two-dimensional representation of the space of real intervals $\mathbb{IR}$ called the *MR-diagram* [1], see Fig. 1(a). In this diagram, an interval is represented by a point with its centred coordinates: midpoint $\check{u}$ and radius $\hat{u}$. Besides midpoint and radius, one can also easily obtain the endpoints $\underline{u}$ and $\overline{u}$ of the interval using the diagonal lines. In this way, the MR-diagram combines conveniently all three common representations of intervals—midpoint-radius, endpoint, and the one-dimensional representation as a segment of the real number line (here on the **Om** axis).

The main diagonals **lb0** and **ub0** constitute a dividing line between intervals *containing zero* (they all lie on or above the diagonals) and those *without zero* (below the diagonals). The *interval axis* **Ou** of the interval u consists of a positive half **Ou**+ going through the interval u, and the negative half **Ou**− through the interval $-u$, see Fig. 1(b). Note how negation (change of sign) of an interval is obtained by reflection in the **Or** axis. All intervals v lying on the interval axis **Ou** of the interval u have the same value of the κ function: $\kappa v = \kappa u$, i.e., have the same relative extent. When $\kappa v \neq \kappa u$, u and v have different axes. An interval v is called *more extended* than u if it lies above the interval axis $\mathbf{O}u$. Symmetric intervals (including 0) are considered more extended than all other intervals. Their axis coincides with the $\mathbf{O}r+$ coordinate axis.

Multiplication of an interval u by a scalar (real number) $m \in \mathbb{R}$ is defined by $m \cdot u = \{m\tilde{u} \,|\, \tilde{u} \in u\} = m\,\check{u} \pm |m|\,\hat{u}$. The interval axis $\mathbf{O}u$ groups all products of the interval u and all real numbers, symbolically: $\mathbf{O}u = \mathbb{R} \cdot u$. To find the product of an interval u and a real number m, it suffices to map appropriately the point on the $\mathbf{O}m$ axis with the coordinate m onto the interval axis $\mathbf{O}u$. The diagrammatic construction for that is shown in Fig. 1(b). The mapping lines are parallel to the lines from the points of value $+1$ and -1 on the $\mathbf{O}m$ axis to u and $-u$, respectively (Fig. 1(b)). It is convenient to define the mapping as a function called *lambda mapping*: $\lambda_u(m) = m \cdot u$. Its inverse allows to find the real number (a point on the $\mathbf{O}m$ axis) by which the interval u has been multiplied to obtain the given point on the axis $\mathbf{O}u$.

4   The Basic Equation and Its Structural Types

The basic one-dimensional equation can be solved diagrammatically. The expression $a \cdot x \lozenge b$ tells us that first we need a representation of all points that are in relation $\lozenge$ to the right-hand side interval b. Thus, we will need a diagrammatic representation of coimages of the coefficient b under the relations defining the solution sets. They are defined in Fig. 2(a), see [3, 5, 8] for more details. Borders of the coimages represent the *border relations* ⌐, ⊐, ⊏, and ⌐ that group intervals one of whose endpoints coincides with one of the endpoints of the coefficient b, as indicated in the figure.

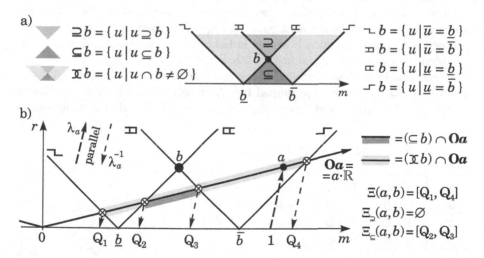

Fig. 2. Coimages of an interval b under interval relations $\mathbb{X}$, $\supseteq$ and $\subseteq$, and the definitions of border relations ⌐, ⊐, ⊏, and ⌐ (a), and diagrammatic solution example of the $a \cdot x = b$ equation (b)

For the given interval a and all possible real numbers x, the set of products $a \cdot x$ coincides with the axis $\mathbf{O}a$. Thus, to find all values of x that fulfill the expression $a \cdot x \Diamond b$ for the given $\Diamond \in \{\sqsubset, \supseteq, \subseteq, =\}$, we must first find the subset of $\mathbf{O}a$ whose member intervals are related to b by the relation $\Diamond$. It is obviously the intersection $\mathbf{O}a \cap (\Diamond b)$ of $\mathbf{O}a$ with the coimage $\Diamond b$. Since $a \cdot x = \lambda_a(x)$, then $x = \lambda_a^{-1}(a \cdot x)$ and the solution set $\Xi_\Diamond$ is the result of the inverse lambda mapping of the said intersection onto the $\mathbf{O}m$ axis, that is, $\Xi_\Diamond = \lambda_a^{-1}(\mathbf{O}a \cap (\Diamond b))$. An example diagrammatic construction for one of the cases is shown in Fig. 2(b), together with the resulting definitions of the three solution sets for this case.

The endpoints of the solution sets are thus given by the points $Q_i = \lambda_a^{-1}(w_i)$, where w_i denotes one of the points of intersection (marked by $\otimes$ in Fig. 2) of $\mathbf{O}a$ with one of the border relations. As it was derived in [3, 5], the points Q_i, called *quotients* of the expression $a \cdot x \Diamond b$, are obtained according to the rule shown in Fig. 3(a), depending on the border relation whose intersection with the $\mathbf{O}a$ axis generates the quotient and its sign (the position of the quotient with respect to the $\mathbf{O}r$ axis in the diagram). The shorthands L, S, Z, and T were chosen on a mnemonic principle, as they mimic the graphical structure formed by dashes and a division operator in the quotient expressions.

When for the given coefficients a and b we sort the quotients Q_i in an increasing order of their numerical values and then list their names $N_i \in \{\text{"L"}, \text{"S"}, \text{"Z"}, \text{"T"}\}$ in the same order $N_{i_1} N_{i_2} N_{i_3} N_{i_4}$, $(N_{i_j} \neq N_{i_k}$ for $j \neq k)$ we obtain the *characteristic quotient sequence* for these coefficients (and hence for the type of the equation with these coefficients). The sequence will be denoted by $\mathbb{Q}(a, b)$. Characteristic quotient sequences are usually augmented by the indications of the position of zero, equality, and special values of some quotients (like infinity or undefined values), see the examples further on.

After arranging quotients in a two-dimensional array as in Fig. 3(b), the sequence can be represented as a *quotient sequence diagram*. Solution sets determined by the given sequence will be indicated with the graphical annotation explained by the two examples in Fig. 3(b).

Diagrammatic analysis sketched above revealed that there are only 16 different basic quotient sequences, grouped into 6 structural types corresponding to different possible configurations of the interval axis $\mathbf{O}b$ and the coefficient a,

a)

border relation	sgn Q_i +	−	$Q_i \in \{L, S, Z, T\}$, $Q^{\alpha\beta} = b^\beta/a^\alpha$:
⌐	S	L	$L = Q^{--} = \underline{b}/\underline{a}$
⌐	L	S	$S = Q^{+-} = \underline{b}/\overline{a}$
⌐	T	Z	$Z = Q^{-+} = \overline{b}/\underline{a}$
⌐	Z	T	$T = Q^{++} = \overline{b}/\overline{a}$

b)

Fig. 3. Notation for quotients of the $a \cdot x \Diamond b$ relation and correspondence between them and intersections with border relations (a); quotient sequence diagram and graphical notation for solution sets (b). It is assumed that $\alpha, \beta \in \{-, +\}$ and $u^- = \underline{u}$, $u^+ = \overline{u}$. Small circles "o" denote the position of zero.

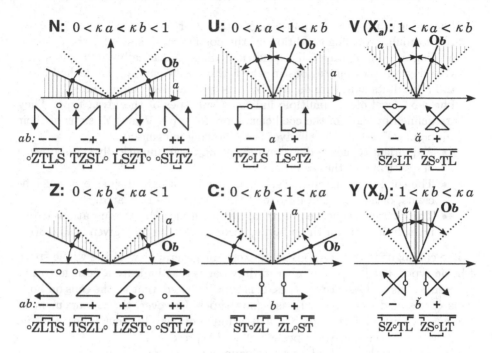

Fig. 4. The catalogue of all basic subtypes of interval equation $a \cdot x = b$. Defining conditions for the (sub)types explicitly exclude intermediate cases.

as shown in the catalogue of basic types in Fig. 4. Letter names for the types were chosen to mimic the shape of the quotient sequence diagram for the type. Concerning the names **V** and **Y**, see the remark in Section 5.

An important property of quotient sequences [3] says that they are invariant with respect to radial moves of the coefficients a and b along their respective positive interval semi-axes (excluding zero), i.e., $\mathbb{Q}(a, b) = \mathbb{Q}(sa, tb)$ for any $s, t \in \mathbb{R}^+$. As a consequence, the solution types (hence their qualitative configurations) depend only on the values of rex a and rex b. Therefore, regions with the same types have the shape of angular wedges in the MR-diagram and are independent of the scale of the diagram. That property allows for convenient diagrammatic representation of conditions for coefficients a and b defining the type of the equation as shown in Fig. 4.

The six basic types (without subtypes) are also explained in [7, 8] in the form of a set of solution diagrams like that for the type **N** in Fig. 2(b).

5   Catalogues of Types

5.1   A Multiple for Basic Types

The catalogue in Fig. 4 is a two-level multiple whose six high-level components (cells) describe individual basic types. Every cell is a hybrid (partially diagrammatic, partially propositional) representation of basic data about the type:

- The name of the type (a bold-face capital letter).
- The formula providing condition on the coefficients a and b for the type.
- A diagram illustrating the condition diagrammatically. It depicts the MR-diagram showing regions in which the coefficients a (the dashed area) and b (with its axis, as indicated by arcs with arrows) should lie.
- The next level of the multiple indicates subtypes of the given type. They are defined by signs of the coefficients, or, for types **V** and **Y**, signs of their midpoints. Every small cell gives the following information:
  - A quotient sequence diagram indicating diagrammatically the sequence of quotients for the subtype.
  - The condition for the subtype, indicated by a pair of signs of the coefficients, or a single sign for one of the coefficients indicated.
  - The quotient sequence in textual form, with graphical annotations defining solution sets for the subtype, according to the rules given in Fig. 3(b).

The catalogue in this form gives an excellent general view of the whole space of basic types, but it has a number of drawbacks that indicate a need for other representations of the space of types. It is type-oriented, that is, the cells on both levels correspond to individual basic (sub)types. However, the arrangement of cells conveys little information about relations between different types, that is, the structure of the space of types is not well represented.

The main problem is that the catalogue contains only basic types (6 main types with 16 subtypes). A considerable number of intermediate and degenerate types is not shown here. These types in principle can be generated from the data provided, but the process is rather tedious and error-prone, and it is easy to miss some of the possibilities. On the other hand, some information, namely quotient sequence diagrams, seems superfluous for the end user. These diagrams encode only part of the data included anyway in the textual quotient sequence below them. They are useful at the stage of enumeration of the basic types, providing a visual classification criterion based on shapes of the diagrams which led to the choice of letter names of the types. Even in this role, these diagrams proved to be somewhat misleading, as they suggested a single type **X** for $\kappa b > 1$. When the classification was extended to multidimensional equations (see [4, 5, 8]) it was found more convenient to split this type into two, as indicated in the diagram (with old type names provided in parentheses).

5.2 Diagrammatic Catalogue of All Types

As a result of these considerations, a new form of the catalogue was developed. It disposes of quotient sequence diagrams and is complete, listing all types, including intermediate and degenerate ones. It combines only diagrammatic representation of conditions for coefficients with textual representation of quotient sequences, augmented by diagrammatic annotations defining solution sets. An additional convention concerning quotient sequences is used here, namely quotients that are equal to zero are omitted. The full diagram occupies a single page; for brevity, in Fig. 5 only a part of it is shown. Note that the original letter labels of cells (c, d, e, f) were retained to provide easy reference to the full catalogue

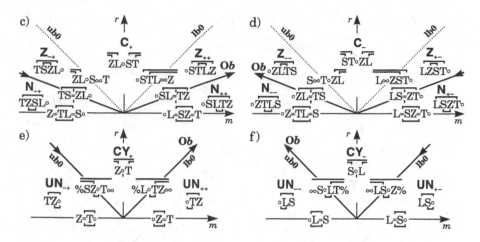

Fig. 5. Part of the diagrammatic catalogue of all subtypes of interval equation $a \cdot x = b$ for $a, b \neq 0$, and b: thick without zero (c, d), and with zero at endpoint (e, f)

of [7]. Two upper cells in this multiple provide data for all 10 basic subtypes of types **N**, **Z**, and **C**, as well as 12 intermediate types connected with them. The other two cells list 14 intermediate types obtained when $\underline{b} = 0$ or $\overline{b} = 0$.

This form of the catalogue is no longer type-oriented like that in Fig. 4. Instead, the cells here correspond to different conditions on the coefficient b. Possible positions of this coefficient in the MR-diagram are indicated by the position of the interval axis **O**b. When the axis is placed within some region, the coefficient b is allowed to vary within the region occupied by the positive semi-axis of the axis **O**b. When it coincides with some characteristic line of the diagram (main diagonal, **O**m or **O**r axes), the coefficient b can vary along the positive semi-axis **O**b only. The regions of the MR-diagram delineated by axes and main diagonals, as well as the axes and diagonals themselves, are labelled by annotated quotient sequences obtained when the coefficient a falls within the indicated region or on the indicated axis. Due to space limitations, only the regions are additionally labelled by names of subtypes. Thus, the basic information provided by this form of the catalogue is the definition of solution sets for all cases. The degenerate types (for $a = 0$ or $b = 0$) are depicted by separate cells of a slightly different design (not shown here). The complete catalogue in this form is provided in [7]; its older versions with different, less convenient layout were published in [3, 5].

A dual a-oriented form of the catalogue can be constructed as well. In it the cells would correspond to appropriate conditions on the coefficient a, while the types would be selected by the position of the coefficient b in the MR-diagram. It may be more convenient for certain purposes. In fact, in the form provided in [7], the degenerate types for $a = 0$ are grouped in a separate cell and represented in that a-oriented form. That, however, violates the b-oriented structure of the whole catalogue. A possible solution, assuring uniformity of the catalogue, is used in the construction of the restricted catalogue in Fig. 6.

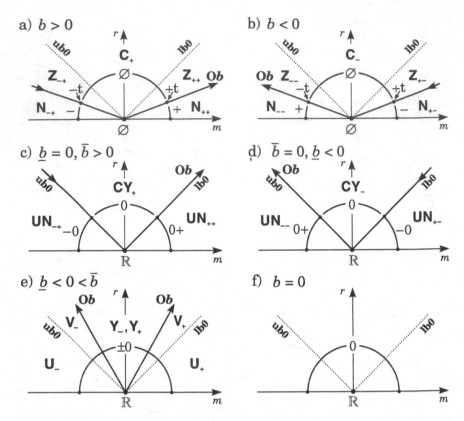

Fig. 6. Diagrammatic catalogue of tolerance solution set configurations for various conditions on the coefficient b

Despite its completeness and compactness, this form of the catalogue has its drawbacks. First, the lack of formulae defining conditions for types is inconvenient in some situations, forcing the user to translate from formulae to the situation in the diagram and back. That can be corrected for the conditions on the coefficient b by adding appropriate formulae as headers of the cells, as in the version in Fig. 6 (see [8] for such improvement). Second, although the representation is aimed at describing solution sets, it does not provide a good view of the distribution of solution set structures depending on positions of the coefficients in the MR-diagram. As information about all basic solution sets is lumped together into annotations of quotient sequences, it is easy to read out locally the particular definition of the solution sets for any single type. E.g., in Fig. 5(e) one may easily read that solution sets for the subtype intermediate between $\mathbf{CY_+}$ and $\mathbf{UN_{++}}$ (with $\underline{a} = 0$), are $\Xi = \mathbb{R}, \Xi_{\supseteq} = [\mathrm{T}, \infty), \Xi_{\subseteq} = [0, \mathrm{T}]$, and thus $\Xi_= = \{\mathrm{T}\}$. However, the overall picture is hard to comprehend.

5.3   Catalogues for Individual Solution Sets

For providing an overall view of possible structures of a single given solution set, useful in finding conditions for occurrence of interesting structures for that set, a specialized catalogues for individual sets may be more useful. Such a catalogue for the tolerance solution set is shown in Fig. 6. Analogous catalogues for control and united solution sets can be found in [7, 8]. The catalogue shown here (in Fig. 6) is additionally augmented with formulae describing conditions on the coefficient b for every cell, as it was discussed above. This catalogue is significantly smaller and simpler than the full catalogue whose part is shown in Fig. 5. This is due to omission of quotient sequences and replacing exact definitions of solution sets by qualitative codes (see [7] for their explanation). That, with the fact that the structure of only a single solution set is represented, allows for aggregation of the cases into a smaller number of cells. The price for that is losing exact definitions of solution sets and losing the possibility to directly compare structures of different sets.

In this version the degenerate cases for $a = 0$ are not depicted with a separate cell of different kind. Instead, descriptors of solution sets for these cases are put at the point $(0, 0)$ in the b-oriented cells. Such a solution can be also adopted in the full catalogue of Fig. 5, though with some difficulties due to a rather large size of quotient sequence descriptors for this case (see [7]).

5.4   Tabular Catalogue

For certain regular structures of information it may be convenient to represent the catalogue in a tabular form. A small part of the table of types, containing 6 intermediate types shown in top parts of Fig. 5(e) and 5(f), from among 73 subtype entries in the complete catalogue, is provided in Table 1. The data in the cells are represented only propositionally. Solution sets are described with both exact definitions in terms of quotients, and with qualitative descriptors used in Fig. 6. The fourth column links table entries to corresponding cells of the

Table 1. Part of a detailed table of descriptions of solution sets for intermediate subtypes

Type	Quotient sequence	Conditions $a \neq 0,\ b \neq 0$	Fig. 5	Solution sets			
				$\Xi_\subseteq$	$\Xi_\supseteq$	$\Xi_=$	Ξ
	...			...			
CY		$\kappa b = 1 < \kappa a$					
–	SoL	$\check b < 0$	f	0	$\pm\infty$:]S, L[	$\emptyset$	$\supseteq$
+	ZoT	$\check b > 0$	e	0	$\pm\infty$:]Z, T[	$\emptyset$	$\supseteq$
	...			...			
CU		$\kappa a = \kappa b = 1$					
– +	%SZoT∞	$\check a < 0,\ \check b > 0$	e	–0: [Z, 0]	$-\infty$:]Z, ∞[	$-t$: Z	$\mathbb{R}$
+ –	∞LSoZ%	$\check a > 0,\ \check b < 0$	f	–0: [S, 0]	$-\infty$:]S, ∞[	$-t$: S	$\mathbb{R}$
– –	∞SoLT%	$\check a < 0,\ \check b < 0$	f	0+: [0, L]	$+\infty$:]∞, L[	$+t$: L	$\mathbb{R}$
++	%LoTZ∞	$\check a > 0,\ \check b > 0$	e	0+: [0, T]	$+\infty$:]∞, T[	$+t$: T	$\mathbb{R}$
	...			...			

diagrammatic catalogue in Fig. 5. Although most convenient for some purposes, these tables have their drawbacks too. First, they tend to be large—the complete catalogue in [7, 8] uses four pages (for 73 subtype entries), compared to the single page of the diagrammatic catalogue in Fig. 5. Second, the overall picture of relations between different types and structures of their solution sets is almost completely lost.

5.5   RR-Diagram Maps and Graphs of Types

New forms of the catalogue can be based on the RR-diagram, introduced in [3, 5]. In this diagram, values of rex a and rex b (or their reciprocals) are put on the coordinate axes. Because types do not change when extent functions of the coefficients a and b do not change, to every point in the RR-diagram corresponds some type. It is unique except when one of the coefficients is thin, because then the value of the extent function is zero, independently of the sign of the interval. Thus, the sign of a thin interval cannot be distinguished by its position in the RR-diagram. Labelling appropriate regions in the diagram by the type of its points we obtain a sort of map, partitioning the diagram into typed regions as in Fig. 7(a) and 7(b). Intermediate and degenerate types (not shown in the figure) correspond to borders and vertices of the regions. Obtaining the complete catalogue in this way is, however, troublesome, as some different intermediate types involving thin intervals of different signs (that includes also all degenerate types) fall on the same points and segments of the rex a and rex b axes.

Representing regions as nodes and neighbourhood relations between them as edges, we can obtain various graphs (networks) of types. Two such graphs are shown in Fig. 7(c) and 7(d). The second graph is useful for enumerating types of multidimensional equations, see [4, 5, 8]. The RR-diagram and graph representations of the catalogue combine the catalogue aspect with another usage type of diagrams, namely showing the structure of the space of types, see [6–Fig. 2(a)], the feature lacking in the multiple-like catalogues.

6   Discussion

Design of a diagrammatic catalogue of types of the basic interval equation $a \cdot x = b$ presented several nontrivial problems, leading to the development of various forms of the catalogue and searching for new ways of structuring them. The nontrivial, though not overwhelming complexity of the catalogue has made it a convenient case study of the problem of designing catalogues of various pieces of mathematical knowledge. The main obstacle here is the lack of general guidelines for designing such catalogues. On the basis of this case study one may try to formulate some preliminary design guidelines.

Catalogues of cases serve as reference databases, but also as research tools for searching patterns of differences and similarities between the cases and their various constituent parameters, in this case especially the structure and definitions of solution sets for every type. As observed by Tufte [15]: "Comparisons must be enforced within the scope of the eyespan." Therefore, the catalogue

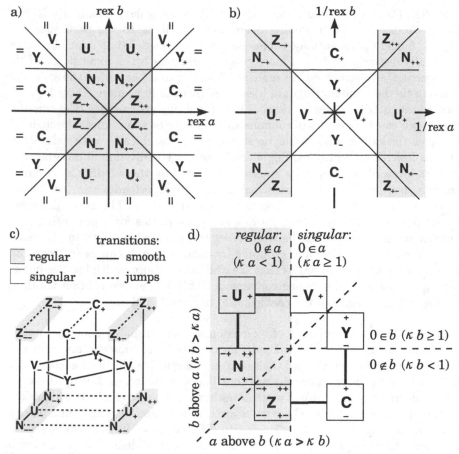

Fig. 7. Type catalogues on the RR-diagram (a), 1/RR-diagram (b); and in the form of graphs (c, d)

should be, if possible, not larger than a single page of paper or a single computer screen. The considerable number of types and their parameters makes such an attempt rather hopeless in this case (and in many others, unfortunately). Either the catalogue becomes too cluttered and unreadable, or it must occupy larger area, or it must omit a substantial amount of information. All these outcomes occurred in the catalogues discussed in the paper. A possible solution is to produce different catalogues for different purposes, differing by the selection of represented data and the form of their presentation. Such catalogues can be generated (semi-)automatically from some underlying complete database, or prepared separately beforehand and then browsed through (as it is currently the case with the catalogues of equation types [7]). Such a solution, however, blocks or makes troublesome some possible comparisons of data, hence other solutions should still be searched for, like interactive catalogues.

The observation of our case shows that there seem to be three basic types of such catalogues: *multiples* (including *tables*), *maps*, and *graphs* (or *networks*).

Multiples. These are regular structures (usually rectangular) of similar cells containing chunks of data (including diagrams) pertaining to the particular case (Figs. 4, 5, and 6). Essentially, tables can be also considered as multiples. They are usually distinguished from more general multiples by an explicit use of the two dimensions to structure the multiple. Namely, cells occupying the same column contain the same type of data (described by column headers, Table 1), while rows contain attribute descriptions of individual objects or cases (the equation types in our case). The roles of columns and rows can be sometimes interchanged. Multiples (including tables) can be structured hierarchically, with cells structured as lower-level multiples, Fig. 4 and Table 1. The division of data between different levels of the hierarchy is usually dictated by the intrinsic structure of the data, but often can be varied depending on the *intended use* (see below) of the catalogue (compare catalogues in Figs. 4 and 5). The main drawback of multiples is that they do not provide adequate means for representing more complex structural relations between cases (represented by data in the cells). The hierarchical grouping and grouping by data type in columns (or rows) are practically the only possibilities that are available in pure multiples. See hierarchical grouping in Fig. 4, typed columns in Table 1, and two-column multiples (corresponding to signs of the coefficient b) in Figs. 5 and 6.

Maps. They are arrangements of regions on a plane (usually; sometimes other arrangements, e.g. three-dimensional, can be used). Regions contain data pertaining to individual cases, and their shapes and relative positions encode additional data about properties and relations between the cases. Multiples can be also considered a special case of maps. Another special case can be distinguished, let us call it *constructions*, where cases are distinguished by diagrammatic constructions placing the results belonging to different cases in different regions of the space. An example is provided by the catalogue of definitions of means in [6–Fig. 4]. The appropriate division of a plane into regions is often obtained with the help of a coordinate system whose coordinates correspond to parameters distinguishing the cases. The maps used within cells of multiples in Figs. 4, 5, and 6 use the MR-diagram midpoint-radius coordinate system, while the RR-diagram based catalogues in Figs. 7(a) and (b) use the values of the function rex for the coefficients a and b. The advantage of maps comes from a richer layout structure that can be used to represent relations between cases, especially when the intrinsic structure of the set of cases conforms well to the structure of the Euclidean plane. Otherwise, the structure must be "planarized", for the price of losing information or introducing information noise. This is the case with our catalogue of types which is essentially at least three-dimensional, see Fig. 7(c).

Graphs (networks). In this form, nodes of a graph represent cases, and edges relations between them. That allows for representation of arbitrary systems of relations between cases, but often for the price of making them hard to comprehend, especially for more complex systems. The proper layout of complex graphs of relations is a nontrivial problem—it has given rise to the whole discipline of graph drawing [11]. Sometimes the proper layout can be obtained by using an appropriate map as a guide for placing the nodes. The graph in Fig. 7(d) is superimposed over a partition of the plane generated by different conditions for the coefficients a and b. Note another hybrid element in this graph—the data within the nodes are arranged as small multiples of subtypes.

Hybrid solutions and user's goals. In practice, as was indicated above, hybrid solutions are used, with different presentation means used for different portions of a catalogue. This includes combining different types of representations, like multiples containing maps (Figs. 4, 5, and 6) or graphs containing multiples and superimposed on maps (Fig. 7(d)). To some extent this depends on the structure of data, but in most part on the intended use (user's goals) of the catalogue. The importance of user's goals for proper design of information presentation has been recognized some time ago (see e.g. [13]). Like for the presentation graphics of quantitative (statistical) data, the design of mathematical diagrams should also be based of the analysis of user's goals and selecting the way of presentation appropriate for them, possibly in a similar way as developed in [13].

Interactive catalogues. The use of many catalogues for different purposes solves some of the problems but makes relating of different pieces of information contained in different catalogues difficult. A possible solution would be to make the set of catalogues interactive. In such a system, selection of certain piece of information in some catalogue may either highlight the corresponding piece of information in another catalogue, or provide that information in a separate small window or "balloon" near the place pointed to. That may not solve all the problems, especially as proper organization of such interaction, when there are several differently structured catalogues, can be a considerable problem in itself.

A useful addition to such a catalogue is an algorithmic component, namely a program producing for any given numerical values of coefficients the type data (including location in the catalogue) and solution sets for the equation with this coefficient. Besides allowing the user of the catalogue to browse the space of possibilities also quantitatively, such a subroutine is a necessary component of any program using solutions of this equation to characterize solution sets of a general multidimensional equation (e.g., calculating radial and parallel cuts through its solution space, see [8]). Such a subroutine was developed and is available for interested users, see [7].

References

1. Z. Kulpa, Diagrammatic representation for a space of intervals. *Machine GRAPHICS & VISION*, **6** (1997) 5–24.
2. Z. Kulpa, Diagrammatic representation for interval arithmetic. *Linear Algebra Appl.*, **324** (2001) 55–80.
3. Z. Kulpa, Diagrammatic analysis of interval linear equations. Part I: Basic notions and the one-dimensional case. *Reliable Computing*, **9** (2003) 1–20.
4. Z. Kulpa, Diagrammatic analysis of interval linear equations. Part II: The two-dimensional case and generalization to n dimensions. *Reliable Computing*, **9** (2003) 205–228.

5. Z. Kulpa, *From Picture Processing to Interval Diagrams*. IFTR Reports 4/2003, Warsaw, 2003. [See http://www.ippt.gov.pl/~zkulpa/diagrams/fpptid.html]
6. Z. Kulpa, On diagrammatic representation of mathematical knowledge. In: A. Asperti, G. Bancerek, A. Trybulec, eds., *Mathematical Knowledge Management* (Third International Conference MKM 2004, Białowieża, Poland, Sept. 19-21, 2004) *Lecture Notes in Computer Science*, vol. 3119, Springer-Verlag, Berlin 2004, pp. 190-204.
7. Z. Kulpa, *Structural Types of Interval Equation $a \cdot x = b$: A Complete Catalogue*. Technical Report B-1/2005, Institute of Fundamental Technological Research, Warsaw 2005. [See http://www.ippt.gov.pl/~zkulpa/quaphys/st1rep.html]
8. Z. Kulpa, *Diagrammatic Interval Analysis with Applications*. IFTR Reports, Warsaw, 2005 (in preparation).
9. S. Markov, On the interval arithmetic in midpoint-radius form. In: *Mathematics and Education in Mathematics* (Proc. 33rd Spring Conference of the Union of Bulgarian Mathematicians, Borovets, April 1–4, 2004), 391–396.
10. A. Neumaier, *Interval Methods for Systems of Equations*. Cambridge University Press, Cambridge 1990.
11. J. Pach (Ed.), *Graph Drawing*. Revised selected papers of the 12th International Symposium GD 2004. Lecture Notes in Computer Science, Vol. 3383, Springer-Verlag, Berlin 2005.
12. J. Rohn, Systems of linear interval equations. *Linear Algebra and Its Applications*, **126** (1989) 39–78.
13. S.F. Roth, J. Mattis, Automating the presentation of information. In: *Proc. IEEE Conf. on Artificial Intelligence Applications*. IEEE Press, 1991.
14. S.P. Shary, A new technique in system analysis under interval uncertainty and ambiguity. *Reliable Computing*, **8** (2002) 321–418.
15. E.R. Tufte, *Envisioning Information*. Graphics Press, Cheshire, CT 1991.
16. E.R. Tufte, *Visual Explanations*. Graphics Press, Cheshire, CT 1997.

Gröbner Bases — Theory Refinement in the Mizar System

Christoph Schwarzweller

Department of Computer Science,
University of Gdańsk, Gdańsk, Poland
schwarzw@math.univ.gda.pl

Abstract. We argue that for building mathematical knowledge repositories a broad development of theories is of major importance. Organizing mathematical knowledge in theories is an obvious approach to cope with the immense number of topics, definitions, theorems, and proofs in a general repository that is not restricted to a special field. However, concrete mathematical objects are often reinterpreted as special instances of a general theory, in this way reusing and refining existing developments. We believe that in order to become widely accepted mathematical knowledge management systems have to adopt this flexibility and to provide collections of well-developed theories.

As an example we describe the Mizar development of the theory of Gröbner bases, a theory which is built upon the theory of polynomials, ring (ideal) theory, and the theory of rewriting systems. Here, polynomials are considered both as ring elements and elements of rewriting systems. Both theories (and polynomials) already have been formalized in Mizar and are therefore refined and reused. Our work also includes a number of theorems that, to our knowledge, have been proved mechanically for the first time.

1 Introduction

One major goal of mathematical knowledge management is to design and construct large repositories containing a wide range of different topics, such as algebra, analysis, topology and many more. To be as broad as possible seems reasonable in order to explore the use of such repositories for distributing mathematics over the internet and extracting introductory courses, among others. On the other hand, to be attractive for professional mathematicians also, more advanced mathematics must be taken into account. As has been pointed out at the last MKM-meetings by Andrzej Trybulec "We should try to reach the research frontier".

Advanced, contemporary mathematics, however, cannot be brought onto the computer by simply choosing one theory and formalizing it "to its end". More advanced mathematics usually uses a number of theories to develop its results. Different theories are reused or combined to get new ones. Moreover, modern mathematics lives from the fact that one and the same object can be considered

M. Kohlhase (Ed.): MKM 2005, LNAI 3863, pp. 299–314, 2006.
© Springer-Verlag Berlin Heidelberg 2006

as a special instance of different theories. For example the integers can be considered as a group (generated by 1), as an Euclidean ring, as an ordered domain or even as (the ring of) coefficients for polynomials rings. In each case the instantiation, or refinement as we shall also call it, of the general theory with the integers allows for both reusing results of the general theory and deducing new results for the particular case.

We believe that mathematical repositories should reflect this way of "working" with mathematical theories. Continuing the work of [GS04] where combination of theories has been investigated, we focus in this paper on theory refinement in the Mizar system [Miz05, RT01]. We consider the theory of Gröbner bases [Buc98] as an example. Gröbner bases are a method to decide among other things the ideal membership problem in polynomial rings: Via computing normal forms of polynomials with respect to a given ideal — a reduction in the sense of rewriting systems — ideal membership can be decided by syntactic equality, if the polynomials generating the ideal form a Gröbner base. We thus have polynomials as basic objects, usually defined as lists of elements from a coefficient ring or as functions from terms into a coefficient ring. Note that the definition of polynomials already uses a theory, the theory of rings. In the theory of Gröbner bases, however, polynomials are also used as special elements for different, more general theories:

1. Polynomials are considered as elements of a ring, that is addition and multiplication of polynomials coincide with ring addition and multiplication.
2. Polynomials are considered as elements of ideals, that is, though almost trivial, polynomials coincide with elements of sets while still obeying their addition and multiplication.
3. Polynomials are considered as elements of a relation, the reduction relation, that is polynomials coincide with the elements of relations.

Not taking into account the second item from above, we thus get the theory structure illustrated in figure 1. Of course one can define polynomials and all the concepts necessary for Gröbner bases from scratch without even employing theories for rings and reduction systems (see for example [The01]), but in repositories for mathematical knowledge management we should — if possible — build new theories by reusing and refining older ones.

The plan of the paper is as follows. After an introduction to the Mizar language we briefly recall polynomials, rings, ideals, and rewriting systems by reviewing their Mizar formalization as done in [RT99, BRS00, Ban95]. Section 4 and 5 describe the development of Gröbner bases based on these theories. In the last two sections we discuss the Mizar approach for refining and reusing theories and compare it to other approaches in the literature. Conclusions for the design of mathematical knowledge repositories are also drawn.

2   The Mizar System

Mizar's [RT01, Miz05] logical basis is classical first order logic extended with so-called schemes. Schemes allow for free second order variables, in this way

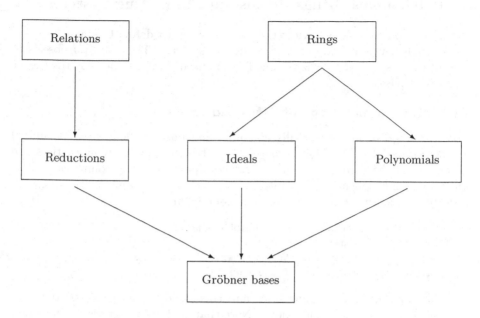

Fig. 1. Theory structure for the development of Gröbner bases

enabling, for example, the definition of induction schemes. The current development of the Mizar Mathematical Library (MML) is based on Tarski-Grothen-
dieck set theory — a variant of Zermelo Fraenkel set theory using Tarski's axiom
on arbitrarily large, strongly inaccessible cardinals [Tar39] which can be used to
prove the axiom of choice —, though in principle the Mizar language allows for
other axiom systems also. Mizar proofs are written in natural deduction style
similar to the calculus of [Jaś34]. The rules of the calculus are connected with
corresponding (English) natural language phrases so that the Mizar language
is close to the one used in mathematical textbooks. The Mizar proof checker
verifies the individual proof steps using the notion of obvious inferences [Dav81]
to shorten the rather long proofs of pure natural deduction.

Mizar objects are typed, the types forming a hierarchy with the fundamental type `set` [Ban03]. New types are constructed using type constructors called
modes. Modes can be decorated with adjectives — given by so-called attribute
definitions — in this way extending the type hierarchy: For example, given the
mode `Ring` and an attribute `commutative` a new mode `commutative Ring` can
be constructed, which obeys all the properties given by the mode `Ring` plus
the ones stated by the attribute `commutative`. Furthermore, a variable of type
`commutative Ring` then is also of type `Ring`, which implies that all notions
defined for `Ring` are available for `commutative Ring`. In addition all theorems
proved for type `Ring` are applicable for objects of type `commutative Ring`; indeed the Mizar checker itself infers subtype relations in order to check whether
theorems are applicable for a given type.

3 Polynomials, Rings, Ideals, and Rewriting Systems

In this section we briefly review the theories used to define Gröbner bases — polynomials, rings and ideals, and reduction systems. The main purpose is to present the basics of their Mizar formalization needed later to develop the theory of Gröbner bases.

3.1 Mizar Formalization of Rings and Ideals

In Mizar rings, or more generally algebraic domains, are defined as attributed structures, see [RST01]. That is, based on a structure mode giving carriers and operations of the domain properties of these operations, e.g. commutativity of addition or multiplication, are introduced by Mizar attributes. For rings the underlying structure mode is called `doubleLoopStr`:

```
struct (LoopStr, multLoopStr_0) doubleLoopStr
  (# carrier      -> set,
     add, mult    -> BinOp of the carrier,
     unity, Zero -> Element of the carrier #);
```

and the mode `Ring` is nothing else than this structure mode decorated with attributes describing the ring axioms. Note that `doubleLoopStr` is a descendant of two other structure modes `LoopStr` and `multLoopStr_0`, and thus a subtype of these. Now given a subset F of a structure mode L one can easily define an attribute describing that F is closed with respect to addition, left- and right-multiplication, for example

```
definition
let L be non empty LoopStr, F be Subset of L;
attr F is add-closed means
  for x,y being Element of L st x in F & y in F holds x+y in F;
end;
```

Combination of these properties then gives the definition of ideals. Note that using the already defined attributes it is trivial to additionally define left and right ideals. Also ideals generated by a subset F can be easily defined as a functor from subsets of the domain into ideals.

```
definition
let L be non empty doubleLoopStr;
mode Ideal of L is
  add-closed left-ideal right-ideal (non empty Subset of L);
end;
```

```
definition
let L be non empty doubleLoopStr, F be Subset of L;
assume F is non empty;
func F-Ideal -> Ideal of L means
  F c= it & for I being Ideal of L st F c= I holds it c= I;
end;
```

Note that both definitions actually do not use the mode `Ring`, but only the underlying structure mode `doubleLoopStr`: The existence of ideals and generated ideals does not depend on algebraic properties of the ring (just take the whole domain). Hence such objects should be defined without using these. Nevertheless, due to Mizar's type system, the so-defined notions of (and theorems proved for) ideals are available for rings in Mizar, because the mode `Ring` is based on and thus a subtype of `doubleLoopStr`.

3.2 Mizar Formalization of Polynomials

Polynomials are defined as functions from terms (called `bags` in Mizar) into coefficients, see [RST01, RT99]. Thus a polynomial is a subtype of `Series of n,L`, where `n` gives the number of indeterminates used to build terms and `L` the structure mode describing the coefficients. The attribute `finite-Support` ensures that the `Support` of a polynomial, that is the set of terms with a non zero coefficient, is finite.

```
definition
let n be Ordinal, L be non empty ZeroStr;
mode Polynomial of n,L is finite-Support Series of n,L;
end;
```

In the theory of Gröbner bases terms are assumed to be ordered. In Mizar we can use the mode `Order of X`, where `X` is an arbitrary set to do so. This will enable us to develop Gröbner bases for arbitrary orderings on terms and is actually another example for reusing theories in Mizar: The set of all terms for a given set of variables — remember that `n` gives the number of variables — is denoted by `Bags n`. Thus we get term orders, that is orders on terms, by just defining

```
definition
let n be set;
mode TermOrder of n is Order of Bags n;
end;
```

A term order is admissible, if the empty term — in Mizar denoted by `Empty Bag n` — is the smallest one and the order respects multiplication of terms, that is if we have for all terms t, t_1, t_2 both `Empty Bag n` $\leq t$ and $t_1 \leq t_2$ implies $t_1 \cdot t \leq t_2 \cdot t$. In Mizar this can be straightforwardly formalized as an attribute `admissible` for the mode `TermOrder`. Thus the Mizar type `admissible TermOrder` describes arbitrary admissible term orders. Note, that an admissible term order is well-founded.

Given an order `T` on terms, the head term of a polynomial `p` is the biggest term in `Support p` with respect to `T`; head coefficients and head monomials are defined analogously. In order to develop Gröbner bases for arbitrary term orders, we defined functors `HT(p,T)`, `HC(p,T)` and `HM(p,T)` taking the (admissible) term order `T` as an additional argument. Note that in order to define head terms the order `T` must be total (called `connected` in Mizar), but not admissible.

3.3 Mizar Formalization of Rewriting Systems

Rewriting systems are an approach to handle structures defined via equivalence relations. The idea is to decide a given equivalence by computing (unique) normal forms for the objects of concern: The objects are reduced until no more rewrite rules are applicable. Then, if the set of rules is "suitable", deciding the equivalence relaton is no more than syntactical comparison. Rewriting systems have various applications in such different fields as specification and verification, algebraic computation or pure theorem proving, see [DJ90]. In the following we recall the basic definitions of general rewriting systems as defined in [Ban95]. Given a relation R a reduction sequence is a sequence over R in which each two neighboured elements are in R. Thus the theory of rewriting systems actually is an extension of the theory of relations.

```
definition
let R be Relation;
mode RedSequence of R -> FinSequence means
  len it > 0 &
  for i being Nat st i in dom it & i+1 in dom it
  holds [it.i, it.(i+1)] in R;
end;
```

Then we have that a can be reduced to b (R reduces a,b), if there exists a reduction sequence of R with a being the first and b the last element in the sequence. Based on these definitions it is straightforward to introduce other basic concepts of rewriting systems, such as confluence, local confluence or the Church-Rosser property, for example

```
definition
let R be Relation;
attr R is locally-confluent means
  for a,b,c being set st [a,b] in R & [a,c] in R
  holds b,c are_convergent_wrt R;
end;
```

where `are_convergent_wrt` means that there exits an element $d \in R$ such that both b and c can be reduced to d. Termination properties such as **strongly-** and **weakly-normalizing** are also introduced in [Ban95]. Finally, a complete rewriting system is a strongly-terminating confluent one, that is a rewriting system in which in particular for every element a unique normal form exists.

4   Reinterpreting Polynomials

In this section we describe how the general theories of rings and rewriting systems are refined with polynomials. This allows to use notations and theorems of these theories for the special case of polynomials as well as further properties of polynomials themselves when later developing the theory Gröbner bases. Note that no special care has to be taken for ideals of polynomials; these are given automatically because ideals have been defined for general rings.

4.1   Polynomials as Rings

To get the theory of rings available for polynomials (the domain and) the operations of polynomials have to be interpreted as the ring (domain and) operations.
This is done by defining a functor `Polynom-Ring(n,L)` into the underlying structure mode of rings (see [RT99]), called `doubleLoopStr` in Mizar. Here, n gives
the number of indeterminates and L the structure mode describing the coefficient domain; note that the number of indeterminates need not be finite. In the
definition the components of the structure, that is the domain and the operations of the ring, are simply identified with polynomials and the corresponding
operations on polynomials.

```
definition
let n be Ordinal,
    L be right_zeroed add-associative right_complementable
         unital distributive non trivial (non empty doubleLoopStr);
func Polynom-Ring(n,L) -> strict non empty doubleLoopStr means
 (for x being set holds
   x in the carrier of it iff x is Polynomial of n, L) &
 (for x,y being Element of it, p,q being Polynomial of n, L
   st x = p & y = q holds x+y = p+q) &
 (for x,y being Element of it, p,q being Polynomial of n, L
   st x = p & y = q holds x*y = p*'q) &
 0.it = 0_(n,L) & 1_ it = 1_(n,L);
end;
```

Now, to apply theorems proved for general rings we need not only that
`Polynom-Ring(n,L)` is a `doubeLoopStr`, but also that the attributes establishing the type `Ring` hold. It turns out that for different attributes to hold different
properties of the coefficient domain are necessary. Therefore each attribute is
proved in a cluster registration stating exactly the properties of the coefficient
ring necessary to prove it (see [RT99]), for example

```
registration
let n be Ordinal,
    L be Abelian right_zeroed add-associative right_complementable
         unital distributive non trivial (non empty doubleLoopStr);
cluster Polynom-Ring(n,L) -> Abelian;
end;
```

The effect is that properties of `Polynom-Ring(n,L)` are automatically bound
to properties of the coefficient domain L: If L obeys the properties stated in
the registration, the Mizar checker itself infers that `Polynom-Ring(n,L)` has the
concluding property, hence is a subtype of `Ring`. This supports reusing theorems
of the general ring theory for the special case of polynomials.

4.2   Polynomial Reduction

Polynomial reduction establishes a generalization of polynomial division for the
univariate case: Each reduction step describes a single step in the division process. Thus a non-zero polynomial f reduces to a polynomial g using polynomial

p by eleminating term (bag) t if there exists a term s such that s · HT(p,T) = t
and g = (f - f.t/HC(p,T)) * s *' p, which in Mizar can be easily defined
as a predicate f reduces_to g,p,b,T. Note that the reduction depends on the
term order T used to define head terms and head monomials.

Now, to introduce polynomial reduction as a special case of general rewriting
we have to define the reduction relation, that is the relation R which contains
all pairs (p₁, p₂) such that p₁ reduces (in one step) to p₂ with respect to a given
set of polynomials P. This is done with a functor PolyRedRel(P,T) returning an
object of type Relation of. Note that the ring of polynomials which is available
according to section 4.1 is used to describe the domain of the relation.

```
definition
let n be Ordinal, T be connected TermOrder of n,
    L be Field, P be Subset of Polynom-Ring(n,L);
func PolyRedRel(P,T) ->
  Relation of (the carrier of Polynom-Ring(n,L)) \ {0_(n,L)},
              the carrier of Polynom-Ring(n,L) means
for p,q being Polynomial of n,L holds [p,q] in it iff p reduces_to q,P,T;
end;
```

Now, PolyRedRel(P,T) being of type Relation — the type Relation of
widens to Relation — allows reuse of the whole theory, that is both notations and theorems developed for rewriting systems, for polynomial reductions.
For example, to show that polynomial reduction is terminating we just use the
attribute strongly-terminating defined for arbitrary reduction systems and
prove in a cluster registration that PolyRedRel(P,T) fulfils it for an arbitrary
set P of polynomials and an arbitrary term order T:

```
registration
let n be Nat, T be connected admissible TermOrder of n,
    L be Field, P be Subset of Polynom-Ring(n,L);
cluster PolyRedRel(P,T) -> strongly-normalizing;
end;
```

Also, showing that polynomial reduction with respect to a set of polynomials
P describes the congruence given by the ideal generated by P — a necessary
precondition to decide ideal membership with reduction techniques — needs no
further preparations: The reflexive symmetric transitive closure of the reduction
relation is given by the predicate are_convertible_wrt from rewriting theory,
whereas generated ideals and their congruences — the functor P-Ideal and the
predicate are_congruent_mod — are reused from general ideal theory. We thus
get the following

```
theorem
for n being Nat, T being admissible connected TermOrder of n,
    L being Field, P being non empty Subset of Polynom-Ring(n,L),
    f,g being Element of Polynom-Ring(n,L)
holds f,g are_congruent_mod P-Ideal
      iff f,g are_convertible_wrt PolyRedRel(P,T);
```

5   Mizar Formalization of Gröbner Bases

We start with a brief introduction to Gröbner bases, see also [BW93, CLO'S96].
Let $K[X_1, \ldots X_n]$ be the ring of polynomials over a field K with n indeterminates. For $P \subseteq K[X_1, \ldots X_n]$ the ideal generated by P — the minimal ideal
including P — is given by $<P> = \{\sum_{i=0}^{n} f_i \cdot p_i \mid f_i \in K[X_1, \ldots X_n],\ p_i \in P\}$.
The basic problem that can be algorithmically solved using Gröbner bases is the
following: Given $f \in K[X_1, \ldots X_n]$ and $P \subseteq K[X_1, \ldots X_n]$, does $f \in <P>$ hold?
Denoting the reduction for polynomials introduced in section 4.3 by $\longrightarrow_P$ it is
easy to show that $f \overset{*}{\longrightarrow}_P 0$ implies $f \in <P>$. The other direction, however,
does not hold in general and can actually serve as a definition for Gröbner bases.
In our formalization we use the equivalent definition, that $G \subseteq K[X_1, \ldots X_n]$ is
a Gröbner base if and only if $\longrightarrow_G$ is locally confluent. Note again, that $\longrightarrow_G$
is terminating.

To check whether a given (finite) set $P \subseteq K[X_1, \ldots X_n]$ is a Gröbner base it
is sufficient to consider the (finite set of) s-polynomials generated by P, that is

$$\mathrm{spoly}(p_1, p_2) = \mathrm{HC}(p_2) \cdot \frac{t}{\mathrm{HT}(p_1)} \cdot p_1 - \mathrm{HC}(p_1) \cdot \frac{t}{\mathrm{HT}(p_2)} \cdot p_2$$

where $t = lcm(\mathrm{HT}(p_1), \mathrm{HT}(p_2))$ for all $p_1, p_2 \in P$: G is a Gröbner base if we have
$\mathrm{spoly}(p_1, p_2) \overset{*}{\longrightarrow}_G 0$ for all $p_1, p_2 \in P$, which in the view of general rewriting
can be interpreted as checking critical pairs. This gives rise to a completion
algorithm: If $\mathrm{spoly}(p_1, p_2)$ not reduces to 0, its normal form is added to P — note
that $p_1, p_2 \in P$ implies $\mathrm{spoly}(p_1, p_2) \in <P>$, so that the generated ideal $<P>$
is not changed — and s-polynomials are recursively computed. This is the basic
version of Buchberger's Algorithm transforming a set $P \subseteq K[X_1, \ldots X_n]$ into a
set $G \subseteq K[X_1, \ldots X_n]$ such that $<P> = <G>$ and $\longrightarrow_G$ is locally confluent.
Further investigations and improvements of the algorithm can be found in the
literature (see for example [Buc79]).

In the following we present the main results of our formalization in Mizar so
far. Besides the definition of Gröbner bases and the usual characterization using
s-polynomials, we also considered other characterizations and the existence of
both ordinary and reduced Gröbner bases.

5.1   Definition and Characterizations

A Gröbner base for a given ideal I is a set G of polynomials such that the induced reduction relation `PolyRedRel(G,T)` is locally confluent (hence a complete
rewriting system) and the ideal generated by G equals I. Note again, that the
term order T is a parameter of the reduction relation.

```
definition
let n be Ordinal, T be connected TermOrder of n,
    L be Field, G,I be Subset of Polynom-Ring(n,L);
pred G is_Groebner_basis_of I,T means
  G-Ideal = I & PolyRedRel(G,T) is locally-confluent;
end;
```

We proved a number of further characterizations of Gröbner bases from [BW93], so for example, that G is a Gröbner base if each polynomial in G-Ideal is top-reducible with respect to G, if each polynomial in G-Ideal is reducible to the zero polynomial 0_(n,L) or if each head term of a polynomial in G-Ideal is divided by a head term of a polynomial in G. The main property of Gröbner bases G — ideal membership of a polynomial p is decidable by reducing p with respect to G — can be formulated as follows.

```
theorem
for n being Nat, T being connected admissible TermOrder of n,
    L being Field, p being Polynomial of n,L,
    G being non empty Subset of Polynom-Ring(n,L)
st G is_Groebner_basis_wrt T
holds p in G-Ideal iff PolyRedRel(G,T) reduces p,0_(n,L);
```

A completely different characterization of Gröbner bases, that is often used to prove more involved theorems, relies on so-called standard representations of polynomials [BW93]. A standard representation of a polynomial p with respect to a set of polynomials P is a linear combination $p = \sum_{i=1}^{k} m_i p_i$ where the m_i are arbitrary monomials, the p_i are from the set P and the head terms of the $m_i p_i$ are bounded by HT(p,T), or more general by a given term t. The concept of linear combinations again can be reused from ring theory [BRS00]:

```
definition
let L be non empty multLoopStr, S be non empty Subset of L;
mode LeftLinearCombination of S -> FinSequence of the carrier of L means
    for i being set st i in dom it
    ex u being Element of L, s being Element of S st it/.i = u * s;
end;
```

A standard representation of a polynomial f is then straightforwardly defined as a LeftLinearCombination of P with the two additional conditions from above. Now one can show that a set G is a Gröbner base if and only if there exists a standard representation for each polynomial in the ideal generated by G. Defining a predicate f has_a_Standard_Representation_of G,t,T with the obvious meaning we thus get

```
theorem
for n being Nat, T being connected admissible TermOrder of n,
    L being Field, G being non empty Subset of Polynom-Ring(n,L)
holds G is_Groebner_basis_wrt T
    iff for p being Polynomial of n,L st p in G-Ideal
        holds p has_a_Standard_Representation_of G,HT(f,T),T;
```

5.2   Construction of Gröbner Bases

The key point in the construction of Gröbner bases is the observation that there exists a finite test to check whether a set of polynomials G is locally confluent, hence a Gröbner base. Critical situations that have to be checked are given by s-polynomials describing the "difference" between two polynomials p1 and p2. The Mizar definition is as follows. Note again that p1,p2 ∈ G implies S-Poly(p1,p2,T) ∈ G-Ideal.

```
definition
let n be Ordinal, T be connected TermOrder of n,
   L be Field, p1,p2 be Polynomial of n,L;
func S-Poly(p1,p2,T) -> Polynomial of n,L equals
  HC(p2,T) * (lcm(HT(p1,T),HT(p2,T))/HT(p1,T)) *' p1 -
  HC(p1,T) * (lcm(HT(p1,T),HT(p2,T))/HT(p2,T)) *' p2;
end;
```

Now, if for a given set G of polynomials we have that PolyRedRel(G,T) reduces S-Poly(g1,g2,T) to 0_(n,L) for all p1,p2 ∈ G, then G is a Gröbner base. Note that if G is finite there exist only finitely many s-polynomials. Using the transition lemma, basically stating that if a polynomial p1-p2 is reducible to 0_(n,L) with respect to G, then there exists a polynomial q such that PolyRedRel(G,T) reduces both p1 and p2 to q, we proved the following

```
theorem
for n being Nat, T being admissible connected TermOrder of n,
   L being Field, G being Subset of Polynom-Ring(n,L)
holds (for p1,p2 being Polynomial of n,L st p1 in G & p2 in G
       holds PolyRedRel(G,T) reduces S-Poly(p1,p2,T),0_(n,L))
implies G is_Groebner_basis_wrt T;
```

This theorem gives rise to a completion algorithm to compute Gröbner bases (see [Buc98]). Note, that the proof of the theorem's opposite direction is almost trivial using the characterizations from the section 5.1.

For the construction of Gröbner bases, however, not all s-polynomials need to be considered, hence detecting such s-polynomials saves a number of reductions in the construction process. In the literature theorems characterizing such situations can be found (see e.g. [Buc79]). We formalized a first theorem into this direction stating that s-polynomials of polynomials p1 and p2 with lcm(HT(p1,T),HT(p2,T)) = HT(p1,T) · HT(p2,T), in other words the head terms of p1 and p2 have no variables in common, need not be considered, they always reduce to the zero polynomial:

```
theorem
for n being Ordinal, T being connected admissible TermOrder of n,
   L being Field, p1,p2 being Polynomial of n,L
st HT(p1,T),HT(p2,T) are_disjoint
holds PolyRedRel({p1,p2},T) reduces S-Poly(p1,p2,T),0_(n,L);
```

5.3   Existence of Gröbner Bases

Finally we consider the existence and uniqueness of Gröbner bases. It is a theoretically interesting fact that a finite Gröbner base exists for any given ideal I; or from a rewriting point of view that there exists a (finite) completion for every set

G of polynomials. Using the characterization of section 5.1 — that G is a Gröbner base if each head term of a polynomial in G-Ideal is divided by a head term of a polynomial in G — and Dickson's lemma from [LR02] it is easy to prove the following

```
theorem
for n being Nat, T being connected admissible TermOrder of n,
    L being Field, I being Ideal of Polynom-Ring(n,L)
ex G being finite Subset of Polynom-Ring(n,L)
st G is_Groebner_basis_of I,T;
```

which actually is another formulation (and another proof) of the Hilbert basis theorem. Note that the theorem states even more, namely that a Gröbner base for a given ideal exists for any total admissible term order T.

We also considered reduced Gröbner bases. In general a Gröbner base is of course not uniquely determined by the ideal I, even if we choose a fixed term order T. However, introducing the concept of reduced Gröbner bases, the situation looks different. A set G of polynomials is called reduced, if every p ∈ G is monic, that is HC(p,T) = 1 for all p ∈ G, and every p ∈ G is irreducible with respect to G\{p}. Note that only the second condition can be reused from rewriting theory, the other being a property of polynomials. Using a predicate is_reduced_wrt we proved the following theorems showing existence and uniqueness of reduced Gröbner bases.

```
theorem
for n being Nat, T being connected admissible TermOrder of n,
    L being Field, I being Ideal of Polynom-Ring(n,L) st I <> {0_(n,L)}
ex G being finite Subset of Polynom-Ring(n,L)
  st G is_Groebner_basis_of I,T & G is_reduced_wrt T;

theorem
for n being Nat, T being connected admissible TermOrder of n,
    L being Field, I being Ideal of Polynom-Ring(n,L),
    G1,G2 being non empty Subset of Polynom-Ring(n,L)
st G1 is_Groebner_basis_of I,T & G1 is_reduced_wrt T &
    G2 is_Groebner_basis_of I,T & G2 is_reduced_wrt T
holds G1 = G2;
```

6   Mathematical Knowledge Repositories

6.1   Mizar Mathematical Library

The Mizar Mathematical Library [Miz05] is a long term project that aims at developing both a comprehensive library of mathematical knowledge and a formal language for doing so. At the time of writing the library consists of 904 articles stating about 40000 theorems and 8000 definitions. Also because of the huge number of covered areas Mizar is well-suited for our experiments concerning building up new developments on existing ones. In the following we discuss some issues of our formalization which we consider of general interest for the development of mathematical repositories.

As a first point, we want to stress that adopting notations is a crucial issue when building mathematical repositories. By "adopting" we mean not only reusing notations in a more specialized situation, but also slightly changing and extending these. We illustrate this with reduced sets of polynomials already mentioned in section 5.3. Irreducibility of sets stems from rewriting and can be defined as follows.

```
definition
let R be Relation, A be set;
pred A is_irreducible_wrt R means
  for a being Element of A holds a is_a_normal_form_wrt R;
end;
```

Reduced sets of polynomials, though based on this notion, are somewhat different: Each polynomial must be irreducible with respect to all other polynomials. Furthermore only monic polynomials are considered here. Hence, the predicate concerning reduction has not only to be refined but also to be extended. In Mizar this can be straightforwardly done as follows.

```
definition
let n be Ordinal, T be connected TermOrder of n,
    L be Field, P be Subset of Polynom-Ring(n,L);
pred P is_reduced_wrt T means
  for p being Polynomial of n,L st p in P
  holds p is_monic_wrt T & {p} is_irreducible_wrt PolyRedRel(P\{p},T);
end;
```

Note, that the reduction relation `PolyRedRel` is used with the argument P\{p} rather than P. We believe, that it is this flexibility that we need to built up large repositories covering not only few theories.

Equivalence proofs are often cyclic, that is actually given by a number of implications. Of course this can be easily mirrored in Mizar (or other repositories) by stating each implication as a theorem. However, using such equivalences then becomes rather tedious, because to get an equivalent formulation more than one theorem is necessary. In [BW93], for example, theorem 5.35 gives 10 equivalent characterizations of Gröbner bases, so that using these equivalences requires up to 9 theorems in Mizar. Here, it might be helpful to extend the language of mathematical repositories to also include "equivalence theorems".

A last point we want to mention concerns the general development of repositories. Most projects are concerned with the formalization of a particular theorem to illustrate the usability of a certain approach. Therefore, for obvious reasons, often parts or theorems actually belonging to the theory considered are ignored just because they are not really necessary to prove the goal. We believe that the development of a general mathematical repository as necessary for mathematical knowledge management has to go another way: Often it turns out that the parts left out would enable a better development beyond the theorem originally chosen. So we should seek for completeness in the sense that when formalizing a theory we should bear alternative characterizations in mind. An example here are standard representations. Of course one can define Gröbner bases and their construction by s-polynomials without using standard representations. To prove more involved results on s-polynomials, however, one finds that in the literature often standard representations are used, which implies that the definition of standard representations cannot be left out.

6.2   Other Formalizations

Gröbner bases and polynomials have been defined in other systems; we first mention [The01], where Buchberger's algorithm is formalized using the Coq proof assistant. From this development an implementation of the algorithm in Ocaml has been extracted. In [MPAR04] a Common Lisp implementation of Buchberger's algorithm is presented that has been verified in Acl2. This is part of a larger project that aims at the computational formalization of polynomial algorithms in the spirit of combining computer algebra and theorem proving. Harrison [Har01] presents a Gröbner base algorithm for complex polynomials in HOL and uses it as a semi-decision procedure for polynomial equations in his work on quantifier elimination. Focusing on the algorithm, however, the notations of both rings and rewriting are not introduced, but defined from scratch for polynomials only.

C-CoRN [CGW04] also includes polynomials as a result of the "Fundamental Theroem of Algebra"-Project. Here a real number structure is used is used to develop polynomials in Coq so that an instantiation with a construction of the real numbers results in a full constructive proof. Polynomials have been defined in other repositories such as for example IMPS [FGT93] and Theorema [Buc01]. However, none of these approaches has been used to develop Gröbner bases so far. It would be interesting to do so and to compare these developments with our experiences in Mizar.

7   Conclusions and Further Work

We believe that mathematical repositories serves at least two goals. Firstly, of course, repositories form the basis for other Mathematical Knowledge Management activities by providing the knowledge to deal with. Secondly, it seems to us that mathematical repositories are also the key for attracting mathematicians and other users: The more knowledge we include in our repositories, the more likely will be the acceptance of both mathematical repositories and its attached software. Therefore developing mathematical repositories should

- be broadly based, that is a large number of different fields of mathematics has to be covered.
- be highly reusable and refinable to impress possible users how easily one can adopt existing developments, in particular basic theories.
- aim at describing (basic) theories as completely as possible — and not only at developing the proof a special theorem — in order to increase the number of possible users.

We also believe that such development techniques will lead to the formalization of contemporary mathematics easier just because a broad basic repository supports — and is necessary for — more involved mathematics.

In this paper we have presented a case study in Mizar to illustrate what such a broad theory development may look like. The Mizar type mechanism, especially the possibility to extend types with adjectives to describe additional properties, elegantly supports the refinement and reuse of existing developments and theories: Adjectives allow not only to refine theories as a whole. In addition theorems itself can be formulated using only properties, that is adjectives, necessary to prove them and can therefore be reused in every theory fulfilling these adjectives. Because this kind of reasoning is present in nearly all areas of mathematics, we claim that such a flexible type system is of major importance for the development of mathematical repositories.

The work presented in this paper can be continued in two ways. Firstly, the theory of Gröbner bases in Mizar should be further developed: Theorems concerning avoiding s-polynomials should be formalized, also to explore the use of "non-standard characterizations", here by standard representations. Also, in the spirit of section 6.1, the characterization of Gröbner bases by division with remainder and of course generalizations of the topic such as for example syzygies are of further interest. Secondly, it would be interesting to transform the formalized material into an introductionary course on Gröbner bases. The main point here would be to take into account both the underlying ring and rewrite theories and the proofs as they have been written in the Mizar language. We think that this would not only give insights in how to use Mizar for generating teaching material, but also — due to the number of theories involved — how to structure courses with larger number of prequisites.

References

[Ban95]   G. Bancerek, Reduction Relations; Formalized Mathematics, 1995, available in JFM from [Miz05].

[Ban03]   G. Bancerek, On the Structure of Mizar Types; in: H. Geuvers and F. Kamareddine (eds.), Proc. of MLC 2003, ENTCS 85(7), 2003.

[BW93]   T. Becker and V. Weispfenning, Gröbner Bases — A Computational Approach to Commutative Algebra; Springer Verlag, 1993.

[BRS00]   J. Backer, P. Rudnicki, and C. Schwarzweller, Ring Ideals; Formalized Mathematics, 2000, available in JFM from [Miz05].

[Buc79]   B. Buchberger, A Criterion for Detecting Unnecessary Reductions in the Construction of Gröbner Bases; in: Proceddings of Eurosam 79, Lecture Notes in Computer Science 72, pp. 3-21, 1979.

[Buc98]   B. Buchberger, Introduction to Gröbner bases; in: B. Buchberger and F. Winkler (eds.), Gröbner Bases and Applications, pp. 3-31, Camebridge University Press, 1998.

[Buc01]   B. Buchberger, Mathematical Knowledge Management in Theorema; in: B. Buchberger, O. Caprotti (eds.), Proceedings of the First International Workshop on Mathematical Knowledge Management, Linz, Austria, 2001.

[CGW04]   L. Cruz-Filipe, H. Geuvers, and F. Wiedijk, C-CoRN, the Constructive Coq Repository at Nijmegen; http://www.cs.kun.nl/~freek/notes/.

[CLO'S96]   D. Cox, J. Little, and D. O'Shea, Ideals, Varieties and Algorithms; Springer Verlag, New York, 2nd edition, 1996.

[Dav81]   M. Davies, Obvious Logical Inferences; in: Proceedings of the 7th International Joint Conference on Artificial Intelligence, pp. 530-531, 1981.

[DJ90]   N. Dershowitz and J.P. Jounnaud, Rewrite Systems; in: J. van Leeuven (ed.), Formal Models and Semantics — Handbook of Theoretical Computer Science, vol. B, Elsevier, 1990.

[FGT92]   W. Farmer, J. Guttman, and F. Thayer, Little Theories; in: D. Kapur (ed.), Automated Deduction – CADE-11, LNCS 607, pp. 567–581, 1992.

[FGT93]   W. Farmer, J. Guttman, and F. Thayer, IMPS – An Interactive Mathematical Proof System; Journal of Automated Reasoning 11, pp. 213–248, 1993.

[GS04]   A. Grabowski and C. Schwarzweller, Rough Concept Analysis — Theory Development in the Mizar System; in: G. Bancerek, A. Asperti, and A. Trybulec (eds.), Proceeding of the Third International Conference on Mathematical Knowledge Management, Lecture Notes in Computer Science 3119, pp. 130-145, 2004.

[Har01]   J. Harrison, Complex Quantifier Elimination in HOL; in: supplementary Proceedings of the 14th International Conference on Theorem Proving in Higher Order Logic, pp. 159-174, 2001.

[Jaś34]   S. Jaśkowski, On the Rules of Suppositon in Formal Logic; in: Studia Logica, vol. 1, 1934.

[LR02]   G. Lee and Piotr Rudnicki, Dickson's Lemma; Formalized Mathematics, 2002, available in JFM from [Miz05].

[Miz05]   The Mizar Home Page, http://mizar.org.

[MPAR04]   J. Medina-Bulo, F. Paloma-Lozano, J. Alonzo-Jiménez, and J.-L. Ruiz-Reina, Verified Computer Algebra in Acl2: Gröbner bases; in: B. Buchberger and J. Campbell (eds.), 7th Conference on Artificial Intelligence and Symbolic Computation (AISC04), Lecture Notes in Computer Science 3249, pp. 171-184.

[RST01]   P. Rudnicki, C. Schwarzweller, and A. Trybulec, Commutative Algebra in the Mizar System; in: Journal of Symbolic Computation vol. 32(1/2), pp. 143-169, 2001.

[RT99]   P. Rudnicki and A. Trybulec, Multivariate polynomials with arbitrary number of variables; Formalized Mathematics, 1999, available in JFM from [Miz05].

[RT01]   P. Rudnicki and A. Trybulec, Mathematical Knowledge Management in Mizar; in: B. Buchberger, O. Caprotti (eds.), Proc. of MKM 2001, Linz, Austria, 2001.

[Tar39]   A. Tarski, On Well-Ordered Subsets of Any Set; in: Fundamenta Mathematicae, vol. 32, pp. 176-183, 1939.

[The01]   L. Théry, A Machine-Checked Implementation of Buchberger's Algorithm; in: Journal of Automated Reasoning 26, pp. 107-137, 2001.

An Interactive Algebra Course with Formalised Proofs and Definitions*

Andrea Asperti[1], Herman Geuvers[2], Iris Loeb[2],
Lionel Elie Mamane[2], and Claudio Sacerdoti Coen[3]

[1] Dept. of Comp. Science, University of Bologna, Italy
[2] ICIS, Radboud University Nijmegen, NL
[3] Project PCRI, CNRS, École Polytechnique, INRIA,
Université Paris-Sud, France

Abstract. We describe a case-study of the application of web-technology (Helm [2]) to create web-based didactic material out of a repository of formal mathematics (C-CoRN [5]), using the structure of an existing course (IDA [4]). The paper discusses the difficulties related to associating notation to a formula, the embedding of formal notions into a document (the "view"), and the rendering of proofs.

1 Introduction

One of the aims of the recently concluded European IST Project MoWGLI was the development of a suitable technology supporting the creation of web-based didactic material out of repositories of formal mathematical knowledge.

In particular, the validation activity reported in this paper consists of the application of the Helm [2] technology, developed at the University of Bologna, to the C-CoRN [5] repository of constructive mathematics of the University of Nijmegen and aiming at the creation of an interactive algebra course.

The Helm system [2] provides tools and techniques for displaying formalised mathematics on the web. It uses XML technology for rendering repositories of formal mathematics, supporting hyperlinks, browsing and querying functionalities, as well as a sophisticated stylesheet mechanism for the notational reconstruction of the symbolic content.

The C-CoRN system [5] is a repository of constructive mathematics, formalised in Coq, covering a considerable body of basic algebra and analysis.

Potentially, Helm and C-CoRN produce a large body of formalised mathematics, which can be viewed and browsed through standard web tools. In order to exploit this potentiality, Helm must be suitably *instantiated* to the particular case of C-CoRN. This instantiation essentially takes place at two levels:

notational defining (directly or indirectly) a set of XSLT transformations providing the required notational rendering for the formal notions coded in C-CoRN;

---

* This research was supported by the European Project IST-33562-MoWGLI.

structural providing the didactic organisation and the natural language glue of the course notes

This paper is a report of the work. The final result is available at `http://helm.cs.unibo.it/`. The structure of the paper is the following. Sect. 2 and 3 introduce respectively the Helm system and the C-CoRN repository; in Sect. 4 we discuss the association of notation to a formula and Sect. 5 is about documents as views; the rendering of proofs is addressed in Sect. 6. Finally we draw some general conclusions about this validation activity.

2   Helm

The process of transforming a Coq proof to a XHTML or MathML-Presentation proof is shown in Fig. 1. The process is essentially split in two parallel pipelines, respectively dealing with *proof objects*, i.e. single mathematical items such as theorems, definitions, examples and so on, and *views* that are structured collections of (links to) objects, possibly intermixed with text and pictures.

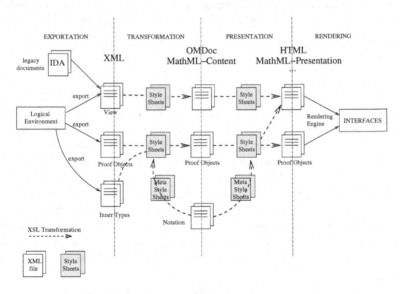

Fig. 1. Transformation process in Helm

From the Coq files, the lambda-term is exported to an XML-language, CICML. Similarly, from scripts we export a sort of minimal, canonical view, that is essentially the index of theorems in the order they have been defined by the user, plus some special comments possibly added to the script. The XML exportation module is currently a standard component of the Coq distribution.

An alternative, simple way to produce a view is by starting from some legacy documents, for instance using a traditional latex to HTML converter (links to the repository have to be added manually). In Section 5 we shall describe in detail this job, adopting, as a view, the HTML material of the Interactive Algebra Course [4] on algebra of the Eindhoven University of Technology (NL).

The raw XML encoding used by Helm to store the information in the repository is transformed into a presentational markup by a suitable set of stylesheets. This transformation process is split, for modularity reasons, in two main parts, passing through an intermediate "content" markup, which is provided by Content-MathML for statements, and OMDoc for Views. The idea is that different foundational dialects may be mapped to a same content description, and on the other side, the same content can be translated to different presentational markups.

The choice of using XSLT for performing the transformation (see also [1]) is mainly motivated by the XML framework of the project (MoWGLI was explicitly conceived as a major validation test for XML technology). However, the limited expressive power of XSLT also combines well with a major philosophical commitment of the Helm project, namely that rendering must be a simple operation, not involving major transformations on the proof. The point is that if the rendering operation is too sophisticated we may loose confidence in what has been actually proved in the machine.

3   The C-CoRN Repository

The C-CoRN repository [5] consists of a Coq formalisation of basic algebra and basic constructive analysis. The algebra covers an algebraic hierarchy consisting of semi-groups, monoids, groups, rings, fields, ordered fields and metric spaces. Apart from algebra, there is a large amount of analysis, including real numbers, complex numbers, polynomials, real valued functions, differentiation / integration and basic results from analysis like the intermediate value theorem, Taylor's theorem up to the fundamental theorem of calculus and the fundamental theorem of algebra (using both algebra and analysis).

4   Mathematical Notation

Part of the descriptive power of mathematics derives from its ability to represent and manipulate ideas in a complex system of two-dimensional symbolic notations refined during centuries of use and experience.

Especially for didactic reasons it is important to adopt a mathematical notation as close as possible to the usual notation in textbooks. As we will see in Sect. 5.2, this notation is much more dynamic and context dependent than it appears at first sight, posing very interesting and challenging problems not yet solved by the current technology.

A friendly and concise mathematical notation is also required during the formal development of the proofs in Coq. In particular, the syntax and the notational expressivity of Coq have greatly changed in the last releases of the system. However, the priorities are different with respect to a course. In particular, Coq requires notation to be perfectly unambiguous in every situation, severely limiting overloading. It also restricts notation to a mono-dimensional language. As a consequence the notation adopted for the development of C-CoRN and the one that is used in the electronic course will not be the same, preventing an automatic translation in the general case.

4.1   Notation in Helm

The process of associating notation to a formula during rendering is made of two phases. The two phases are both XSLT transformations, since the formula, expressed as a term of the logic of Coq, the Calculus of (Co)Inductive Construction (CIC), is stored in XML. The first transformation is a semantically lossy operation that maps the formula to a MathML Content expression that captures its intended meaning. The second one maps MathML Content to either XHTML or MathML Presentation, only the latter giving access to the most complex bidimensional notations.

The stylesheet to MathML Presentation is quite standard, but for the handling of the layout. Indeed the stylesheet automatically breaks long formulae on multiple lines, exploiting the MathML Content expression to decide where to break the lines and how to indent the expression. Greatly suboptimal with respect to the most natural layout a human can provide, our strategy is superior to the trivial algorithms implemented in the browsers that cannot exploit the content expression. However, layouting greatly increases the complexity of the stylesheet: every time a notation can be applied a template is invoked to estimate the size of the subexpressions and decide if line breaking and consequent indentation are necessary. For instance, detecting and rendering the "less than" relation requires respectively 7, 139 and 142 lines of XSLT for the CIC to MathML Content, MathML Content to XHTML and MathML Content to MathML Presentation transformations.

Providing a new notation, especially in those cases where the notation is quite standard, should not require more than a few seconds, nor any major knowledge from the user, who is not supposed to write about 300 lines of XSLT for each mathematical operator. Since the stylesheets have a very simple and repetitive structure it is possible to automatically generate them from a concise description of the operator, including, for instance, its arity, associativity and type (infix/prefix/postfix) for non binding operators that have a mono-dimensional notation.

To automatically generate stylesheets, Helm provides two intermediate XML languages to describe notation and a set of transformations (in XSLT) which translate the first simplified language in the second, and the second to the three stylesheets CIC to MathML Content, MathML Content to XHTML and MathML Content to MathML Presentation (see [7]). Moreover, links are automatically added from one level — say, XHTML — to the generating expression at the previous level — MathML Content —, and from the occurrence of the operator (in MathML Presentation or XHTML) to the CIC document where it is defined. Describing the notation for the "less than" operator in the simplified language simply amounts in giving: 1) its arity; 2) its type; 3) its associativity; 4) its URI (unique identifiers) of the operator at the CIC level; 5) the name of the MathML Content element that it must be mapped to; 6) the Unicode symbol that it must be mapped to for presentation. This information can easily be provided by the user by directly editing the XML file. Trusted Logic has also implemented a tool to generate this information from the corresponding one used by Coq to describe notations [6]. The simplified language also recognises binding operators, operators that have arguments that must be left implicit and operators that are rendered as the negation of other operators.

When the simplified language is not expressive enough to describe the notation, the user can directly use the intermediate language, writing the expected representation directly in MathML Presentation (or HTML) extended with special macros to suggest desired positions for line breaking, to insert the required links, to process recursively the sub expressions and so on.

At first the automatic generation of notational stylesheets seems to definitively solve in an elegant way the problem of defining new notations. However, in Sect. 5.2 we will consider the problem of making notations evolve within a document, facing the current limitations of the Helm technology.

4.2   Notation in Natural Language for Predicates

Automatic stylesheet generation has been conceived to translate CIC expressions to their usual mathematical symbolic rendering, for instance to replace "plus" by "+". The same technology can be exploited to provide a natural language like notation for predicates. For instance, a generated stylesheet can transform "(commutes plus nat)" into "+ is commutative on ℕ", with links from "+" to the definition of "plus", from "ℕ" to the definition of "nat" and from "is", "commutative" and "on" to the definition of "commutes".

Even if the descriptions generated for large formulae are far from being as fluent as those that a human would write, the results are pretty satisfactory and the statements are much easier to understand. The Mizar language has already successfully adopted a similar notation for predicates for years.

From the point of view of the user who must provide the notation, treating words in the same way as symbols is rather onerous. Moreover, the algorithm that automatically breaks a formula into several lines does not produce the expected output for natural language notations, where we would expect long sentences to simply continue on the next line. An improvement of the Helm technology is required to handle this kind of notation in a completely successful way.

5   Documents as Views

A course is much more than a set of definitions and theorems. It has its own buildup, which is directed by the rules of didactics and clarity of mathematical exposition. It may sometimes sacrifice formality to provide intuitions in the most direct way, or it may provide the intuitions first and the formal counterpart later on. It contains plenty of examples, exercises and rhetorical text, while omitting technical lemmas and results that will not be necessary or that will be presented only when used.

Thus, to develop our course notes from C-CoRN, it is not a reasonable idea to start from the script, the Coq development where definitions and proofs are given in their order of definition, and integrate it with examples and the rest. On the contrary we decided to start from the buildup of the course, seen as a huge hypertextual electronic "document" (divided into chapters, sections and the like) in which we embed formal definitions, lemmas, exercises and references to definitions and lemmas in the middle of informal sentences.

The embedding does not need to be done statically. On the contrary we may imagine a document — called *view* in the Helm terminology — where special elements are put where formal notions need to be embedded. The view can be written in any XML language we like, and XML namespaces are used to avoid confusion with the special elements. To visualise a view, the document is processed on the fly by the Helm processor and mapped to a standard XHTML page (eventually embedding MathML Presentation islands and made dynamic by means of JavaScript). All the special elements are expanded by the processor which embeds in the document the requested rendering of the formal notions.

To speed up the development time and to be sure of the didactic quality of the course obtained we decided to exploit the course organisation, the rhetorical text and all the exercises and examples (the latter to be first formalised in C-CoRN that used to lack examples completely) from IDA.

IDA [4] is an interactive course on algebra, which has been developed at the Eindhoven University of Technology (NL), for first year mathematics students. The IDA course notes have been developed over the years in the context of a first year class on algebra. It consists of a book with a CD-ROM. (Before the book + CD-ROM were published by Springer, β-versions of the material were available through the web.) The material on the CD-ROM is the same as in the book, but there is additional functionality:

- Alternative ways of browsing through the book.
- Hyperlinks to definitions and lemmas. However, since the hyperlinks have been inserted manually, many of them are missing, in particular in the textual flow.
- Applets that show algorithms (e.g. Euclid's algorithm).
- Multiple choice exercises.

IDA does not provide a formal treatment of proofs. Proofs in IDA are like proofs in ordinary mathematical text books: plain text, so we couldn't reuse any proof from IDA. However it is very instructive to compare the proofs in IDA, targeted at an audience of students (and developed in classical mathematics), with the formalised (constructive) ones in C-CoRN. Because the structures in IDA, like monoids and groups, are not as rich as the ones in C-CoRN, which include also an apartness relation, the C-CoRN proofs are in general a bit more involved than the original IDA ones. However, we have only come across one theorem that was false constructively.

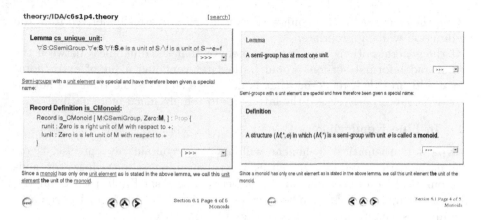

Fig. 2. Comparison between our course notes and IDA

Concretely, our goal was to obtain electronic course notes from the C-CoRN repository having the same look and feel as IDA, using Helm-tools to render mathematical objects (definitions, statements and their proofs) and to create hyperlinks between them.

Fig. 2 compares the page that defines a monoid in our course notes (on the left hand side) with the corresponding page in IDA (on the right hand side). The two pages are similar and the interactivity of the IDA page has been preserved: the lower frame is used to navigate in the notes and the drop down boxes give access to examples and exercises which are the same in the two courses. Our page is richer in functionality. First of all it adds another navigation frame on top which provides a breadcrumb trail and a link to the Whelp search engine. Whelp [3] is a search engine developed in MoWGLI to index and retrieve proofs and definitions in a formal library by means of powerful queries over the mathematical expressions in the statements. As IDA does not have any search or index facilities, this is a pure gain from the formalisation. We also observe that our course offers many more links to definitions than IDA. Indeed every occurrence of a formal concept is given its own link, both in the formal statements and definitions that are automatically generated from C-CoRN and in the free text that comes from IDA.

Finally we notice remarkable differences between C-CoRN and IDA for the statement of uniqueness of a unit and the definition of a monoid. One of the reasons for the increased verbosity of the formal definitions is that the formal library lacks or does not exploit a few notions that help to make the statements more concise. For instance, it would be possible to define the notion of uniqueness over a type T and a property over T, using it to state the lemma of uniqueness of the unit element. This is not normally done when working in Coq since Coq unification does not automatically expand this kind of general notions, making proofs much more cumbersome for the user. The same phenomenon appears in the definition of a monoid. Instead of relying on the notion of unit, in C-CoRN it is preferred to explicitly state that $\mathtt{Zero}$ is both a left and a right unit, since in Coq the constituents of a conjunction are not automatically derived from the conjunction when appropriate.

If the general notions were used, we believe that the differences between the formal and informal versions would be less relevant and a bit of extra notation would provide a rendering that is not more cumbersome than the human provided counterpart (even if the latter would remain much more natural).

5.1   In-Line Rendering

A typical mathematical document will not only contain raw, precise, rigorous mathematics, but also sentences whose meaning is more fuzzy or non-mathematical, somewhere in the range between normal English text and rigorous mathematics. This "free text" can contain for example historical remarks or remarks such as (this is an actual excerpt from IDA)

> Most binary operations in which we are interested distinguish themselves from arbitrary ones in that they have the following property.

The meaning of "in which we are interested" cannot be expressed in a theorem prover and escapes its checking framework, but that kind of text is still an important part of a mathematical document. However, it still speaks about and refers to formally defined mathematical notions ("binary operations"). It is thus desirable to be able to have free-form (non proof assistant checked) text that nevertheless uses names and notations of the proof assistant checked "library of mathematics" the document is about.

The initial implementation of the Helm system only allowed to refer to a mathematical notion by embedding its whole definition, as a separate paragraph, right there in the document. To complete our task we have implemented the possibility to have a rendering that

- integrates in a sentence, as a word or phrase, instead of being its own paragraph. We call this the *in-line* rendering.
- removes (or adds) some parts of the definition of the object that are included (or not included) by default, such as the name of the object, the body of the definition or the notation for the object.

These choices must be made by specifying attributes to the XML element that is used to provide a link to the formal notion and to ask for the embedding of its rendering in place of the element. Even if in principle all these choices are orthogonal, not every combination makes sense and gives a reasonable result.

To represent the free text of IDA in our course notes we have also sometimes explicitly chosen to avoid references to formal objects even if they were available. This was done in cases where some particular translation of the formal expression was necessary for the general flow of the document, to rise from the level of "sequence of mathematical statements" to the level of "document that tells a story". For example, the very first sentence of the chapter we have treated:

> The map that takes an element of $\mathbb{Z}$ to its negative is a unary operation on $\mathbb{Z}$, while addition and multiplication are binary operations on $\mathbb{Z}$ in the following sense.

Mathematically, this says the same as "$-_Z$ is a unary operation on $\mathbb{Z}$ and $+_Z$ is a binary operation on $\mathbb{Z}$ and $*_Z$ is a binary operation on $\mathbb{Z}$" (which is approximately how Helm would have rendered the corresponding formal mathematical statement). For a human reader, however, "while" does not just express a conjunction, as far as quality (e.g. ease of reading, clarity of point) of the document is concerned, even though "while" and "and" are equivalent mathematically. A document making absolutely no use of this kind of subtleties of language would be quite "dry" and hard to read for a human. Also notice the contraction of "$+_Z$ is a binary operation on $\mathbb{Z}$ and $*_Z$ is a binary operation on $\mathbb{Z}$" into "$+_Z$ and $*_Z$ are binary operations on $\mathbb{Z}$".

5.2   Context Depending Rendering

In Sect. 4 we have described the Helm facilities to provide mathematical notation easily, and we claimed that at first sight— when considering the rendering of a single definition or theorem at a time in a sort of vacuum — the machinery seems to be expressive enough. As soon as we started to consider the rendering of views, however, we realised that mathematical notation is much more dynamic and context dependent than what it seems to be at first sight. We will examine a few examples where the current context-free machinery of Helm is not sufficient.

Notations Depending on Lemmas. Stating a lemma can implicitly change the notation used in the rest of the view. The most well known example is proving the associativity of a binary operation. Until associativity is proved, every expression involving the operator must be fully parenthesised. However, once associativity is known, parentheses are omitted. Since "most of the time" associativity is known (i.e. the majority of the theorems depend logically on the lemma of associativity), we could have decided to always omit parentheses. However, this solution is not satisfactory since we sometimes face statements that in this way are reduced to trivialities. For instance, the lemma of associativity for $\circ$ would become $\forall x, y, z. x \circ y \circ z = x \circ y \circ z$.

Another similar annoying example can be found in the theory of semi-groups where first the notion of *being a unit* is defined and only after that the uniqueness of the unit in semi-groups is proved. So when we state the uniqueness of the unit, we would like to speak about "*a* unit", because the uniqueness has not yet been established:

Lemma 1. $\forall S{:}CSemiGroup.\forall e{:}S.\forall f{:}S.$ *e is a unit of $S \wedge f$ is a unit of $S \rightarrow e{=}f$*

But after we have proved the uniqueness, we would like to see "*the* unit" in all following uses.

It has also been argued that this is not really a problem, because differences in the English language — like "a" and "the" here — reflect also a difference in the mathematical meaning: "a unit" would denote a relation and "the unit" would denote a function. So, it has been proposed to use "a" whenever we see the relation, and to use "the" whenever we see the function. However, it is not clear that the correspondence between the English language and the mathematical meaning is as straightforward as suggested here. It is not a mistake to use the relation even after uniqueness has been established, but to use the indefinite article in English while we know the uniqueness, seems highly unusual.

Overloaded Notations. It is pretty common in mathematics to overload notations. However, sometimes the two semantics attached to a notation are needed at the same time and the context may not be sufficient to disambiguate. Then, the notation for at least one of the semantics must be changed, for example by further qualification. In Helm we can easily overload a notation and we can also provide qualified notations. However, the choice of using a qualified notation depends heavily on the context and cannot be made automatically.

Multiple Notations for (Instances of) Abstract Notions. When defining an abstract notion, say a group, a default notation is also usually defined to state the general theorems about the abstract notion. For instance, we could choose a multiplicative notation as the default for an abstract group. Later on abstract notions are instantiated to concrete ones and different notations should be applied to the different instances. For instance, for particular groups we often prefer an additive notation.

In the formalisation, a group becomes a record (a tuple with named projections) whose first element is the carrier and whose second element is the operation on the carrier. Thus an occurrence of the operation of a group is formally represented as the projection π_2 applied to the group G: $(\pi_2\ G)$. The operation applied to two arguments x and y becomes $(\pi_2\ G\ x\ y)$. In Helm notations must be associated to constants and are independent of the arguments the constants are applied to. In this case we can associate a notation to π_2 — say the additive notation — by saying that π_2 is a binary infix operator with an implicit argument. Thus $(\pi_2\ G\ x\ y)$ is represented as $x + y$. We could make the implicit argument explicit, representing the previous expression as $x +_G y$ equally easy (notice, however, that G can be a huge expression). However, we have no way of saying that $+$ must be replaced by $*$ for some particular groups G.

Having seen a few examples where the notation depends on the context, we will now discuss a possible enhancement of Helm to allow dynamic notation.

5.3   Extending Helm with Context Dependent Notation

As we have seen in the previous sections, there are several different kinds of dependencies on the context. Here we discuss the ways these could possibly be dealt with in Helm.

The first one is the case of temporary dependencies on lemmas and definitions, where a lemma or definition that occurs *in the view*, i.e. in a single page in our electronic notes, activates or changes a notation. Since the order of appearance in a view is different from "logical causality" (i.e. the partial order induced when a lemma refers to previously defined lemmas and definitions), we need to search for a solution by associating extra information to a view. In particular we may apply a batch process to a view that collects for each formal notion referred in the view all the formal notions that occur before it. Notice that a view can follow other views in the intentions of the authors, i.e. all the pages that precede it. Thus the batch process should visit all the views in their order of dependencies to collect the information to associate to objects. Once the information has been collected, we can identify for each object the notations that are active when presenting it and store this as metadata in the Helm database, associating it to the occurrence of an object in a particular view. The notational stylesheets can retrieve from the database the metadata (in RDF format) and apply only the templates supposed to be active.

Even if the previous solution seems satisfactory at first, it induces a new problem, e.g. when the user follows a hyperlink from an occurrence of a notion in a view to the definition of the notion. This definition exists in a vacuum with respect to the view that contains the hyperlink: a notion is not required to be defined in the views it is used in and it can be used in multiple views unrelated to the current one. This precludes any contextual metadata from being used in the decision of which notation to apply. Theoretically, the solution consists in considering the vacuum as a new view generated on the fly that inherits its metadata from the view the user is coming from. Practically, the Helm tools cannot handle this operation right now and implementing the new functionality requires a significant overhaul of the design of the tools.

The second case of contextual dependency, overloading, can be handled in a similar way. Indeed we can just give the user the possibility to associate to each lemma or definition metadata that says that some overloaded notation needs to be qualified. The main difference with the previous case is that the metadata is likely to be associated to the object in any possible view, and not to the occurrence of the object in a particular view.

The third case is the dependency of the notation on the arguments of the operator, which is much harder. Currently we have no practical solution. The problem becomes particularly complex when the user follows a hyperlink to the proof of a general statement. Since the statement was applied to an argument in the current view, its notation is supposed to depend on the argument. However, in the new window that shows the statement the actual arguments are universally quantified and the notation is likely to change accordingly to a default value (e.g. from additive to multiplicative). While this behaviour can be confusing, the situation on this point is exactly the same as with classical paper documents: The reference text book about a notion may use a notation that is different from a particular instance of the notion in another document. One would hope, however, that semantically rich electronic documents would improve the situation here.

6   Proofs

6.1   What Is a Formal Proof?

Rendering of formal proofs is a complex task. The first issue that arises when rendering formal proofs is: what kind of "object" is a formalised proof? This very much depends on the proof assistant used. The user interacts with a proof assistant via a so called *proof script*, the sequence of input commands that a user enters to prove a result. An important difference between proof assistants lies in the style of the language used in these scripts. Roughly speaking there exists two styles: *procedural* and *declarative*. In the procedural style (Coq, NuPRL, HOL, Isabelle), a user inputs *tactics*, commands that *modify the state of the system* (e.g. by applying a logical rule or a hypothesis or calling an automation procedure). So in a procedural proof style the user tells the system what to *do*. In the declarative style (Mizar, Isabelle/Isar) a proof script is essentially a sequence of intermediate results that *tell the system what our knowledge state is*. These go together with hints (known lemmas) on why the alleged intermediate result should hold. In a declarative proof style, the user tells the system where we *are*, and the built-in automation should verify that that's indeed the case.

In procedural proof assistants there may be another notion of proof, which is a *proof object* or proof term. This is a mathematical object (typically a typed lambda term) composed of only the very primitive logical rules. A tactic script creates a proof object, which is usually quite big, but has the advantage that it is directly verifiable by a small and trusted kernel. The Coq system, which we use in C-CoRN, has these proof objects.[1]

6.2   What Proof Do We Store and What Proof Do We Render?

The problem of procedural scripts is that the syntax and semantics of tactics is very system dependent and changes rapidly together with the overall system evolution. Declarative scripts are more robust, but similarly changes in the automation engine of the system can break a declarative script. As a consequence, scripts cannot be reasonably used for long term preservation of the information, and are also hardly reusable by any external, third party tool. Proof objects have a very clean and formal semantics, defined by the underlying foundational system (that is usually quite stable, even along the evolution of the tool). The main drawback of proof objects is that they are quite verbose and could be far away from the original proof of the user (which is usually better reflected by the tactic macrosteps of the proof script).

The Helm Project is particularly focused on the long term preservation and management of repositories of formal knowledge. To this aim, as explained above, the most relevant information is provided by proof objects. In the framework of the MoWGLI project we have developed an exportation module able to extract

---

[1] In systems like HOL and Isabelle no proof object is created, but in principle one could let the tactics create a proof object on the fly. In Mizar, the declarative proof itself is seen as *the* proof.

from the Coq proof assistant the raw lambda terms encoded into a suitable XML dialect[2]. Helm also provides stylesheets attempting a natural language reconstruction of the proof object; since the lambda term is isomorphic to a natural deduction proof, the result is, if not really appealing, surely readable.

In line with to the philosophy described in section 2, we make no major transformation on the input lambda term (the proof object) before presenting it. It is obvious that with a little effort we could easily improve the presentation by suppressing (or removing) a lot of detail. A typical example are proof objects that are found by an automatic decision procedure: these are usually extremely verbose and complex proofs that humans don't want to read. Not showing these subproofs' details by default would probably work well. However, in the proof object as such we can't trace the application of an automation tactic, so this would require a major transformation of the term before rendering it. A particularly fortunate case of an automation tactic in Coq is that of a *reflexive tactic*. This is not the place to go into detail about the nature of reflexive tactics, but the crucial point is that it does not create a huge proof term. Instead it encodes the goal to be proved as a syntactic object and then applies a generic lemma to an argument that solves the goal by mere computation. (And in Coq, computation does not require a proof.) So, to detect a reflexive tactic it is sufficient to recognise (in the stylesheets) an application of the generic lemma and hide its rendering.

6.3   The Actual Rendering of the Proofs

Thanks to the Helm exportation module we have produced proof terms for each C-CoRN theorem and we have use the standard Helm technology to render in our course the natural language generated from them. This uses the standard well-known transformation of typed lambda terms to natural deduction proofs. If rendering the proof term has not required major changes in the Helm technology, on the contrary we had to sensibly augment our library to match the textual flow of IDA. Below we give an example of a proof in C-CoRN, as rendered by the Helm tools. We observe that the C-CoRN proof is much more verbose than the one from IDA, which just reads

$$e = e + f = f.$$

On the other side, the C-CoRN proof gives the full formal details. Some of the details are "hidden" under a green link: clicking this link unfolds the details on request. In the black-and-white printout: these links are the crossed box before "Assumptions", which unfolds explicitly the local assumptions of the lemma, and the "Proof of $e = e + f$", which gives an explicit proof of this fact.

---

[2] We have also attempted to define a similar exportation functionality for Coq *proof trees*, which integrate information from the proof script with the proof objects, but we eventually failed. This failure was due to the complexity of the data structures (e.g. handling of bound variables) and the fact that the proof trees are much more subject to changes from one version of the system to the next.

```
cic:/CoRN/algebra/CSemiGroups/cs_unique_unit.con        [search]

DEFINITION cs_unique_unit()
TYPE =
    ∀S:CSemiGroup.∀e:S.∀f:S.e is a unit of S/\f is a unit of S→e=f
BODY =
    ▣Assumptions
    we must prove e is a unit of S/\f is a unit of S→e=f
    or equivalently (∀a:S.e+a=a/\a+e=a)/\(∀a:S.f+a=a/\a+f=a)→e=f
    suppose H: (∀a:S.e+a=a/\a+e=a)/\(∀a:S.f+a=a/\a+f=a)
        consider H
        we have:
        (H0) ∀a:S.e+a=a/\a+e=a
        (H1) ∀a:S.f+a=a/\a+f=a
        by (H0 .)
        we have:
        (H2) e+f=f
        (H3) f+e=f
        by (H1 .)
        we have:
        (H4) f+e=e
        (H5) e+f=e
        we have the following chain of (in-)equalities:
          e
            by
              Proof of  e=e+f
          = e+f
            by  H2
          = f
        we proved e=f
    we proved (∀a:S.e+a=a/\a+e=a)/\(∀a:S.f+a=a/\a+f=a)→e=f
    that is equivalent to e is a unit of S/\f is a unit of S→e=f
    we proved ∀S:CSemiGroup.∀e:S.∀f:S.e is a unit of S/\f is a unit of S→e=f
```

7   Conclusions

This research has been a test case for the use of a formal library, especially C-CoRN, in a mathematical document. It has also been a test case for the Helm tools in supporting the creation of such a document. This has led to interesting refinements of the Helm tools, as described in Section 5.

The use of the mathematics in the formal library has some huge advantages: one can use everything that is already in the library. In our case, not much from what we needed was present in C-CoRN when we started: only about 20% of the definitions, lemmas and examples we have used in formal form in our final document. Considering that the document covered about ten pages of basic mathematics and that we have not replaced every definition, lemma or example, that seems a bit disappointing. But it has been very useful to us anyway, because many items that were missing in the library were fairly easy to formalise from lemmas that were present. And by adding items to the library, we have made a contribution to future use. So, in short, it is beneficial to use a library because it is *economical*: if something has been defined or proved, it can be used over and over again. Nothing has to be done more than once.

It should also be noted that, using a formal library within an existing mathematical document (IDA) by putting references to it does note guarantee the *coherence* of the document. A mathematical document aims at satisfying a strict requirement: objects and lemmas are not to be used before they have been defined or proved. Of course, sometimes a lemma is used and is only proved later at a more suitable place, but this is mostly announced in the surrounding text and this can be seen as the use of a local assumption, instead of premature use of a lemma. But using an object or a lemma that is unfamiliar to the reader without any explanation, can be seen as a mistake. Because we use an existing library (C-CoRN) and in the (IDA) document we just put references to the library, we have paid no attention to this kind of logical coherence in the IDA document. It could well be that the order of the references does not meet the logical order of the corresponding items in the library. The fear for logical mistakes *between* items, as opposed to logical mistakes *within* items (proofs and definitions), is not imaginary. In the ten pages of IDA that we have looked at, the logical coherence has been violated several times. A positive aspect of our development is that the Helm-links from IDA provide the possibility to track the formal dependencies (inside C-CoRN) of the statements that we find in IDA.

The order in which the Helm system forces the user to make the document, i.e. first formalise the mathematics and only after that describe its rendering can sometimes be inconvenient. In Helm the link from the informal description to the formal object is made by the user, who writes (in the XML description) explicitly where the object can be found. This means that whenever an object in the library moves, the user has to alter its XML description. And objects do move, because the library is not a static object: theory files are reorganised and are being split up when they become very large. One may expect these libraries to stabilise at a certain point, but C-CoRN is still very much a dynamic library.

Acknowledgements. We thank the anonymous referees for their useful comments.

References

1. A. Asperti, L. Padovani, C. Sacerdoti Coen, I. Schena. *XML, Stylesheets and the re-mathematization of Formal Content.* Proceedings of "Extreme Markup Languages 2001 Conference", August 12-17, 2001, Montreal, Canada.
2. A. Asperti, L. Padovani, C. Sacerdoti Coen, F. Guidi, I. Schena. Mathematical Knowledge Management in HELM. Annals of Mathematics and Artificial Intelligence, 38(1): 27–46; May 2003.
3. A. Asperti, F. Guidi, C. Sacerdoti Coen, E. Tassi, S. Zacchiroli. A content based mathematical search engine: whelp. Submitted for publication.
4. A. Cohen, H. Cuypers, H. Sterk, *Algebra Interactive!*, Springer 1999.
5. L. Cruz-Filipe, H. Geuvers and F. Wiedijk, C-CoRN, the Constructive Coq Repository at Nijmegen. In: A. Asperti, G. Bancerek, A. Trybulec (eds.), *Proceedings of MKM 2004*, Springer LNCS 3119, 88-103, 2004.
6. E. Gimenez, *Validation 2: Smart Card Security*, MoWGLI deliverable, 2005.
7. P. Di Lena, *Generazione automatica di stylesheet per notazione matematica*, Master's thesis, University of Bologna, 2002.

Interactive Learning and Mathematical Calculus*

Arjeh M. Cohen, Hans Cuypers, Dorina Jibetean, and Mark Spanbroek

Technische Universiteit Eindhoven,
P.O. Box 513, 5600 MB Eindhoven, The Netherlands

Abstract. A variety of problems in mathematical calculus can be solved by recursively applying a finite number of rules. Often, a generic solving strategy can be extracted and an interactive exercise system that emulates a tutor can be implemented.

In this paper we show how software developed by us can be used to realize this interactivity. In particular, an implementation of a generic exercise for computing the derivative of elementary functions is presented.

1   Introduction

This paper deals with mathematical interactive learning. In most implementations of multi-step exercises for interactive learning the exercises are completely authored for each instance of the problem. Here we discuss an implementation of *generic* exercises for specific mathematical problems. This means that the exercise system provided will depend on the type of problem and not on the particular instance of the problem. In our example, the generic problem is *computing derivatives of elementary functions* while a particular instance of the problem is *computing the derivative of* $x \mapsto x^2 + 3$. (From here on, we shall omit the binding operation '$x \mapsto$' when referring to a function in x.)

For the design of automated exercises we use mathematical knowledge coming from two sources. Firstly, many problems in mathematical calculus can be solved by applying recursively a finite number of rules. The rules used for solving mathematical problems in a particular domain will be called *domain rules*. In our example, these will be the differentiation rules of composed functions (sum, product, quotient, chain rule) and the formulas for derivation of elementary non-composed functions (for instance, the derivative of sin equals cos).

Secondly, we need information on the particular instance of the problem, like the type of mathematical object we deal with and its definition. This information can be immediately deduced from its OpenMath [10] representation and it is typically sufficient to infer the domain rule(s) to be used for solving the problem in this particular instance. In fact, what we need is the map assigning to a mathematical (OpenMath) object the domain rule which needs to be applied. We will call this map a *domain reasoner*.

---

* Work carried out within the LeActiveMath project.

M. Kohlhase (Ed.): MKM 2005, LNAI 3863, pp. 330–345, 2006.
© Springer-Verlag Berlin Heidelberg 2006

In an interactive exercise, the student is required to solve a problem. Our strategy is to decompose the original problem into simpler sub-problems, so as to obtain a multi-step exercise. An interactive exercise can be seen as a collection of problems together with the order in which they are executed. An exercise always has a first step, corresponding to the original question. According to the student's answer and a predefined strategy, the next step is selected. In this way, the student is guided in solving the initial question. The correctness of the student's answer is evaluated by the use of a computer algebra system (CAS) connected to the exercise. There are quite a few interactive exercise systems which check automatically the correctness of the student's answer (such as [11], [7], [5]) but they are not as concerned with tutoring. These exercises consist of a single evaluation step. Some systems (like [7]) also provide automatically generated hints.

The novelty of our approach is that we produce generic exercises for certain mathematical problems by using the semantic information encoded in the Open-Math expression. Our approach can be applied to any mathematical problem for which a complete set of domain rules and a map from the OpenMath expression to the set of domain rules can be defined. We believe that this is possible for many problems in mathematical calculus.

The structure of the paper is as follows. In section 2 we present our general approach to interactive exercises, exemplified for the computation of derivatives. It shows how we use the domain rules and the OpenMath tree structure of the instance to generate the interactive exercise. Sections 3 and 4 contain details on the design of the interactive exercises, respectively on the set-up of the system running the exercise and some of the tools developed for this type of interactive exercise. In Section 5 we show our implementation for computing the derivative of an arbitrary (differentiable) univariate function. Section 6 discusses the applicability of the method to other problems in mathematical calculus. Section 7 presents ways to enhance the interactive exercise into detecting misconceptions of the student and other errors. Conclusions are presented in Section 8.

2   Interactive Exercises

Let us analyze the problem of computing the inverse of a given matrix using Cramer's rule. Here, a possible sub-problem is computing the determinant of the matrix. This helps for example to determine whether the inverse exists or not. Another sub-problem involves computing the elements of the inverse using Cramer's rule, as the fraction of two determinants. This exercise will consist of several steps, namely the original question, the problem of computing the determinant of a matrix, the problem of computing an element of the inverse (repeated n^2 times, where n is the size of the matrix) and, at last, an acknowledgment of the result.

There are however more complicated examples of calculus problems in which, although there is a finite number of procedures that can be applied, the order in which they are executed is not the same for any instance of the problem. In case few alternatives are possible, an interactive exercise should allow the students to choose the course of action they want to pursue.

The choice of the rules to apply depends in general on the particular instance of the problem. For example, in differentiating $\sin(x^2 + 1)$ one applies the chain rule and the elementary differentiation rules for sin and polynomials, while differentiating $x + \log(x)$ requires applying the sum rule and the elementary differentiation rules for the logarithm and the identity.

Still, a generic exercise can be implemented in such cases. Using the Open-Math tree structure of the expression of the function, the system can recognize the rule to be applied. Here we concentrate on derivatives, although in Section 6 we discuss the applicability of the method to other fields in mathematical calculus.

Derivatives are the *case study* considered by the European project LeActive-Math [2] for interactive, user-adapted e-learning. Much of the work reported in the present paper is carried out within the context of this project.

2.1 Derivatives

The domain rules in the case of derivatives are easy to derive. They are the differentiation rules of composed functions and the formulas for differentiation of elementary non-composed functions.

- sum: $(f + g)' = f' + g'$.
- product: $(fg)' = f'g + g'f$.
- quotient: $(f/g)' = (f'g - g'f)/g^2$.
- chain rule: $(f \circ g)' = (f' \circ g)\, g'$.
- $c' = 0$, where c is a constant.
- $id' = 1$, where id denotes the identity function.
- elementary non-composed functions: $\sin' = \cos$, $\exp' = \exp$, etc.

The OpenMath expression of a function contains all necessary information for computing the derivative of the respective function according to the above rules. One can determine whether the elementary function is composed or not, by the fact that a composed function has at least two OMAs in its OpenMath expression. In case the function is composed one can easily decide which one of the rules for composed functions needs to be applied. Let us consider for example the OpenMath expression of $\sin(x^2 + 1)$.

```
<OMOBJ>
  <OMA><OMS cd='transc1' name='sin'/>
    <OMA><OMS cd='arith1' name='plus'/>
      <OMA><OMS cd='arith1' name='power'/>
        <OMV name='x'/> <OMI>2</OMI>
      </OMA>
      <OMI>1</OMI>
    </OMA>
  </OMA>
</OMOBJ>
```

By looking at the root operator (the first OMS) it is clear that we either need to apply the chain rule if the function is composed or, otherwise, a differentiation rule for elementary functions. For $x+\log(x)$ the root of the OpenMath expression is the 'plus' symbol which suggests applying the sum rule. In this way we define a map matching the root of the OpenMath tree to the particular domain rule.

In Section 5, we discuss this generic exercise in greater detail.

3   The Design of an Interactive Exercise

This section explains the structure of an interactive exercise. Each exercise corresponds to a particular area in mathematical calculus. Specifically, we will have one interactive exercise for computing derivatives, one for computing limits, etc. Such an exercise is applicable to any instance of the problem (under some general assumptions) and it is in this sense generic.

We decompose a given problem into several sub-problems. Then we construct a directed graph having a source (start node) and a sink (end node). Each node of the graph corresponds to a sub-problem. Each arc shows a possible succession from one sub-problem to another one. The graph is constructed in such a way that each path from the source to the sink is a possible solution of the original problem. Therefore, we call the graph a solution graph and a path from the source to the sink a solution-path.

Each sub-problem has a text field, containing for example a question, and a user-input field, in which the student types the answer. The only exception is the sink which contains only the text field, to acknowledge the result. According to the input of the user or to the state in which the system finds itself, an arc is chosen and implicitly the next step/sub-problem to be solved.

An abstract solution graph is depicted in Figure 1. Each node represents a subproblem. The source, denoted by 1 contains the original question. In case condition $c1$ is satisfied, the student is directed to a a different question 2. In case another condition is satisfied, $c1'$, the student is redirected for example to 1. In case neither $c1$ nor $c1'$ are satisfied, the student is directed to the sink, represented by n.

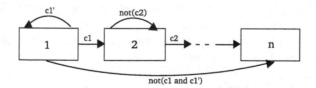

Fig. 1 An abstract solution graph

The arcs leaving a node must represent exclusive conditions and must cover all possibilities. The conditions can be imposed for example on the student input, leading for example to a specific sub-problem if the answer to a question is correct and to another one if the answer is incorrect. The conditions can be imposed on other parameters as well, like a maximum allowed number of trials, previous performance of the student, etc.

The solution graph is characteristic for a mathematical problem. It is designed to take into consideration the domain reasoner of the problem and aspects of a teaching methodology. There is no unique design for the solution graph of a specific mathematical problem.

4   Implementation of the Interactive Exercises

This section explains the implementation of the interactive exercises. More details on the general set-up and tools necessary to run the interactive exercises can be found in [6].

The exercises are implemented as a web-application. They are written in an XML based language that offers mark-up support for interactivity enriched with OpenMath for handling mathematics.

4.1   XML Representation of the Solution Graph

XML trees are ideal for capturing the structure of the solution graph of an exercise. Our examples use the following principal XML tags: **step**, **message**, **userinput** and **choice**. Each **step** corresponds to one sub-problem in the solution graph. The step/sub-problem has an attribute **id** by which it can be called. Each step has the following components: **message**, **userinput** and **choice**. The only exception is the last step, the sink, which has only **message**. The message contains the text or mathematical question posed to the student. That is, for example in Figure 2, *Please input a function in x whose derivative you would like to compute*. In Figure 2, the user input is a text-box having the label **Function**. The tag **choice** implements the *arcs* of the solution graph, and redirects to the next step. The choice and redirect option are further explained in the following section. The choice is hidden from the user.

4.2   Variables and Flow Control

The interactivity is catered by a Java based server (e.g. JSP [4]) which also provides a basic programming-like language for setting and retrieving variables, performing tests, flow control, etc. The interactivity consists of walking through the solution graph and the exchange of information with the user. The path one takes is determined by various parameters which can be determined by queries but also by previously set variables.

4.3   Mathematical Queries

Mathematics is represented according to the OpenMath standard ([10]). Open-Math encoded objects can be displayed by browsers (through conversion to MathML) and interpreted by computer algebra systems via OpenMath phrasebooks. Moreover, we use here the OpenMath expression of a mathematical object for interactivity in order to determine the domain rule to be applied (see Sections 2.1 and 6).

Within the frame provided by the Java based server we construct custom tags. Examples are provided below.

Query to a CAS. It is necessary for the automatic evaluation of the student's input. A standard for mathematical queries was defined by MONET (see [8]). See also the OpenMath webpage, Software and Tools, for some implementations of query services ([10]).

Query to a taglib. The following queries to the OpenMath tree structure of a mathematical expression turn out to be very useful in our interactive examples: extraction of a node, extraction of the root (operator) of a node, navigation in the tree (e.g., move to parent/next sibling/first child), rewriting the tree (e.g., mathematically $x + x^2$ equals $x(x + 1)$, but the corresponding OpenMath-expressions are different), keeping track of the current node (a query of the type *What is the parent of x in* $(1 + x)^2/x$ is ambiguous). The use of these operations will be made clear in Section 5.

The list of queries above is not exclusive and may need to be extended in case an interactive exercise for a different mathematical problem is considered.

5   The Example for Computing Derivatives

We present here an implementation of a generic exercise for computing the derivative of a function, see [3]. As a back-engine for checking the student's answer we use Mathematica. Expressions like $(x - 1)/(x^2 - 1)$ and $1/(x + 1)$ are therefore considered equal.

Students are allowed to choose a (univariate) function. Then they are asked to compute the derivative. In case they give a wrong answer they are guided through decomposing the function into simpler ones and then apply differentiation rules. In this implementation the next step is determined in interaction with the student. The student receives hints automatically, as shown in Step 3.

We present below snapshots taken while running the exercise. The exercise consists of a finite set of interactions, one asking the student to compute the derivative of a function, one asking the student to decompose the function, etc. This 'modularization' of the exercise allows us to reuse parts of the exercise. Note that each module is largely humanly authored but that parts of it are automatically updated by means of variables as described in Section 4. Note that the path taken for solving the exercise is decided in interaction with the student.

We describe below a possible scenario.

Step 1. The student introduces a function on which they want to practice. Alternatively, the function can be drawn from a database or randomly drawn from a particular class of functions such as polynomials, trigonometric, transcendental or any composition of functions belonging to these classes.

Step 2. The student is asked to submit the elementary function that is the derivative of the function.

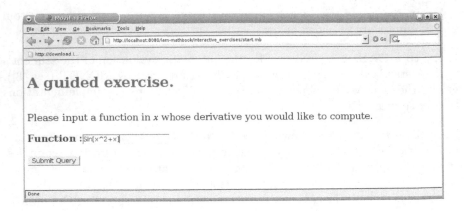

Fig. 2

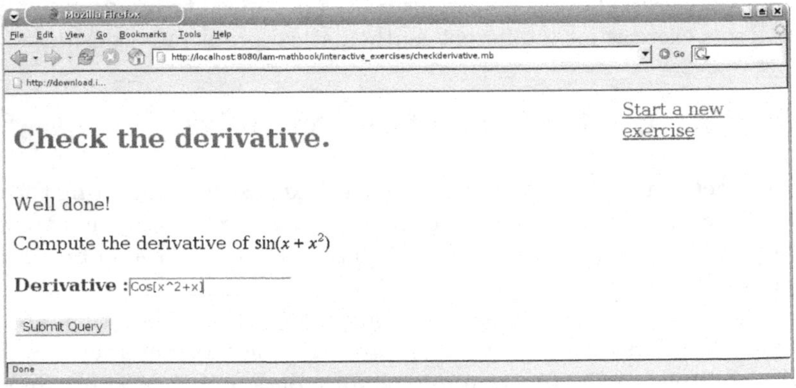

Fig. 3

Step 3. The student's answer is checked by Mathematica. If the student's answer is not correct, they are asked to choose a differentiation rule they want to apply. The button Hint, when pressed, displays the text: 'Use the chain rule'. Here the word *chain* is determined automatically by the system by analyzing the OpenMath expression of the function whose derivative we are computing at this step. The system finds the rule to be applied by using the special tools described in Section 4, namely the extraction of the root operator.

Step 4. After having typed the rule, the student is redirected accordingly.

If the rule chosen by the student matches the rule corresponding to the main operator, the system can also suggest the two functions g and h. This is not implemented in the current version.

Step 5. The student introduces the two functions and the CAS verifies that by composing the two, one obtains indeed the original function. If that is correct, the student is directed to the first function and asked for its derivative. The expression of the first child is obtained using the tools of Section 4, namely the

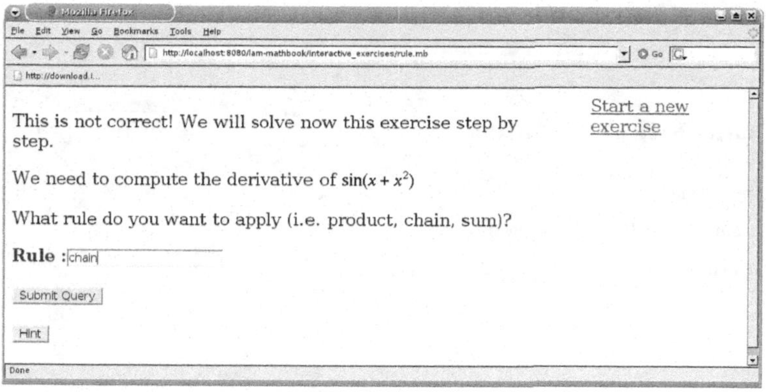

Fig. 4

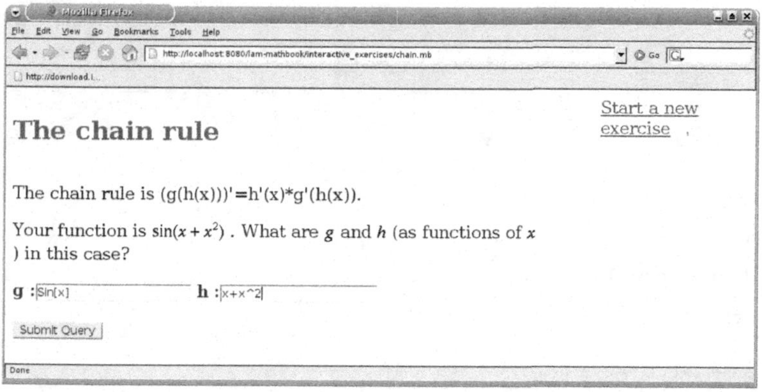

Fig. 5

navigation in the tree and the extraction of a node. Note that we reuse here the interaction from Step 2, for a different function.

Step 6. Since the answer at Step 5 is correct, we compute the derivative of the next sibling (reusing interaction from Steps 2 and 5). The next sibling is also found using the tools of Section 4.

Step 7. Since the second answer is correct and there are no more siblings we are redirected to compute the derivative of the parent (interaction of Steps 2, 5 and 6). If the answer were not correct at this point the student would have been asked to decompose further $x^2 + x$. (Students are allowed to decompose it both as a sum of x and x^2 and as a product of x and $x + 1$).

Note that the hint, when pressed, displays the sum rule for derivation.

Step 8. The answer is correct and this is acknowledged in the last page.

The complete solution graph of this problem is illustrated in Figure 10. In the graph, the subproblem 1 corresponds to Figure 2. From subproblem 1 we are always redirected to subproblem 2 which corresponds to Figure 3. Here

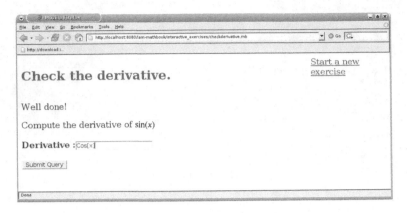

Fig. 6

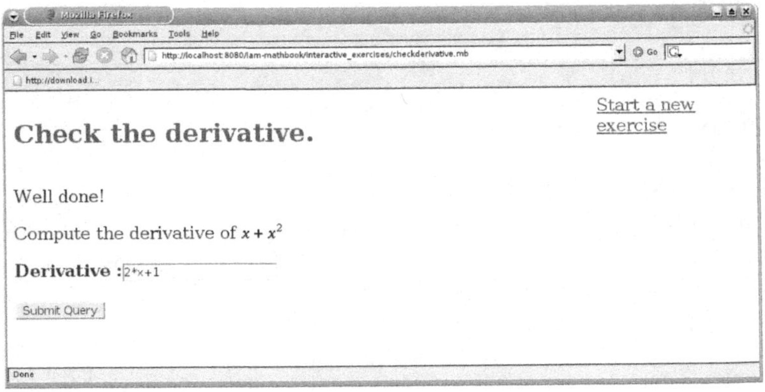

Fig. 7

depending on the student's answer and on previous knowledge we choose the next step. Let us explain the arcs. C_{11} is taken if the answer is incorrect and the derivative for the children are known. C_{12} is taken when the answer is correct and there is a next sibling, as it happens for example at Step 6. In this case, the current function is also updated. C_{13} is taken if the answer is incorrect and the derivatives of the children have not yet been computed. C_{14} is taken in case the answer is correct and there is no parent of the current node as it happens in Step 8. C_{15} is taken if the answer is correct, the derivatives of all children have been computed and the current node has a parent. This situation occurs at Step 7.

The subproblem of 3 corresponds to Figure 4. The subproblem of 4 corresponds to Figure 5, while subproblem 5 corresponds to a different rule. The arcs C_{21}, C_{22} correspond to the choice the student is making. Arc C_{31} corresponds to an incorrect decomposition of the function into simpler ones, while the arc C_{32} corresponds to a correct decomposition. Note that if the choice is correct, the function

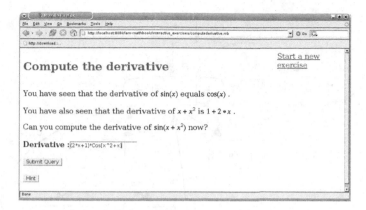

Fig. 8

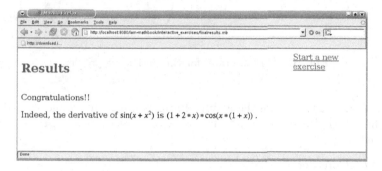

Fig. 9

we work with (the current node) is updated and takes the value of the first child as it happens at Step 5. The sub-problem 7 corresponds to Figure 8. Arc C_{51} corresponds to a correct answer. In this case the function we work with (current node) is updated to the parent. Arc C_{52} corresponds to an incorrect answer.

6   Other Applications

In this section we discuss the domain reasoners for the problem of computing limits, respectively the problem of computing indefinite integrals of elementary functions. Recall that the domain reasoner is the map from the mathematical knowledge on the instance of the problem (typically available in its OpenMath expression) to the set of domain rules.

6.1   Limits

The limit of a univariate elementary function when its argument goes to a specified value is in general obtained by taking the limits of its parts. In other words, taking the limit commutes with the composition of continuous functions. There

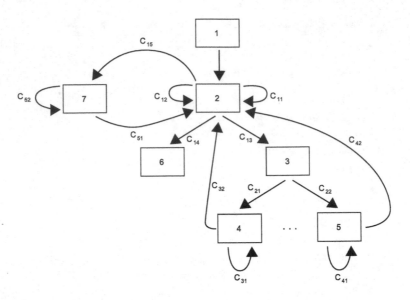

Fig. 10. solution graph for the derivative problem

are a few exceptions: in case the result of applying the above mentioned rule is one of the undecided cases $0^0, 1^\infty, 0^\infty, \infty^0, 0/0, 0 \times \infty, \infty - \infty$, rewriting rules are applied.

Example 1.   Consider

$$\lim_{x \to 1} \left(\frac{1}{x}\right)^{\frac{1}{1-x}}$$

which results in the undecided situation 1^∞. To solve the problem we reduce it to the undetermined case $0/0$ by applying the rewriting rule

$$\left(\frac{1}{x}\right)^{\frac{1}{1-x}} = \exp\log\left(\frac{1}{x}\right)^{\frac{1}{1-x}} = \exp\frac{-\log x}{1-x},$$

and then apply l'Hôpital rule; we conclude that the original will equal

$$\exp^{\lim_{x \to 1} \frac{1}{x}} = e.$$

L'Hôpital rule says that in the undecided cases where $\lim f(x)/g(x)$ is either $0/0$ or $\pm\infty/\pm\infty$ and $\lim f'(x)/g'(x)$ exists, we have $\lim f(x)/g(x) = \lim f'(x)/g'(x)$. L'Hôpital rule can be applied repeatedly and is equivalent to the method based on Taylor series expansions. Other undecided cases can be reduced to the $0/0$, $\infty/\pm\infty$ cases using rewriting rules. For example, the 'power' cases $0^0, 0^{\pm\infty}, 1^{\pm\infty}$ are rewritten using the equivalence $f(x) = \exp(\log(f(x)))$ while for the 'difference' cases $\infty - \infty$ we can use $f(x) = \log(\exp(f(x)))$. However, when the expression of $f(x)$ contains radicals, a different rewriting rule may need to be applied. As shown by the example, $\lim_{x \to \infty} x/\sqrt{x^2 + 1}$, l'Hôpital's rule occasionally fails to yield useful results. In certain cases, the limit of a 'part' of the function does not exist, although the limit of the whole function exists.

Example 2. Consider

$$\lim_{x \to \infty} \frac{\sin(x)}{x}.$$

In such a case the *sandwich* rule is applied:

$$-\frac{1}{x} \leq \frac{\sin(x)}{x} \leq \frac{1}{x} \quad \text{implies} \quad \lim_{x \to \infty} \frac{\sin(x)}{x} = 0.$$

The domain rules for computing limits are the rewriting rules, l'Hôpital/Taylor series rule and the sandwich rule.

- l'Hôpital rule: if $\lim_{x \to a} f(x)$ and $\lim_{x \to a} g(x)$ are both 0 or are both $\pm\infty$ and $\lim f'(x)/g'(x)$ exists, then $\lim_{x \to a} f(x)/g(x) = \lim_{x \to a} f'(x)/g'(x)$.
- sandwich rule: if $g(x) \leq f(x) \leq h(x)$, for all x in $(a - \epsilon, a + \epsilon)$ (x in (M, ∞), x in $(-\infty, -M)$, when $a = \infty$, respectively $a = -\infty$ and $M > 0$ large) and $\lim_{x \to a} g(x) = \lim_{x \to a} h(x) = L$, then $\lim_{x \to a} f(x) = L$.
- rewriting rules: $f(x) = \exp(\log(f(x)))$, $f(x) = \log(\exp(f(x)))$, rewriting rules for radicals, etc.

Computing the limit of a composed function is based on computing the limits of its sub-functions. Again, the OpenMath expression of a function is very useful for identifying the sub functions. Let us consider the OpenMath expression of $\lim_{x \to 1}(\frac{1}{x})^{\frac{1}{1-x}}$ (Example 1).

```
<OMOBJ>
  <OMA><OMS cd="limit1" name="limit"/>
    <OMI> 1 </OMI>
    <OMS cd="limit1" name="both_sides"/>
    <OMBIND><OMS cd="fns1" name="lambda"/>
      <OMBVAR><OMV name="x"/></OMBVAR>
        <OMA><OMS name='power' cd='arith1'/>
          <OMA><OMS name='divide' cd='arith1'/>
            <OMI>1</OMI><OMV name='x'/>
          </OMA>
          <OMA><OMS name='divide' cd='arith1'/>
            <OMI>1</OMI>
            <OMA><OMS name='minus' cd='arith1'/>
              <OMI>1</OMI><OMV name='x'/>
            </OMA>
          </OMA>
        </OMA>
    </OMBIND>
  </OMA>
</OMOBJ>
```

As in the case of differentiation, we can start by asking the student to compute the limit of the function. In case of a wrong answer we ask the student to decompose the function. The hint is given according to the OpenMath expression of the function. By doing this recursively, at the end of this procedure we have either computed the limit or identified the undecided case.

According to the undecided case we find, a rewriting rule (from the domain rule) is applied in order to bring the function to a case in which l'Hôpital can be applied as described in Section 6.1.

As remarked in Example 2, it is possible that evaluating a sub-function at some point we obtain a whole 'interval'. In this case, the rule (of the domain rule) to be applied is the sandwich rule.

In general the mathematical knowledge contained in the OpenMath expression is sufficient for defining the domain reasoner. An exception is the computation of limits in which knowledge on the range of the function is necessary for applying the sandwich rule.

6.2   Indefinite Integrals

Computing indefinite integrals (also called primitives) of elementary functions is in general much harder than computing derivatives or even limits. However a lot of progress has been made in the area of symbolic integration (see e.g. [1]) and many computer algebra systems have an integrated module for symbolic integration. Powerful algorithms have been developed which compute the indefinite integral of an elementary function in case it exists or prove that there is no elementary primitive.

Nevertheless our problem is in some sense simpler. We do not want to compute the primitive of a given function or to establish whether the primitive exists. To make sure that a primitive of f exists, we will randomly generate an elementary function F and compute its derivative f. Knowing F, we can guide the student through finding F as shown below.

Example 3. Consider the function $f : \mathbf{R} \to \mathbf{R}$ given by $f(x) = 1/(1 + 2x + 2x^2)$ and a primitive $F(x) = \arctan(x/(x+1))$ of it. From the OpenMath expression of $F(x)$ we see that F equals $\arctan(y)$ for some function y of x. Knowing that $(\arctan(y))' = y'/(y^2 + 1)$, a helpful suggestion to the student is:

Can you write the function f as $y'/(y^2 + 1)$ for a suitably chosen function y of x?

Note that the answer is not unique. Besides F, other primitives of f are

$$F_1(x) = -\arctan\left(\frac{x+1}{x}\right) \text{ obtained by rewriting } f(x) \text{ as } -\frac{(1/x)'}{1 + (1/x)^2}$$

and

$$F_2(x) = -\arctan(1 + 2x) \text{ obtained by rewriting } f(x) \text{ as } \frac{(2x+1)'}{1 + (2x+1)^2}.$$

All answers are correct and it can be checked that F, F_1 and F_2 all differ from each other by a constant.

There are several differences between our problem, which is teaching rules for computing integrals and the problem of computing integrals using state-of-the-art algorithms for symbolic integration described for example in [1]. We know the answer and want to deduce the steps that were taken in order to solve the problem. In some cases, as shown in Example 3, this is trivial. However other rules, like the integration by parts, are hard to recognize. Also, the algorithms described in [1] can be very different from the methods taught in schools, like the sum, chain, partial fractions, integration by parts rules and the elementary rules for integration.

The first implementations of algorithms for computing indefinite integrals are closer to the methods taught in school. To make our job easier, we can simply take over the domain reasoner of such a computer algebra system, for example the one described in [9]. There, the domain reasoner is based on 'pattern recognition' of a sub-function like sin, exp, etc., which is trivial from the OpenMath expression of the given function and uses heuristics for determining the rule to be applied. A multi-step interactive exercise for symbolic integration will have to implement the domain reasoner of a computer algebra system, for example of [9].

The alternative for us would be to write a domain reasoner ourselves, using the knowledge we have on the primitive F and on the function f, as suggested in Example 3. However we do not pursue this idea here.

7   Enhanced Interactive Exercises

The work presented in the previous sections can be easily extended in a few directions.

7.1   Detecting Errors and Misconceptions

A common error the student can make is forgetting brackets. If the number of opened brackets does not equal the number of closed brackets, the error can be detected at the moment of parsing the answer to OpenMath. Otherwise, forgotten brackets can still be detected in simple expressions, by extensive trials, although that would involve a number of computations which increases exponentially with the size of the expression.

Another interesting way of using the computer power for educational purposes, namely for detecting the misconceptions of the student, is by introducing the so-called *buggy rules*. Experienced teachers know what are the most frequent errors that a student makes when solving a particular type of problem. For example, a common mistake (buggy rule) when computing derivatives is $(x^n)' = x^{n-1}$. This knowledge can be formalized and the system can compare the student's answer against a list of buggy rules, in case the answer is not correct. In this way, the computer algebra system can make a guess as to what the student did wrong.

7.2 Simplifications

The student may often give a correct result which is however not simplified, as in $(x - 1)/(x^2 - 1)$. Most computer algebra systems have implemented simplification procedures to bring an expression to a normal form. At the moment we only distinguish between two cases, namely between a correct answer and a wrong answer. For pedagogical purposes, a third case would be interesting, namely 'correct but not fully simplified'. For this we would have to compare the student's answer with the normal form(s) available in the computer algebra system. However, this is a tricky issue since a normal form is not uniquely defined (e.g., both forms $x(x + 1)$ and $x^2 + x$ are acceptable). Also different computer algebra systems will have different implementations of simplification procedures, leading to different normal forms. Nevertheless, computer algebra systems often have a complexity measure of an expression which may be used to detect very elaborate answers.

7.3 Solution Generators

In this paper we have proposed implementing the domain reasoner for each particular problem in order to identify a possible next step. An alternative would be to use already existing tools, such as [12], which generate a detailed solution for a particular instance of the problem. For example [12] can display the solution of practically any instance of a variety of calculus problems among which we find computing derivatives and computing limits. This is achieved by using the domain reasoner implemented in Mathematica. The fact that [12] is an extension of Mathematica is in our view a disadvantage since this may restrict its use. At the moment the only interactivity [12] provides is in choosing the problem to be solved, hence it cannot be directly applied for interactive exercises. However, it is an innovative use of computer algebra systems and it may turn out to be very useful for interactive exercises.

8 Conclusion

The paper describes a method for implementing multi-step interactive exercises for certain problems in mathematical calculus. This is possible by exploiting the mathematical knowledge available in the OpenMath expression of the mathematical objects we deal with, by deriving the domain rules corresponding to the problem and constructing a domain reasoner for solving the problem.

As a first approximation, in the case of differentiation, the necessary properties of the mathematical object are deducible from its OpenMath expression. However, we have noticed already an exception in Example 2 where extra knowledge (about bounds of the function) is necessary in order to be able to compute certain limits.

Our set-up for the particular case of computing derivatives for functions is a first attempt, made possible within the LeActiveMath project [2].

References

1. M. Bronstein, Symbolic Integration I, Transcendental Functions, Springer, 1997.
2. LeActiveMath, http://www.leactivemath.org.
3. LeActiveMath exercises, http://www.riaca.win.tue.nl:8080/leam-exercises.
4. The J2EE Trademarked 1.4 Tutorial, http://java.sun.com/j2ee/1.4/docs/tutorial/doc/index.html.
5. MapleTA, http://www.maplesoft.com/products/mapleta/index.aspx.
6. A.M. Cohen, H. Cuypers, E. Reinaldo Barreiro, MathDox: Mathematical Documents on the Web, Contribution to OMDoc book, http://www.win.tue.nl/~hansc/mathdox3.pdf
7. Metric, Imperial College London, http://metric.ma.imperial.ac.uk/new/html/index.html.
8. Monet (Mathematics on net), http://monet.nag.co.uk/cocoon/monet/index.html.
9. Joel Moses, Symbolic integration: the stormy decade, http://portal.acm.org/citation.cfm?id=362651
10. OpenMath, http://www.openmath.org/cocoon/openmath/index.html.
11. Stack (System for Teaching and Assessment using a Computer algebra Kernel), http://eee595.bham.ac.uk/~stack/index.html.
12. webSolutions Detailed and Dynamic Mathematical Solutions, http://www.webmath.ch/.

XML-izing Mizar: Making Semantic Processing and Presentation of MML Easy

Josef Urban

Dept. of Theoretical Computer Science, Charles University,
Malostranské nám. 2/25, Praha, Czech Republic
urban@kti.ms.mff.cuni.cz

Abstract. Since version 7.2 the Mizar system produces quite detailed XML-based semantic description of Mizar articles during their verification. This format is now used natively for most of the processing done by Mizar, e.g., also for the whole Mizar internal database. The main motivation for switching to this XML-based representation is to make semantic communication with Mizar and presentation of the MML more accessible to a variety of external tools and systems. This article briefly describes this format and its current implementation, and shows examples of its usage. These examples include presentation of linked Mizar articles in modern XML-capable browsers like Mozilla, authoring assistance in the Mizar mode for Emacs, and experiments with XML-based querying languages like XQuery over the Mizar Mathematical Library loaded into a native XML databases like eXist. This work makes the currently largest repository of formal mathematics available to many kinds of presentational, data-mining, and automated reasoning applications and experiments, and the goal of this article is also to encourage, facilitate and provide recipes for the development of such applications.

1 Motivation

The Mizar[1] [Rud92, RT99] system is today probably the longest living proof checker of the world. There have been several re-implementations of this system during the previous decades [MR04], the current version is Mizar 7. These re-implementations were usually triggered by changes in the Mizar language or in the core proof-checking algorithms, coming from Mizar-based research on formalization of mathematics. Unlike these changes, the XML-ization of Mizar described here was triggered and also largely made possible by external influences. In this section we state the reasons that motivated this work.

1.1 Standard Format for External Semantic Processing of MML

First, a number of Mizar-external tools working with the semantic content of the Mizar Mathematical Library (MML) have been implemented in the recent years: MML Query [BR03], MoMM [Urb05b], MPTP and Mizar Proof Advisor [Urb04],

---

[1] http://mizar.uwb.edu.pl

M. Kohlhase (Ed.): MKM 2005, LNAI 3863, pp. 346–360, 2006.

semantic browsing in Emacs [BU04, Urb05a], etc. Such projects have shown the general interestingness and usefulness of Mizar-external semantic processing of the MML. However, since the Mizar language attempts to be very close to mathematical vernacular, the raw Mizar articles are difficult to process without a good knowledge of the Mizar parser. That is why all the above mentioned systems are not completely Mizar-external. They usually have to include some kind of a *Mizar exporter*, i.e. a tool written on top of the Mizar parser that exports various parts of the Mizar texts into the format used by these systems. Having to write such a tool can however be a "showstopper" for a number of people interested in the semantic processing of MML. It already requires some effort to get to know the Mizar language, and it requires even more effort to gain some working knowledge of the Mizar implementation. The obvious remedy is to export the semantic content of MML into an easily parsable and well-described format. The first effort that largely succeeded in this was the ILF project [Dah98], for which Czeslaw Bylinski wrote a tool exporting Mizar articles into a well-described Prolog format. Today, XML seems more suitable for such an export, because of its large-scale adoption throughout the software industry, resulting in the availability of a number of standards and tools for its description, validation, parsing, transforming, presentation, storage and indexing, etc.

1.2   Better Format for Internal Mizar Processing

The second reason for XML-izing Mizar comes from the Mizar implementation itself. The fast progress in the computer industry has exponentially increased the amounts of memory and storage space available to current systems, when compared with the situation thirty years ago. It is no longer necessary to devise as succinct formats as possible for storing internal information. More attention can be paid to the maintainability, extendability, standardization, and ease of use of such formats. Modern computer languages have gone as far as having default procedures for storage/loading complex data in XML.

The old Mizar format for internal data storage suffered from the deficiencies mentioned above. It was so succinct that it was quite hard to extend, and therefore changes to the Mizar language usually meant painful low-level work on data-storage. Only as little information as necessary for the basic verification was often stored. It meant that utilities needing more information were hard to implement, or entirely impossible without a large-scale duplication of code. While the verbosity has to be controlled even today, we can afford to have much more complete information at any time of the Mizar processing with no significant impact on its speed, and thus allow an easy implementation of a number of new Mizar utilities.

1.3   Putting It Together

The distinction between external and internal reasons for this work is not perfect. It has become quite clear during the years of Mizar development that various special-purpose Mizar-exporting tools written for external processing of MML usually become very quickly outdated. The only working way how to avoid it is to maintain one common and well-described format both for internal Mizar processing and for the external tools. Additionally, the distinction between internal and Mizar-external tools is not perfect either. Some tools (e.g., the MizarMode [Urb05a]) eventually became a standard part of the Mizar distribution, while some other Mizar utilities are no longer distributed. It is quite realistic to predict that in some time e.g. more searching tools might be distributed with Mizar, especially now that the internal Mizar database is already queryable using XML standards like XPath and XQuery.

2   Description of the Mizar XML Implementation

The code-base of the Mizar kernel is quite large (ca. 50000 lines). Since Mizar is a "production" system used by many people around the world, the switch had to be done in a conservative way, introducing as little user-visible changes to Mizar as possible. The work has been done in three smaller phases, making the transition smoother and better controlled. After each phase, regression testing on both the MML and randomly created incorrect articles was carried out before releasing the new Mizar version. The first two phases done in 2004 dealt with the old information passing in Mizar. As noted above, these old formats were devised in times when memory and storage space were expensive, and therefore a lot of useful information was not present in them. Additionally, different I/O procedures were used for the same data-structures in different contexts, again in order to save memory and storage. In the third phase the actual switch to XML format was carried out.

2.1   Providing Semantic Form of Mizar Articles

In the first phase the interface between semantic analysis and proof checking was completely rewritten. The output of semantic analysis is now a complete semantically disambiguated form of a Mizar article, while earlier it included only the information that was absolutely necessary for proof checking. The additional information includes, e.g., the description of the Jaskowski-style natural deduction steps used in the proofs, the modified thesis after each deduction step, definitional macros (definientia) used for silent unfolding of thesis, etc. The semantic analysis and proof checking became completely separated, and they can now even be run as separate programs. Though this is not very important for Mizar at the moment, it documents well the conceptual separation that was carried out: on one hand we have the parser for the natural-language-like Mizar articles, and on the other hand we have the Mizar verifier and other utilities working on the easily machine-processable format produced by the parser. It is interesting to observe that such a split is becoming more and more common in many "authoring" areas, and various human-friendly compact representations (e.g. wiki for HTML, compact syntax for RELAX NG, etc.) are increasingly used to produce richly annotated machine-processable data. In this aspect, the Mizar language might be thought of as a similar compact syntax for human-authored formal mathematics, and the detailed output produced by the Mizar parser (including semantic analysis) as its machine-processable equivalent.

This first step has already allowed to rewrite more cleanly a number of Mizar utilities, and to produce some interesting new utilities. It is now possible to collect statistics about the usage of definitional macros in MML, or to easily compute all the suppositions done along the Jaskowski-style proof path to some lemma. The latter can be used e.g. in the MoMM data-miner [Urb05b] as another method of exporting internal lemmas from their proof context. As noted in the Motivation section, a big incentive for this step was also the need for processing of complete Mizar proofs in external systems like MPTP, MML Query, Mizar Proof Advisor and others. However once the first step was finished, it became clear that the complete XML-ization is really feasible in quite a short time, and work on these external systems was largely postponed after that.

2.2   XML-izing Mizar

Mizar can be used on a number of computer architectures, and can be compiled with all major objective Pascal compilers (FPC, GPC, Delphi, Kylix). This variety makes it quite difficult to use some external XML-parsing library for Mizar. So the starting point that triggered further work on XML in Mizar was the implementation of a native "Tiny XML" Pascal parser by Czeslaw Bylinski. This fast parser now implements just a subset of XML needed by Mizar, more advanced features like parsing of XML entities are not yet supported (and probably not necessary for our purpose at all). One thing that still prevented the switch to XML was the number of memory-optimized versions of essentially the same data-structure in Mizar. During the second phase a large unification of such Mizar data-structures was done and the Mizar codebase thus simplified quite a lot (about 3000 lines were pruned). The effects of both the first and this second phase on the speed, memory, and storage requirements of Mizar processing were (according to current standards) really negligible. For instance, the slowdown caused by having the detailed information after semantic analysis was about 1-2 percent. Most of the Mizar processing time is today used for the proof checking, and various modifications of the proof checking module usually cause much higher deviations (in both directions) than this number.

In the third phase, an XML representation was suggested for most of the Mizar data-structures, and most of the I/O procedures have been rewritten for this representation. The initial task of the XML format[2] was just to allow the switch to XML, and it will probably go through some more changes in the future. The applications that we describe below are already giving quite a lot of feedback about the usability of this format for Mizar-external purposes, and some of that feedback already resulted in changes in the XML format. The general policy concerning such changes is to incrementally improve the clarity of the

---

[2] Since Mizar version 7.4, the Mizar XML documentation is distributed with Mizar. It can be also viewed at http://lipa.ms.mff.cuni.cz/~urban/Mizar.html

XML representation, so that it is more useful for various external applications, however this has to be done without serious influence on the Mizar processing. A lot of information can now be added to the XML representation by external processing via simple XSLT stylesheets (we give an example below). If a really rich format is required by some external tool, the developers of such a tool should use XSLT to produce such a rich format from the basic Mizar XML. On the other hand, most of the Mizar XML texts is now occupied by Mizar terms, formulas and types, so adding markup to other Mizar constructs is usually quite cheap and with negligible influence on the speed, memory and storage requirements of the Mizar processing.

The description of the Mizar XML format is (following Knuth ideas on literate programming) now given directly in the Mizar code, as specially tagged Pascal comments. They are written in the RELAX NG compact syntax[3] which is (unlike other XML-based specification formats[4]) natural, human-friendly, and actually useful even as an internal description of the Pascal data-structures. It would not be feasible to fully produce the I/O routines automatically from the RELAX NG descriptions (or vice-versa), however, at least a part of the Mizar XML parsing code is now produced automatically from them. This, together with validation by the resulting schemata, should reasonably guarantee that the descriptions will be maintained when Mizar changes, and will always faithfully represent the Mizar XML format. These descriptions are then very simply (by a Perl one-liner) collected into the file Mizar.rnc, which is now distributed with Mizar. The particular RELAX NG schemata for each XML file produced by Mizar are then created just by stating the starting element, and including Mizar.rnc. For instance, the file `article.rnc` describing the complete semantic content of Mizar articles looks this way:

```
include "Mizar.rnc"
start = Article
```

2.3 Speed and Space Considerations

The switch to XML had its effects on the speed and storage requirements of Mizar processing. The following table 1 compares the speed of verification and accommodation[5] on complete MML and on average article before and after XML-ization. The verification time of MML increased from ca. 47 minutes to ca. 78 minutes, i.e., about 1.67 times. This is the most important number for Mizar users, because verification is carried out very frequently during author-ing Mizar articles. The accommodation time grew quite considerably, from ca. 2 minutes to about 15 minutes. Accommodation is however quite rare operation,

---

[3] http://www.oasis-open.org/committees/relax-ng/compact-20021121.html

[4] "XML combines all the inefficiency of text-based formats with most of the unread-ability of binary formats." – Oren Tirosh, comp.lang.python.

[5] Accommodation extracts from the internal Mizar database the Mizar items used for processing a new article, and creates a local environment for that article.

Table 1. Comparison of Mizar processing times (in seconds) before and after XML

Action	MML bef. XML	MML after XML	avrg. bef. XML	avrg. after XML
verification	2832	4723	3.2	5.34
accommodation	145	906	0.16	1.02

Table 2. Comparison of the sizes of exported Mizar files (in Kb) before and after XML

MML bef. XML	MML after XML	avrg. bef. XML	avrg. after XML
12270	78080	14	89

and in absolute numbers it is still much below the verification times. Since accommodation spends most of its time in I/O procedures, its time increase corresponds quite faithfully to the growth in the size of processed data, i.e. to the size of the internal database. These sizes are compared in table 2. This increase is considerable (636 percent), and it is obviously caused by the greater verbosity of the XML representation. It would be possible to compress the XML format quite a lot (and therefore also decrease the processing times) by choosing much shorter names of elements and attributes, but it would make the XML format less understandable. Today (spring 2005), one gigabyte of storage costs about 0.75 USD, so from this point of view, an increase by tens of megabytes is really negligible, and should pose no problems to users. The most sensitive value is therefore the verification time, and this is the reason why the verbosity of the XML format needs to be controlled. On the other hand, the Moore's law will take care of the current verification time increase in one year, and as mentioned above, the deviations caused by modifications of the proof checking module are quite frequent and at about the same scale.

2.4 Short Description of the Mizar XML Format

Long before XML and XML-based mathematical formats were devised, Mizar had strict internal division between the presentation layer and the semantic layer. The current Mizar XML format captures only the semantic content of Mizar articles, which is used for Mizar verification and also in practically all the above mentioned Mizar-external tools. As in e.g. MathML, it is possible to construct the presentation from the Mizar semantic layer, and e.g. MML Query has been doing this for several years. However, such reconstructed presentation often differs from the original presentation used by authors of the articles. So the current XML format cannot be used yet for exact presentation purposes, like the Journal of Formalized Mathematics[6]. Therefore the Mizar presentation layer will probably also be XML-ized in the future, it has been already discussed among Mizar developers several times.

---

[6] http://mizar.uwb.edu.pl/JFM/

As noted above, most of the markup is now occupied by Mizar terms, formulas and types. Terms and formulas are now just named patterns representing the various elements:

```
Term =
    ( Var | LocusVar | FreeVar | LambdaVar | Const | InfConst | Num
    | Func | PrivFunc | Fraenkel | QuaTrm | It | ErrorTrm)


Formula =
    ( Not | And | For | Pred | PrivPred | Is | Verum | ErrorFrm )
```

These particular term and formula elements are described in more detail in the RELAX NG documentation[7]. Note that the logical connectives used in the Mizar semantic layer are only `Not`, `And` and `For`, all other connectives are transformed to them. The markup for Mizar types looks this way:

```
## Adjective is a possibly negated (and paramaterized) attribute
Adjective = element Adjective {
    attribute nr { xsd:integer },
    attribute value { xsd:boolean }?,
    Term*}


## Cluster of adjectives
Cluster = element Cluster {Adjective*}


## Parameterized type - either mode or structure
## First goes the LowerCluster, then UpperCluster
Typ = element Typ {
    attribute kind { "M" | "G" | "errortyp" },
    attribute nr { xsd:integer }?,
    Cluster*, Term* }
```

The building blocks of Mizar proofs are Jaskowski-style [Jas34, Pel99] natural deduction steps[8]. Those modifying the currently proved *thesis* are `SkeletonItems` and the rest are called `AuxiliaryItems`. Again, these are just named patterns, the particular elements are described in the documentation:

```
SkeletonItem =
    ( ( Let | Conclusion | Assume | Given | Take | TakeAsVar ), Thesis? )


AuxiliaryItem =
    (JustifiedProposition | Consider | Set |Reconsider | DefFunc | DefPred)
```

The top-level `Article` element describes the semantic content of a whole Mizar article. It contains various Mizar blocks, introducing definitions, theorems, schemes, etc.:

---

[7] http://lipa.ms.mff.cuni.cz/~urban/Mizar.html

[8] Actually, a new name like *Mizar-Jaskowski* or *Jaskowski-Trybulec ND style* might be more proper, since the Jaskowski style was reinvented and augmented in Mizar.

```
## The complete article after analyzer, aid specifies its name in uppercase.
  Article = element Article {
    attribute aid { xsd:string },
    ( DefinitionBlock | RegistrationBlock |
      NotationBlock | Reservation | SchemeBlock |
      JustifiedTheorem | DefTheorem | Definiens |
      Canceled | Pattern | AuxiliaryItem )* }
```

This semantic description of an article is now automatically constructed during the verification, so the way to produce such description for the whole MML from installed Mizar distribution is e.g. following[9]:

```
mkdir tmp; cp -r $MIZFILES/mml tmp; cd tmp/mml
for i in `cat $MIZFILES/mml.lar`; do accom $i; verifier $i; done
```

The Mizar distribution also includes the Mizar internal database (directory $MIZFILES/prel), which is now also fully XML-ized. This database contains all the MML items that can be re-used in other articles, i.e., theorems, definientia, schemes, notations, registrations and constructors[10]. Again, all these files can be validated by setting a suitable starting element (i.e., Theorems, Definientia, Schemes, Notations, Registrations and Constructors) and including the basic schema Mizar.rnc. Despite the large-scale code unification described in 2.2, the markup used for higher-level structures in the internal database is not yet completely the same as the markup used for the description of whole articles. E.g., the Scheme element used in the internal database differs quite considerably from the scheme description used in whole articles inside SchemeBlocks. Complete reconciliation of all such differences would be too time-consuming, and the plan is to reconcile them gradually in the future, also using the feedback obtained from various applications building on the XML format.

3   Applications

3.1   Stylesheet Processing

XSL Transformations (XSLT[11]) is a high-level declarative language used today very frequently for transforming XML documents. A number of fast XSLT processors are now publicly available, e.g., xsltproc[12] or xalan[13]. XSLT is often used for producing customized HTML from XML documents, and state-of-the-art browsers like Mozilla and Firefox are shipped with built-in XSLT processors. Useful compact syntaxes exist for human-friendly authoring of XSLT stylesheets, the XSLTXT[14] compact syntax is used to produce the stylesheets presented here.

---

[9] This will take ca. 1-2 hours on standard hardware.

[10] Just for completeness, the XML-ized form of the file $MIZFILES/mml.vct containing MML symbols will be probably added to Mizar distribution.

[11] http://www.w3.org/TR/xslt

[12] http://xmlsoft.org/XSLT/xsltproc2.html

[13] http://xml.apache.org/xalan-c/

[14] http://www.zanthan.com/ajm/xsltxt/

Some of these stylesheets are now distributed with Mizar, others are available from the author's site[15].

A typical task easily achieved by XSLT is to produce a richer XML description of an article from the description supplied by the Mizar parser. During accommodation, Mizar collects from the internal database various resources required for a new Mizar article. Such resources (theorems, schemes, constructors, etc.) are organized into arrays for fast lookup, and the current XML format only uses their positions in these arrays (one integer) instead of their absolute MML address. Using the absolute MML addresses makes the resulting XML files almost twice as big, and as described above, this has negative effect on the speed of Mizar verification and accommodation. The shortened and the absolute description of the basic Mizar type `set` (with no adjectives) are shown below:

```
<Typ kind="M" nr="1">
  <Cluster/>
</Typ>
```

```
<Typ kind="M" nr="1" aid="HIDDEN" absnr="1">
  <Cluster/>
</Typ>
```

The additional attributes `aid` and `absnr` encode (together with the attribute `kind`) the absolute MML address of this resource, i.e. it is the first mode type introduced in article `HIDDEN`.

On the other hand, having the absolute addresses directly inside the XML is often very useful, and it is indispensable e.g. when one wants to uniformly process multiple articles and their resources. For instance, provided that we use the absolute addresses, the following XPath expression will return all adjectives defined in article `ORDINAL1` used in generalizations (`Let`-items) in articles `CARD_1` and `CARD_2`.

```
/Article[(@aid="CARD_1")or(@aid="CARD_2")]//Let//Adjective[@aid="ORDINAL1"]
```

Addition of the absolute addresses e.g. to Mizar adjectives can be done using the following XSLTXT piece of code:

```
tpl [Adjective] { <Adjective
    { copy-of '@*'; abs(#k="V", #nr='@nr'); apply; }}
```

The XSLT function `abs(#k,#nr)` just looks up the absolute address (in the file containing the lookup table) using the serial number as a key. The complete transformation adding absolute addresses to all Mizar resources takes about 100 lines of XSLTXT[16].

3.2   Browsing Semantic Form of Articles in XSLT-Capable Browsers

One of the original design goals of XML was to "become a better HTML". Several popular internet browsers are able to apply XSLT stylesheets to a given XML file

---

[15] `http://kti.ms.mff.cuni.cz/cgi-bin/viewcvs.cgi/xsl4mizar/`

[16] `http://kti.ms.mff.cuni.cz/cgi-bin/viewcvs.cgi/xsl4mizar/addabsrefs.xsltxt`

and display the result as if it was a normal HTML document. Since the XML is now produced after each verification run, direct browsing of the semantic form of the currently authored Mizar article is just a matter of specifying a suitable stylesheet. If such a stylesheet is present, no other XML tools need to be installed, the only necessary step is to open the XML file in a browser like Mozilla or Firefox after the verification.

An initial stylesheet `miz.xsltxt` serving this purpose has been developed[17], and since Mizar version 7.4 it is also distributed with Mizar. The default installation location of this stylesheet is written to the XML output of the Mizar parser, which means that the XML output can be displayed locally, without internet connection. The default linking of Mizar resources is however done to the internet address of MML Query, so for looking up these resources in the default way an internet connection is necessary. There are several other experimental linking possibilities, including linking to "self", i.e. to the collection of XML files produced by the verification of MML. Another good linking possibility would arise if we (the Mizar developers) decided to distribute a richer version of the internal Mizar database, containing the absolute MML addresses as described above. This version is about twice as big (ca. 100 MB), which would cause further slowdown of accommodation, but the Mizar distribution would then also directly serve as a self-sufficient HTML documentation of the semantic form of MML abstracts, similar to the MML Query generated abstracts [BU04].

All the HTML files produced by applying the default `miz.xsltxt` stylesheet to the XML description of MML articles can be now browsed at the author's site[18]. Adding number of presentational features like e.g. JavaScript hiding/showing of subproofs to `miz.xsltxt` should be very easy, and users and developers are encouraged to use `miz.xsltxt` as a building block for such richer presentations.

3.3   Disambiguating in MizarMode

External tools are not the only uses of the semantic representation of Mizar articles. In more advanced mathematical domains, notation from several other domains are combined, sometimes causing ambiguities. It is often nontrivial to guess how certain terms and formulas are parsed by Mizar, especially in more complex Mizar articles.

In [BU04, Urb05a] we describe a solution (called *Constructor Explanations*) to this problem in the Emacs authoring environment for Mizar (MizarMode), using the MML Query generated abstracts and the old-style Mizar parser output. As noted above, the old output contained only as little information as was absolutely necessary for proof checking, and also complete parsing of its custom format would require detailed rewrite of the Pascal I/O procedures into Emacs Lisp. Both these reasons had lead to a middle-way approach: only Mizar propositions with *SimpleJustification* (XML element By) were parsed by Emacs from the Mizar parser output and made available for semantic disambiguation.

---

[17] http://kti.ms.mff.cuni.cz/cgi-bin/viewcvs.cgi/xsl4mizar/miz.xsltxt
[18] http://lipa.ms.mff.cuni.cz/~urban/xmlmml/

Both these reasons have now disappeared: a complete semantic representation of an article is now available, and it is in XML. Emacs is distributed with a default XML-parsing module since version 21.1. The *Constructor Explanations* are now therefore applicable virtually to any Mizar construct. In practice, the *Constructor Explanations* however use the mechanism of Emacs text properties, to couple pieces of text with their disambiguated form. This coupling is easily done for all XML constructs that remember their original position in the article, so it now covers all Mizar Propositions. Many other XML constructs remember their position too, so *Constructor Explanations* can be easily extended to them. The only additional work needed for this is to specify presentation of the XML fragments in a stylesheet-like manner. Displaying the whole semantic representation inside Emacs will be also feasible (exactly as the MML Query generated abstracts), and it can be done with very similar methods as the HTML presentation.

3.4   Using the XML Technology for Indexing and Querying

The large-scale adoption of XML in the software industry has another benefit for formal mathematics: it brings new tools for storage and indexing of tree-like structures which typically represent mathematical proofs, formulas and terms. Storing such structures directly e.g. in relational databases is often cumbersome, relying e.g. on full-text indexing of formulas represented as strings, or explicit Parent/Children tables. Several native XML databases centered around tree-like structures are being developed today, e.g. X-Hive[19], Berkeley DB XML[20] or eXist[21]. Berkeley DB XML and eXist are probably the most mature freely available systems.

For experimenting with Mizar XML, eXist was chosen. It provides a wider functionality, mainly a working implementation of structural auto-indexing of XML [Mei02] which makes XPath queries efficient without the necessity of low-level manual work on indexing. A very experimental eXist server accepting XQuery queries has been set up[22] for the Mizar XML data. Due to limited resources, only the internal Mizar database (with added absolute MML addresses) is now available there. This basically means that all proofs are omitted. The auto-indexing in eXist is based on a numbering scheme which assigns unique identifiers to all nodes in an XML tree. This has to be done in a regular way, so that Ancestor/Descendant and other XPath axes between two nodes (expressed as numbers) were quickly computable. For that purpose the XML tree needs to be "regularized", i.e. some identifiers are spent on nonexistent nodes. On large or irregular XML documents this can cause that eXist runs out of node identifiers. This limitation made direct eXist storage of practically all XML files with Mizar theorems impossible. The Mizar internal database XML files were therefore split before inserting into eXist, and a separate document was created for each Mizar

---

[19] http://www.x-hive.com/
[20] http://www.sleepycat.com/products/xml.shtml
[21] http://exist.sourceforge.net/
[22] http://lipa.ms.mff.cuni.cz/~urban/existdemo.html

resource (theorem, scheme, constructor, etc.). About 90000 smaller XML documents were thus created and loaded into eXist quite successfully: only 332 of these documents were still too irregular for the current eXist auto-indexing. The loading (and indexing which is done on the fly) takes about 30 minutes on an Intel Pentium 4 3.4GHz computer. The size of the resulting database files is about 0.5 GB.

Preliminary experiments show that the Mizar database is still quite large for the eXist server. Particularly, even with the XPath auto-indexing, the path joins done for frequent elements can require portions of memory which are not available on our server. Queries like `//Func` (yields 731154 elements) or `//Typ` (yields 281998 elements) are very fast, and even queries like `//Func/Func` (yielding 327007 elements) and `//Typ//Func` (308399 elements) are fast (below 1 second) and executable within the current memory limit, but e.g. `//Func//Func` is already aborted because of the memory limit. While part of these problems can be solved by having a dedicated server with several GB of memory, generally the eXist system would also have to be augmented to trade some of its speed for lower memory usage in very large XML collections.

Native storage and auto-indexing of XML is a very young field, and while it may be already fully usable for smaller mathematical libraries than MML, custom solutions like MML Query will probably still be necessary for several following years for MML. The functionality provided by languages like XQuery is quite different from the current MML Query language, which has much lower support for tree-like queries, but adds qualifiers like *atmost* and *exactly*. The combination of both should be useful, and building custom search engines like MML Query on top of a native XML database should make a lot of the low-level work on such custom system unnecessary. On the other hand, for very specialized mathematical operations like matching and unification, the ATP-based indexing methods [RSV01] like discrimination trees (used now in the MoMM system) will probably remain superior to the more general XML indexing schemes.

3.5   Data-Mining Queries

Here we give examples of simple XML processing pieces of code that can be used to extract input for various AI tools from MML. First the XSLTXT and then the XQuery version are shown. XSLT and XQuery will be usable for most of these task, and obviously they can be complemented or embedded into other programming languages for more complex tasks.

a) Get sequences of top-level proof steps for various natural-language-like (e.g. Markov model) learnings:

```
tpl [Proof] { <Proof { for-each [*] { name(); } } }

for $i in //Proof
return <Proof> { for $j in $i/* return name($j) } </Proof>
```

This will yield sequences of rule names like:

```
<Proof>Let Let Assume Consider Take IterEquality Conclusion</Proof>
```

b) For each theorem get all theorems and definitions used in its proof, (should be executed on the richer XML with absolute MML addresses):

```
tpl [JustifiedTheorem] { <Refs
    { for-each [.//Ref[@kind]] { '@aid'; ":"; '@kind'; '@absnr'; }}}
```

```
for $i in //JustifiedTheorem return
  <Refs>
  { for $j in $i//Ref[@kind] return
    concat(string($j/@aid),":",string($j/@kind),string($j/@absnr))}
  </Refs>
```

This will yield sequences of references like:

```
<Refs>CARD_1:T64 ORDINAL1:T19 ORDINAL1:T23 ORDINAL1:T26</Refs>
```

c) For each theorem get all constructors in it (should be executed on the richer XML with absolute MML addresses):

```
tpl [JustifiedTheorem/Proposition] { <Constrs
    { for-each [.//(Pred|Func|Typ|Adjective)]
      { '@aid'; ":"; '@kind'; '@absnr'; }}}
```

```
for $i in //JustifiedTheorem/Proposition return
  <Constrs>
  { for $j in $i//(Pred|Func|Typ|Adjective) return
    concat(string($j/@aid),":",string($j/@kind),string($j/@absnr))}
  </Constrs>
```

This will yield sequences of constructors like:

```
<Constrs>CARD_1:V1 HIDDEN:M1 HIDDEN:R2 ORDINAL2:K5</Constrs>
```

The previous two queries put together are actually the input to the Mizar Proof Advisor training (now a Bayesian classifier implemented by the SNoW [CCRR99] system).

4   Conclusions and Future Work

We believe that the switch to XML representation in Mizar was carried out at about the right time. Though still advancing very fast, the XML and derived technologies like XSLT have matured, and have been implemented in many mainstream applications like internet browsers. Standard XML parsers exist for a number of programming languages, implementations of XPath and XQuery standards are becoming quite common, and both relational and native XML storage and indexing models are being developed quite actively. We have described above several initial Mizar-related applications taking advantage of this new technology. For many simple applications, XSLT processing of Mizar data will be sufficient, and writing a couple of lines in a high-level declarative language like XSLTXT should be much easier than building on the Pascal implementation of Mizar. Quite an important difference between this work and earlier Mizar-exporting tools is that the XML format is now also used internally by Mizar. It means that the XML format will follow the changes introduced during the fast development of Mizar. Third-party developers will therefore no longer be stuck with outdated exporting tools that no longer work on the newest MML, they can easily detect recent XML changes with tools like xmldiff, and keep their external tools as much up-to-date as they wish. The user-visible increase in the verification time is not negligible, however it is neither dramatic, especially in the view of the drop that occurred in the past years thanks to the general growth of computing power.

The largest future extension of the work presented here should be the above mentioned inclusion of the Mizar presentation layer. This should allow simple reimplementation of presentation tools like the Journal of Formalized Mathematics, and improve the current content-derived presentation methods used for direct browsing or in MML Query. Another possible extension is detailed reporting about the smallest verification steps done by Mizar, e.g., in a way similar to the Otter [McC94] proof objects. Some preliminary work has been done in this direction, however this practically amounts to a large-scale reimplementation of the Mizar proof-checking module. A cheaper (though less elegant) alternative how to do this can be e.g. using external automated theorem provers like Otter to re-prove the Mizar verification steps, and get these detailed proof objects from them. This would have the additional cross-verification benefit, and is planned as a part of the MPTP project.

The Mizar XML format is now very fresh, and is likely to go through some more changes, particularly in order to unify the remaining redundancies and to make it more compact. We hope that this article will also stimulate feedback from potential users of the Mizar library, which is a unique source for a number of data-mining, automated reasoning, presentational, and information retrieval experiments.

Acknowledgments

The idea of using XML for a number of Mizar-related tasks has been thoroughly discussed within the Mizar team, naming at least Andrzej Trybulec, Czeslaw Bylinski and Grzegorz Bancerek. The final step that triggered the work described here was the implementation of a Tiny-XML Pascal parser for Mizar by Czeslaw Bylinski. The recent versions of the HTML presentation contain many improvements suggested by Jiří Vyskočil. This work was partially supported by the CALCULEMUS Research Training Network (HPRN-CT-2000-00102) and Charles University research grants (205-03/2060985, 205-10/203336).

References

[BR03]   Grzegorz Bancerek and Piotr Rudnicki. Information retrieval in MML. In
            MKM, volume 2594 of *Lecture Notes in Computer Science*, pages 119–132.
            Springer, 2003.

[BU04]   Grzegorz Bancerek and Josef Urban. Integrated semantic browsing of
            the Mizar Mathematical Library for authoring Mizar articles. In *MKM*,
            volume 3119 of *Lecture Notes in Computer Science*, pages 44–57. Springer,
            2004.

[CCRR99]   A. J. Carlson, C. M. Cumby, J. L. Rosen, and D. Roth. Snow user's guide.
            Technical Report UIUC-DCS-R-99-210, UIUC, 1999.

[Dah98]   Ingo Dahn. Interpretation of a Mizar-like logic in first-order logic. In
            FTP (LNCS Selection), pages 137–151, 1998.

[Jas34]   S. Jaskowski. On the rules of suppositions. *Studia Logica*, 1, 1934.

[McC94]   W. W. McCune. Otter 3.0 reference manual and guide. Technical Report
            ANL-94/6, Argonne National Laboratory, Argonne, Illinois, 1994.

[Mei02]   Wolfgang Meier. eXist: An open source native XML database. In *Web,
            Web-Services, and Database Systems*, volume 2593 of *Lecture Notes in
            Computer Science*, pages 169–183. Springer, 2002.

[MR04]   Roman Matuszewski and Piotr Rudnicki. Mizar: the first 30 years. In
            Grzegorz Bancerek, editor, *MKM Workshop on 30 Years of Mizar*, 2004.

[Pel99]   F. J. Pelletier. A brief history of natural deduction. *History and Philos-
            ophy of Logic*, 20:1 – 31, 1999.

[RSV01]   I.V. Ramakrishnan, R. Sekar, and A. Voronkov. *Handbook of Automated
            Reasoning*, chapter Term Indexing, pages 1853–1964. Elsevier and MIT
            Press, 2001.

[RT99]   Piotr Rudnicki and Andrzej Trybulec. On equivalents of well-foundedness.
            J. Autom. Reasoning, 23(3-4):197–234, 1999.

[Rud92]   P. Rudnicki. An overview of the Mizar project. In *1992 Workshop on
            Types for Proofs and Programs*, pages 311–332. Chalmers University of
            Technology, Bastad, 1992.

[Urb04]   Josef Urban. MPTP - motivation, implementation, first experiments.
            Journal of Automated Reasoning, 33(3-4):319–339, 2004.

[Urb05a]   Josef Urban. MizarMode - an integrated proof assistance tool for
            the Mizar way of formalizing mathematics. *Journal of Applied Logic*,
            2005. forthcoming, available online at http://ktiml.mff.cuni.cz/~urban/
            mizmode.ps.

[Urb05b]   Josef Urban. MoMM - fast interreduction and retrieval in large
            libraries of formalized mathematics. *International Journal on Ar-
            tificial Intelligence Tools*, 2005. forthcoming, available online at
            http://ktiml.mff.cuni.cz/~urban/MoMM/momm.ps.

Determining Empirical Characteristics of Mathematical Expression Use

Clare M. So and Stephen M. Watt

Ontario Research Centre for Computer Algebra,
Department of Computer Science,
University of Western Ontario,
London Ontario, Canada N6A 5B7
{clare, watt}@orcca.on.ca

Abstract. Many processes in mathematical computing try to use knowledge of the most desired forms of mathematical expressions. This occurs, for example, in symbolic computation systems, when expressions are simplified, or mathematical document recognition, when formula layout is analyzed. The decision about which forms are the most desired, however, has typically been left to the guess-work or prejudices of a small number of system designers.

This paper observes that, on a domain by domain basis, certain expressions are actually used much more frequently than others. On the hypothesis that actual usage is the best measure of desirability, this papers begins to quantify empirically the use of common expressions in the mathematical literature. We analyze all 20,000 mathematical documents from the mathematical arXiv server from 2000-2004, the period corresponding to the new mathematical subject classification. We report on the process by which these documents are analyzed, through conversion to MathML, and present first empirical results on the most common aspects of mathematical expressions by subject classification. We use the notion of a weighted dictionary to record the relative frequency of subexpressions, and explore how this information may be used for further processes, including deriving common patterns of expressions and probability measures for symbol sequences.

1 Introduction

Most software that deals with symbolic mathematical information have some pre-defined notion of when expressions are well-formed and, of the well-formed expressions, which are the most desirable. Which forms are deemed most desirable is usually decided by the software system designers, through their experience or preference, and hard-coded into the application's logic. This has made symbolic mathematical software more natural to use in some areas than others, depending on the compatibility of the system designer's choices with the user's needs. As we move toward more sophisticated, knowledge-based mathematical software, this methodology becomes increasingly problematic.

M. Kohlhase (Ed.): MKM 2005, LNAI 3863, pp. 361–375, 2006.

In this paper we argue that it is important to understand what forms of expressions are deemed most desirable in the actual practice of mathematics. We believe that empirical knowledge of which forms of expressions are used most often will lead to more effective mathematical software. For example, this information could be used to guide simplification in computer algebra systems, or to provide disambiguation criteria in mathematical document recognition.

Our initial motivation for this work comes from the area of mathematical handwriting recognition. We note that today's acceptable recognition rates for natural language handwriting is achieved with the aid of dictionary-based methods. For example, if the feature analysis of a stroke could yield either `Hdb` or `Hello`, then `Hello` is chosen because it is in the dictionary. At first consideration, such an approach is not suitable for mathematical handwriting recognition for several reasons: Mathematical expressions are trees, not strings. There is no fixed vocabulary from which to build a dictionary. The set of symbols alone is insufficient, and the set of possible expressions is infinite.

Nevertheless, any mathematically sophisticated person can take an arbitrary volume from a mathematical library, leaf through the pages, and, in a few seconds, have a very good idea of the precise mathematical subject area, in part, simply be noticing some characteristics of the formulae. We therefore claim that there is, in fact, usage knowledge that can and should be used by mathematical software packages. In the mathematical handwriting recognition case, this knowledge could be used to disambiguate between $\sin \omega t$ and $\sin wt$, since the former occurs much more often in practice. In the computer algebra case, this knowledge could be used to order one polynomial as $x^2 + 1$ and another as $1 + \epsilon^2$.

The goals of this present line of work are to understand how

- to capture and represent empirical mathematical usage information
- to employ this information in mathematical software packages
- to analyze and organize this knowledge so as to be most useful.

We report here on our initial results toward these long-term goals. As stated earlier, we see immediate applicability to mathematical handwriting recognition and to symbolic mathematical computing. Other potential applications include mathematical searching, automated classification of mathematical documents, and mathematical data mining.

The contributions of this work are

- the identification of empirical mathematical usage as an important source of information for mathematical software design
- an approach to empirical analysis of mathematical expressions
- specific findings on symbol usage, on a subject-by-subject basis
- specific findings on most common expression usage
- methods to derive pattern expressions, and symbol-sequence Markov chains, based on analysis of instances.

The rest of the paper is organized as follows: We present the methodology of the current study in Section 2. As part of this study, we rely on a TEX to

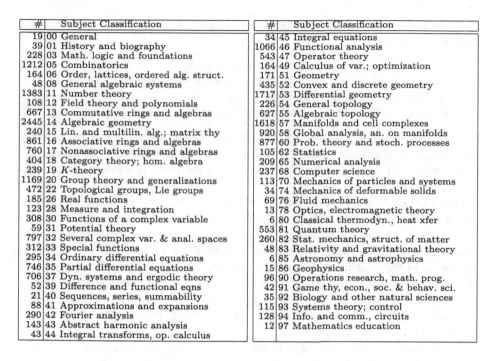

#	Subject Classification		#	Subject Classification
19	00 General		34	45 Integral equations
39	01 History and biography		1066	46 Functional analysis
228	03 Math. logic and foundations		543	47 Operator theory
1212	05 Combinatorics		164	49 Calculus of var.; optimization
164	06 Order, lattices, ordered alg. struct.		171	51 Geometry
48	08 General algebraic systems		435	52 Convex and discrete geometry
1383	11 Number theory		1717	53 Differential geometry
108	12 Field theory and polynomials		226	54 General topology
667	13 Commutative rings and algebras		627	55 Algebraic topology
2445	14 Algebraic geometry		1618	57 Manifolds and cell complexes
240	15 Lin. and multilin. alg.; matrix thy		920	58 Global analysis, an. on manifolds
861	16 Associative rings and algebras		877	60 Prob. theory and stoch. processes
760	17 Nonassociative rings and algebras		105	62 Statistics
404	18 Category theory; hom. algebra		209	65 Numerical analysis
239	19 K-theory		237	68 Computer science
1169	20 Group theory and generalizations		113	70 Mechanics of particles and systems
472	22 Topological groups, Lie groups		34	74 Mechanics of deformable solids
185	26 Real functions		69	76 Fluid mechanics
123	28 Measure and integration		13	78 Optics, electromagnetic theory
308	30 Functions of a complex variable		6	80 Classical thermodyn., heat xfer
59	31 Potential theory		553	81 Quantum theory
797	32 Several complex var. & anal. spaces		260	82 Stat. mechanics, struct. of matter
312	33 Special functions		48	83 Relativity and gravitational theory
295	34 Ordinary differential equations		6	85 Astronomy and astrophysics
746	35 Partial differential equations		15	86 Geophysics
706	37 Dyn. systems and ergodic theory		96	90 Operations research, math. prog.
52	39 Difference and functional eqns		42	91 Game thy, econ., soc. & behav. sci.
21	40 Sequences, series, summability		35	92 Biology and other natural sciences
88	41 Approximations and expansions		115	93 Systems theory; control
290	42 Fourier analysis		128	94 Info. and comm., circuits
143	43 Abstract harmonic analysis		12	97 Mathematics education
43	44 Integral transforms, op. calculus			

Fig. 1. Count of articles by MR Subject Classification

MathML conversion. Section 3 describes this process and extensions we have had to make for the present work. Results on frequency of symbols, as identifiers and operators, are reported in Sections 4 and 5. We present some initial results on expression analysis in Section 6. Section 7 concludes the paper.

2 Methodology

To study the empirical usage of mathematical expressions, the first step was to identify a suitable source of mathematical input. A number of possibilities existed, including

- to use logged input from a software system, such as Maple,
- to use a collection of documents from a set of cooperative authors,
- to use the articles from a particular journal

Although any of these avenues would have been easy to follow, each had its own problems: Logged input from a software system would heavily influenced by the characteristics of the system, and thus be riddled with artifacts. Articles from a small set of authors, or from a particular journal, would likely be heavily slanted in their usage and could not be taken as representative.

Instead, we chose to use the collection of articles available on the widely used, public e-Print server, arXiv.org [2], as our corpus of mathematical usage.

This has the advantage of broad coverage by mathematical area. It also has the disadvantages that:

- Some areas are disproportionately represented.
- The mathematical material is at a research level, and this may not be representative of usage at more elementary levels.
- The material is relatively new, and is not representative of historical usage.

Bearing this in mind, we decided that the collection of articles was sufficiently representative of current mathematical usage to be useful, and that developing a collection that was more balanced by area, level, historical period, *etc*, was a long-term project.

One of the attractive properties of `arXiv.org` is its organization of articles according to the Mathematics Subject Classification, which is used to categorize items covered by the two reviewing databases, Mathematical Reviews (MR) and Zentralblatt MATH (Zbl). The current classification system, MSC 2000 [3], is a revision of the classification scheme that had been used previously by these databases. It consists of more than 5,000 two-, three-, and five-character classifications, corresponding to increasingly finely defined disciplines of mathematics. For example, "11" represents Number theory; "11B" Sequences and sets, and "11B05" Density, gaps, topology.

We followed the following steps to obtain our corpus of expressions to analyze:

The first step was to obtain all articles from `arXiv.org` from the five year period 2000–2004. This data range contained all articles since the new subject classification was introduced. To understand area-specific usage patterns, while having a sufficient number of articles in each category, we grouped articles according to their top-level, two-digit MSC classification. The count by classification of articles considered is shown in Figure 1. Altogether 22,289 articles were accessed. Of these 21,677 came with TeX source. This comprised 4.65GB of PDF files and 794 MB of TeX source.

The second step was to extract mathematical expressions from the articles. It was helpful that the articles had TeX source, but this was not usable directly for our analysis. The problems with TeX source include:

- Mathematical expressions typically use author-defined macros.
- Mathematical expressions my be hidden in macros, and not be visible in the source text.
- TeX expressions typically have only as much structure as is needed to give proper visual grouping. For example $(ad-bc)^2$ consists of a single row of 7 items, (, a, d, -, b, c and)^2. Note that there is no notion that ad and bc are subexpressions, while $d - b$ is not, and note that it is only the closing parenthesis that is squared.

We used our TeX to MathML [1] converter, described in [8], to resolve these difficulties, and performed our analysis on the resulting MathML expressions. The benefit of this approach was that the expressions treated were (for the most part) complete, well formed, and grouped appropriately. The difficulty with the approach was that not all the complexities of TEX were handled, and some expressions were incorrectly translated. However, since we are interested in the most frequently occurring expressions, the incomplete handling of infrequently occurring expressions is not, in principle, a problem. We describe the conversion process in more detail in Section 3. The overall conversion process required about three days of computer time on a personal workstation.

The third step was to examine the MathML expressions for each area, and to build three frequency tables. The first two tables contained counts of all identifier symbols (typically single letter operands) and all operator symbols. The third table counted the number of occurrences in the classification of each sub-expression. These tables were built using syntactic comparison of XML elements. For example, `<mrow><mo>(</mo><mi>a</mi><mo>)</mo></mrow>` would be treated as inequivalent to `<mfenced><mi>a</mi></mfenced>`. We therefore preprocessed the MathML to remove multiple representations for what would appear as *syntactically equivalent* mathematical expressions. This consisted of a number of simple conversions, including

- for `<mi>` and `<mo>`, normalizing the use of the `mathvariant` attribute
- for `<mfrac>`, eliminating any non-zero `linethickness` attribute
- for `<mfenced>`, convert to `<mrow>` with explicit open and close operators
- for trivial `<mmultiscripts>`, convert to `<msub>` or `<msup>`
- elimination of a number of attributes and elements related to presentation, such as spacing.

3   TEX to MathML Conversion

The conversion of TEX to MathML is not a straightforward process. There is not yet a standard tool that completely solves this problem. TEX documents are, in general, programs with the computational power of a Turing machine. In practice, TEX macros are usually used to perform simple substitutions, with a smaller number performing heavy computations and transformations.

There are two principal approaches to TEX to MathML conversion: The first approach is to use alternative style files with modified definitions for the standard mathematical macros. These modified macros leave special markers in the generated `dvi` file, which are then used to generate the MathML. This approach has the advantage that all TEX files can be handled. The disadvantage is that all the high-level structure implicit in the TEX markup is discarded. This is the approach taken by `TeX4ht` [10] and the Hermes project [4].

The second approach is have a (partial) implementation of a TEX processor handle the input, and to generate MathML from the higher-level TEX operators. This has the advantage that implicit semantics in TEX markup (e.g. grouping information from braces, "{" and "}") is available to the MathML generation. The disadvantage is that, in principle, a complete TEX re-implementation is needed.

For this study, we used a TEX to MathML converter, developed within the ORCCA research group. This converter adopts the second approach. It has a partial implementation of the TEX programming language sufficient to expand the macros of interest in mathematics. Source for a TEX document may be given as a single file, or as a tree of files and using external macro packages. The correspondences between TEX and MathML are given by a set of bi-directional mapping files. These mapping files are intended to allow high-level semantic mappings between TEX and XSLT style sheets [8]. Because complex TEX macros are almost always given in style files, rather than being specified at top-level by authors, the mapping files may almost always be used to eliminate any short-comings arising from the incomplete implementation of TEX. This translator is available on-line [5].

The conversion of all TEX source documents in the five year `arXiv.org` collection served as heavy test for the MathML converter, and a number of problems were encountered. Initially only 14,354 of the 21,677 articles could be handled automatically. First, we discovered that there were a number of TEX constructs that were not handled by the converter. The most important of these were (1) the handling of explicit positioning commands, e.g. for kerning symbols, and (2) the ability to handle arbitrary external macro packages from a search path. Dealing with these difficulties proved to be fairly easy.

The second major difficulty in the TEX to MathML translation was that a significant number of the TEX source files did not contain valid TEX. The TEX converter had been constructed assuming valid input, the idea being that an author would first produce a correct file by debugging with TEX and then, possibly long afterward, generate MathML. This assumption proved invalid — authors do not always correct their TEX errors if TEX's error recovery gives a desired output. We therefore were required to extend the TEX to MathML converter to simulate TEX error handling.

With user error handling in place, we were able to process 19,137 of the articles automatically. Of these, 19,063 were able to have their MathML canonicalized, and it is from these that we have extracted the expressions for analysis.

4   Identifiers

Our first analysis determines the most frequently occurring symbols used as identifiers in mathematical expressions. By this we mean letter-like symbols that occur as operands or function names, rather than as operators.

We counted all symbols occurring in expressions and recorded the results both for the global analysis and independently for each category. The first observation is that in each classification some symbols occur much more frequently than others, and which symbols are the most frequent differs from classification to classification.

Figure 2 shows the most frequently occurring identifiers for all the classifications taken together, as well as the most frequently occurring identifiers for three typical classifications, Logic, Number Theory and Partial Differential Equations. For detailed information on all classifications see [7].

All			03			11			35		
Ucode	Id	Freq	Ucode	Id	Freq	Ucode	Id	Freq	Ucode	Id	Freq
006E	n	48,150	0069	i	51,565	006E	n	58,186	0078	x	51,773
0069	i	43,280	006E	n	48,239	0070	p	40,302	0074	t	49,859
0078	x	36,240	0078	x	41,042	006B	k	38,230	0075	u	39,841
006B	k	32,060	0058	X	33,862	0078	x	35,294	006E	n	35,705
0074	t	25,967	0041	A	29,845	0069	i	35,100	006B	k	29,924
0058	X	23,369	0070	p	26,292	0061	a	25,301	0069	i	28,941
006A	j	23,038	03B1	α	24,604	006D	m	23,642	0073	s	25,234
0070	p	22,832	006B	k	24,374	0064	d	22,302	006A	j	24,968
0041	A	22,791	0066	f	22,671	0071	q	21,797	0064	d	24,095
0061	a	21,435	0061	a	22,030	0073	s	21,319	004C	L	21,094
0064	d	19,457	0047	G	21,983	006A	j	21,153	03B5	ϵ	20,740
006D	m	19,263	006D	m	19,893	0072	r	19,695	03BB	λ	20,189
0066	f	18,235	006A	j	18,062	0074	t	19,654	0070	p	19,107
004D	M	18,135	03C9	ω	18,015	0047	G	19,620	0043	C	17,450
0073	s	17,659	004D	M	17,256	0058	X	19,535	03B1	α	17,087
0072	r	17,248	0053	S	17,122	0041	A	19,107	0072	r	16,834
0043	C	16,915	0043	C	17,107	004B	K	18,905	0076	v	16,820
0053	S	16,487	0046	F	16,773	0066	f	18,126	0061	a	15,931
0047	G	16,074	0079	y	16,764	0046	F	16,524	0079	y	15,920
03B1	α	15,943	0074	t	15,693	004C	L	15,921	0066	f	15,215

Fig. 2. The most frequent identifiers (per million) in all classifications (All), Logic (03), Number Theory (11) and Partial Differential Equations (35)

03			11			35		
Ucode	Id	Freq	Ucode	Id	Freq	Ucode	Id	Freq
03C9	ω	18,015	0071	q	21,797	0075	u	39,841
0046	F	16,773	004B	K	18,905	004C	L	21,094
0079	y	16,764	0046	F	16,524	03B5	ϵ	20,740
0054	T	15,605	004C	L	15,921	03BB	λ	20,189
0062	b	15,270	004E	N	15,537	0076	v	16,820
004B	K	15,144	0076	v	14,380	0079	y	15,920
0042	B	15,002	0054	T	14,126	03BE	ξ	15,154
0063	c	14,586	0067	g	13,683	007A	z	14,459
0050	P	14,582	0050	P	13,479	0054	T	14,333
03BA	κ	13,285	007A	z	13,333	004E	N	13,906
004C	L	13,280	0079	y	12,880	0048	H	13,575
0056	V	12,004	0063	c	12,383	0052	R	12,421
0055	U	11,916	0048	H	12,238	0068	h	12,392
0048	H	11,452	0044	D	12,056	03A9	Ω	12,305
0071	q	11,385	0062	b	11,867	0077	w	11,562
03B2	β	11,305	0045	E	11,714	03B4	δ	11,120
0068	h	10,369	03C0	π	11,348	0067	g	10,933
03B3	γ	10,196	0068	h	10,550	0044	D	10,809
0067	g	10,104	0042	B	10,309	0071	q	10,380
0059	Y	9,918	0075	u	10,291	03BC	μ	10,356

Fig. 3. Most frequent identifiers (per million) in Logic (03), Number Theory (11) and Partial Differential Equations (35), after excluding the 20 globally most frequent

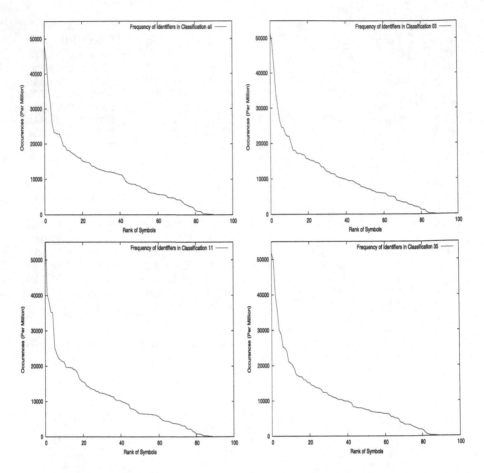

Fig. 4. Most frequent identifiers in all expressions (upper left), Logic (upper right), Number Theory (lower left), and Partial Differential Equations (lower right). The horizontal axis gives the symbol (from most to least frequent), and the vertical axis gives the number of occurrences per million symbols in the classification.

This information could be used for disambiguation in mathematical handwriting recognition. In Number Theory, for example, we see that the letter n occurs more than twice as frequently as the letter r. By feature analysis alone, these two letters are difficult to distinguish. This frequency information is therefore useful in disambiguation.

We have arrived at a generalization of the dictionary used for disambiguation in handwriting recognition: we have constructed here, with symbols (and, in Section 6, with expressions) a *weighted dictionary*. This structure carries information about the vocabulary of potential results, together with empirically determined weights.

Figure 3 shows the most frequently occurring identifiers for the same classifications after excluding the 20 identifiers that appear most frequently in all classifications together. We see these lists are less similar than those of Figure 2. We might use this information to aid in automatic document classification, together with word-frequency and citation analysis. Information such as this could also be used by an interactive system as a heuristic aid to determine the mathematical area in which a user is working.

Figure 4 shows, for the same classifications, the number of occurrences of identifier symbols, with the symbols ordered from most frequent to least frequent. While this will obviously be a monotonically decreasing curve, it is remarkable the degree of similarity in the shapes of these curves. We observe that although *which* symbols are used most varies quite a bit from mathematical area to area, the distribution of use of symbols is remarkably similar. In particular, after the 10% most popular identifiers, the frequency of appearance ordered by identifier decays approximately linearly.

Although, for space reasons, we have presented here the tabular results and graphs for only three classifications, and for the aggregate, the overall picture is similar for the other classifications.

5   Operators

An analogous analysis to that for identifiers was performed for operator symbols. We counted as operators anything occurring in an `<mo>` element, excluding the characters "(", ")", "[", "]", "{", "}", thinspace and underscore. We excluded the

All			03			11			35		
Ucode	Op	Freq	Ucode	Op	Freq	Ucode	Op	Freq	Ucode	Op	Freq
003D	=	128,715	003D	=	121,806	003D	=	130,735	002D	−	138,603
002D	−	116,064	2061		115,262	002D	−	128,330	002C	,	111,176
002C	,	112,818	002C	,	100,880	2061		112,484	2061		103,527
2061		103,090	2208	∋	77,021	002C	,	104,964	003D	=	103,376
002B	+	79,404	002D	−	60,732	002B	+	94,172	002B	+	97,579
2208	∋	43,942	002B	+	60,121	002F	/	40,239	2208	∋	38,370
002A	*	29,210	002A	*	32,796	2208	∋	39,319	2264	≤	34,575
2192	→	23,818	003C	<	28,345	2211	∑	20,165	2202	∂	28,815
002F	/	23,405	02C9	¯	25,805	2264	≤	19,574	002F	/	25,985
2264	≤	20,088	2192	→	24,370	2192	→	18,481	221E	∞	23,460
02DC	~	16,875	2264	≤	24,242	002A	*	17,757	222B	∫	23,196
2297	⊗	14,242	002F	/	14,626	00AF	¯	14,708	02DC	~	19,545
2211	∑	13,560	2026	…	13,495	221E	∞	14,627	003C	<	16,453
003E	>	13,528	222A	∪	12,654	003E	>	12,926	2207	∇	15,387
221E	∞	13,138	2229	∩	12,483	22EF	⋯	12,358	003E	>	15,256
00AF	¯	12,451	2286	⊆	12,330	02DC	~	12,209	002A	*	14,470
003C	<	12,058	003E	>	11,784	2265	≥	11,963	2192	→	14,381
22EF	⋯	12,005	2223	∣	9,883	2113	ℓ	10,997	22C5	·	12,669
2202	∂	11,940	22EF	⋯	9,781	003C	<	10,151	2211	∑	12,394
00D7	×	11,294	02DC	~	9,428	00D7	×	10,144	2265	≥	11,531

Fig. 5. The most frequent operators (per million) in all classifications (All), Logic (03), Number Theory (11) and Partial Differential Equations (35). The Unicode point 2061 is the invisible "ApplyFunction" operator.

bracket forms because they were so frequent their occurrence masked the details of the other operators. Thinspace is often used for adjusting appearance, and underscores were an artifact of incomplete TEX translation. With this, Figure 5 shows the most frequently occurring operators for the same classifications as for the identifiers. Figure 6 shows the most frequently occurring operators, excluding from each category the 20 most globally common operators.

Figure 7 shows the count of operator symbols, by category, sorted from most to least popular. We note that the shape of the operator distribution is roughly similar among categories, although there are some evident differences, and even though it is different operators that are occurring most frequently. The shape of the distribution is quite different from the distribution for identifiers: generally, a few operators are used very frequently.

We see that in all areas there are a few (1-5) operator symbols that occur very frequently followed by a rapid decay in use. In particular see that more than half the symbol occurrences are from the top 10% most popular operators, and almost all occurrences are from the top 40% most popular operators.

We note that the shape of the distribution for the most popular operators varies by category. For example, in Number Theory and Partial Differential Equations, the first few most popular operators occur with similar frequency, followed by a sharp drop, whereas in Logic there there is a more gradual decline in frequency of use.

03		
Ucode	**Op**	**Freq**
02C9	‾	25,805
2026	...	13,495
222A	∪	12,654
2229	∩	12,483
2286	⊆	12,330
2223	|	9,883
2218	∘	8,894
2265	≥	8,252
2329	⟨	7,348
232A	⟩	7,072
2260	≠	6,885
2200	∀	6,390
0022	"	6,177
2227	∧	5,978
02C6	^	5,825
2282	⊂	5,552
2113	ℓ	5,467
2216	\	5,282
2203	∃	4,990
22C5	·	4,745

11		
Ucode	**Op**	**Freq**
2265	≥	11,963
2113	ℓ	10,997
2223	|	9,474
02C9	‾	8,750
2026	...	7,829
22C5	·	7,728
02C6	^	7,464
222B	∫	5,719
220F	∏	5,287
2282	⊂	4,938
2032	′	4,681
2260	≠	4,626
224D	≍	4,534
2229	∩	4,238
0021	!	3,692
2218	∘	3,550
2295	⊕	3,062
0022	"	2,849
00B1	±	2,796
226A	≪	2,644

35		
Ucode	**Op**	**Freq**
222B	∫	23,196
2207	∇	15,387
22C5	·	12,669
2265	≥	11,531
02C9	‾	9,349
02C6	^	8,170
2223	|	6,379
2113	ℓ	6,074
232A	⟩	5,583
2329	⟨	5,559
00B1	±	4,556
2282	⊂	4,130
2229	∩	3,728
2272	≲	3,635
002E	.	3,375
2216	\	3,239
2260	≠	2,843
0022	"	2,767
2026	...	2,397
2032	′	2,328

Fig. 6. Most frequent operators (per million) in Logic (03), Number Theory (11) and Partial Differential Equations (35), after excluding the 20 globally most frequent

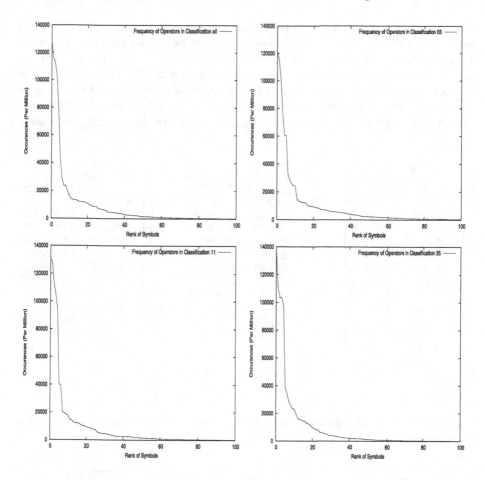

Fig. 7. Most frequent operators in all expressions (upper left), Logic (upper right), Number Theory (lower left), and Partial Differential Equations (lower right). The horizontal axis gives the symbol (from most to least frequent), and the vertical axis gives the number of occurrences per million symbols in the classification.

6   Expressions

We have performed a similar analysis for non-trivial subexpressions, counting the number of times each distinct subexpression occurs in each subject classification. The analysis of the results is more complex, however.

A large subexpression that occurs a certain number of times is more significant than an smaller subexpression that occurs as often, for two reasons. The first reason is that, in absolute terms, there tend to be fewer subexpressions of large size. The second reason is that there are exponentially more potential expressions of the larger size.

With the idea that the size of an expression should be part of determining the significance of its occurrences, we have analyzed each subject classification for the number of expressions. The results for subject classifications 03, 11 and 35 are shown in Figure 8. For each of these classifications and for each size, the figure shows (i) the number of subexpressions of that size that occurred in the articles, (ii) the number of *distinct* subexpressions occurring of that size, and (iii) the number of distinct subexpressions as a percentage of the number of expressions. We measure the size of an expression as the count of the following MathML tags that produce output, as opposed to providing structure. In our case, because of

03				11				35			
Sz	#	distinct	%	Sz	#	distinct	%	Sz	#	distinct	%
2	169,302	10,612	6	2	1,118,829	39,729	3	2	844,311	24,032	2
3	63,952	11,868	18	3	513,041	73,095	14	3	398,139	48,962	12
4	31,485	8,850	28	4	253,320	69,937	27	4	188,363	46,826	24
5	16,688	6,377	38	5	169,908	59,286	34	5	132,814	40,389	30
6	9,726	4,035	41	6	87,256	42,085	48	6	73,267	30,059	41
7	6,271	3,258	51	7	68,645	34,438	50	7	58,787	26,036	44
8	3,878	2,082	53	8	40,907	23,445	57	8	36,283	18,593	51
9	3,395	1,887	55	9	33,558	19,803	59	9	28,560	15,960	55
10	2,196	1,255	57	10	22,060	14,254	64	10	19,930	11,894	59
11	1,977	1,222	61	11	18,614	12,562	67	11	16,770	10,287	61
12	1,229	798	64	12	13,216	9,341	70	12	12,159	8,084	66
13	1,049	782	74	13	11,692	8,430	72	13	10,811	7,284	67
14	868	652	75	14	8,690	6,533	75	14	8,407	5,883	69
15	776	577	74	15	7,857	5,969	75	15	7,056	5,268	74
all	318,000	58,555	18	all	2,436,305	480,023	19	all	1,908,150	363,746	19

Fig. 8. Number of subexpressions and of distinct subexpressions by classification and by subexpression size

Size 2			Size 4			Size 7	
#	expression		#	expression		#	expression
19,752	-1		2,111	$-1/2$		445	$\frac{n+2}{n-2}$
14,565	L^2		955	$-\frac{1}{2}$		194	$\frac{n+4}{n-4}$
8,098	dx		660	$\frac{d}{dt}$		110	(x', ξ', μ)
5,634	t_0		427	$W^{1,p}$		96	$p-1, q-1$
4,735	x_0		397	$\int_{-\infty}^{\infty}$		90	$-(a+1)p+c$
4,628	∂_t		394	$-3/2$		88	$\sum_{i,j=1}^{n}$
4,607	ij		358	$\frac{\partial}{\partial t}$		75	$j_1, j_2 \geq 0$
4,572	u_0		352	S^{n-1}		75	$(g(t), K(t))$
4,183	dt		340	$\delta_{x,\lambda}$		70	$u^{\frac{2n}{n-2}}$
4,142	(t,x)		337	H^{N-1}		69	$(t, x; \tau, \xi)$

Fig. 9. Most frequent subexpressions of size 2, of size 4 and of size 7 in subject classi-fication classification Partial Differential Equations (35)

the nature of the TeX to MathML conversion, the tags we counted were `<mi>`, `<mo>`, `<mn>`, `<mroot>`, `<msqrt>`, `<mfrac>`, `<menclose>`, and `<ms>`.

We observe two phenomena: First, as expected, the number of expressions occurring decreases as size increases. There are many more small expressions than large expressions. Secondly, we note that as expressions become larger, the fraction that are distinct increases. The proportion of unique subexpressions seems to depend strongly on the classification.

This analysis provides a weighted dictionary for each subject classification, providing the frequency that expressions occur in each subject classification. Space limitations preclude giving a detailed accounting of the particular expressions which occur most frequently in each classification, but we give a sample from the classification 35, Partial Differential Equations. These are shown in Figure 9. More details are available in [7].

An important question is how dependent is the weighted dictionary of subexpressions on the choice of TeX to MathML converter. Since each conversion program will have its own choices for MathML output idioms, there is a clearly dependency. However, for each expression there is a well defined collection of symbols and an intended grouping. Provided the TeX to MathML converter is consistent and provided it correctly identifies the intended groupings, the distribution of entries in the weighted dictionary should be stable under choice of converter. The application using the entries must be aware, however, that the choice of the exact way to represent a particular expression may be arbitrary.

The information in these weighted dictionaries may be used directly by applications, or may be used for further analysis. Two such directions of further analysis are deriving expression patterns, and deriving common writing sequences. We foresee many additional uses of this kind of empirical data on expression frequency.

Expression Patterns

We note that very similar subexpressions may occur frequently, for example $\sqrt{A^2 + B^2}$ and $\sqrt{x^2 + y^2}$. While it is possible to maintain a weighted dictionary keeping track of both of these expressions, it would be more desirable to determine that $\sqrt{\alpha^2 + \beta^2}$ was a frequently occurring pattern, with suitable choices of α and β.

"Antiunification" provides an elegant framework to define such patterns. Antiunification is a process dual to unification. Rather than taking expressions and determining the most general expression to which they all can be specialized, antiunification takes a number of instance expressions and finds the least general expression which may be specialized to each instance expression. The syntactic form of antiunification has been studied since the 1970s [6].

We may determine the set of patterns from a weighted dictionary by considering all pairs of expressions. Each pair will give an antiunifier. We then consider all pairs of antiunifiers with expressions from the dictionary. These may give more antiunifiers, which are added to the set of antiunifiers. We continue to consider pairs of antiunifiers with expressions until no new antiunifiers are generated.

Since antiunification is associative, this generates a complete set of antiunifiers for the dictionary. For each antiunification, we may use the one pass algorithm of [9].

We may associate weights with these patterns simply: for each antiunifier, attempt a unification with each expression in the weighted dictionary. Then the weight of the antiunifier is the sum of the weights of the expressions with which it unifies. We note that since we are interested in syntactic expressions, this entire process of antiunification and unification is syntactic. An empirically derived, weighted dictionary of antiunifiers would provide an interesting measure to select among possibilities for "simplified" forms in a computer algebra system.

Tree-Order Symbol Sequencers

The second direction we wish to discuss for deriving expression patterns is the use of ordered tree traversals. We examine this in support of mathematical handwriting recognition. For each type of tree node, we define a traversal order corresponding to the most common writing order. For example, with

$$\sum_{i=0}^{\infty} i^2$$

the summation sign is usually written first, followed by the equation $i = 0$, then ∞, and finally i^2. Ideally the information on writing order for each node type should be determined with user experiments. Without these experiments, it is still possible to have writer-specific traversal order.

Given one or more traversal orders for each node type, we may then examine the weighted dictionary of expressions, traversing each expression, to determine Markov chains for symbol sequences. If the expression $\sum_{i=0} \ldots$ occurs twice as frequently as $\sum_{j=0} \ldots$, then the symbol sequence $\langle \Sigma, i \rangle$ gets twice the weight of $\langle \Sigma, j \rangle$. If there is not a unique traversal order for a node type, then the alternatives may be weighted.

7   Conclusions

We have proposed the idea of empirical analysis of mathematical literature as a new technique to be used in the design of sophisticated mathematical software. This is a break from the tradition of system designers using their own preferences or prejudices in determining which forms of expressions will be deemed most preferable by their systems.

We have taken presented an approach to performing empirical analysis of a body of mathematical literature. We have developed a suite of tools to convert raw TeX source to well-formed MathML, and to build weighted dictionaries of symbols and expressions.

We have made an analysis of all articles from arXiv.org since the new MSC 2000 subject classification. From this, we have observed that the use of mathematical symbols varies considerably from area to area and have produced usage frequency tables for all MSC 2000 classification areas. We have observed that, while the specifics of *which* symbols are most used varies from area to area, the overall *distribution* of symbol use is very similar between areas. This is true both for symbols used as identifiers (function names and arguments), and as operators. We have also analyzed the collection of subexpressions present in the arXiv.org data. As well as developing a weighted dictionary for each classification area, we have observed some general properties of the frequency of distinct expressions. We are currently investigating how to best make these dictionaries available to other research projects.

Beyond these practical experiments, we have explored the potential use of information derived from symbol and expression weighted dictionaries. These have included particular applications to computer algebra, mathematical handwriting recognition and document analysis. We have also shown how weighted expression dictionaries may be used to determine further useful information, including weighted pattern dictionaries (by antiunification) and Markov chains for symbols in writing-order traversal of expression trees.

The applicability of these results depends on how representative the empirical data is. It is likely that different tables would be obtained from high-school mathematics texts, for example. Therefore, the overall approach we have taken is just as important as the specific results for this particular mathematical database.

We are excited and hopeful that the use of empirically gained knowledge may make mathematical software systems more powerful and more natural to use.

References

1. David Carlisle, Patrick Ion, Robert Miner, Nico Poppelier, Editors. Mathematical Markup Language (MathML) Version 2.0 (Second Edition). W3C Recommendation. http://www.w3.org/TR/2003/REC-MathML2-20031021/. October 21, 2003.
2. ArXiv e-Print Archive. http://xxx.lanl.gov
3. Mathematical Subject Classification (2000). American Mathematical Society. http://www.ams.org/msc
4. The Hermes Project. http://alphaserv3.aei.mpg.de/hermes
5. Ontario Research Centre for Computer Algebra. On-line TeX to MathML translator. http://www.orcca.on.ca/MathML/texmml/textomml.html (2002)
6. Gordon D. Plotkin. A Note on Inductive Generalization. Machine Intelligence 5 153–163 (1970).
7. Clare M. So. An Analysis of Mathematical Expressions Used in Practice. MSc. Thesis. University of Western Ontario. (2005)
8. Stephen M. Watt. Implicit Mathematical Semantics in Conversion between TeX and MathML, TUGBoat, Vol 23, No 1 (2002)
9. Cosmin Oancea, Clare So and Stephen M. Watt. Generalization in Maple, pp 377-382, Maple Conference 2005, Maplesoft.
10. TeX4ht: LaTeX and TeX for Hypertext, http://www.cse.ohio-state.edu/~gurari/TeX4ht

Transformations of MML Database's Elements

Robert Milewski

Institute of Computer Science, University of Bialystok,
Sosnowa 64, Bialystok, Poland
milewski@math.uwb.edu.pl

Abstract. The main goal of most systems for formalizing mathematics is creating the database of formalized knowledge. Every system uses its own way of storing that knowledge. Pieces of the information stored in such a database are something more than just elements of that database – we may also look at it from another perspective. Namely, the information is the reflection of some text and that text is the representation of some logical reasoning. Therefore every such piece of information may be treated as an element of some set and then we may analyze the relationships between all these elements. This article describes such an approach to one of the systems for formalizing mathematics.

1 Motivation and Main Ideas

One of the oldest systems for computer-aided formalization of mathematics is the MIZAR system created by Andrzej Trybulec [1], [9], [3], [4], [5], [6]. The knowledge which is formalized with this system is stored in the *Mizar Mathematical Library* (MML). This database has been developed since 1989, and currently it contains more than 900 articles, with over 39 000 theorems, over 7 500 definitions and over 700 schemes. The articles are authored by over 150 people from more than 10 countries. All articles stored in the MML as text files occupy more than 65 Mb of disk space. Freek Wiedijk in his paper [7] devoted to 15 main systems for formalizing mathematics stated that the MIZAR system has one of the biggest databases (MML) among all of the analyzed systems.

Mizar Mathematical Library, like the databases of other systems for formalizing mathematics, is usually treated mainly as a whole structure storing formalized knowledge. But there is another way of looking at this database. The formalized texts may be considered as individual elements – this allows for analyzing the MML database (or the set of formalized texts in some other system) on the basis of mutual comparisons of its elements, and searching for relationships between them. Determining these relationships allows us to compare the results produced by applications which modify formal texts with their original form. Such comparisons allow us to classify these applications. Examining the applications, MML maintainers can decide how useful the application is for database maintenance tasks. This was the main reason for defining relationships between formalized texts and analyzing the levels of equivalence of the applications mentioned above.

M. Kohlhase (Ed.): MKM 2005, LNAI 3863, pp. 376–388, 2006.

2   The Relationships Between MIZAR Texts

When we treat MIZAR articles as elements of such a set of MIZAR texts, it is obvious that the first defined relationship should be some kind of equivalence between the articles (whatever it means). The research on the data-mining and the robustness of the MIZAR system, which started a few years ago [2] and gave rise to many utilities modifying articles in the MIZAR database, confirmed the need to define these relationships to check whether a given application does not change the original text significantly.

But there appears the problem of proper understanding the concept of the equivalence between MIZAR articles. That equivalence may have many meanings, from total identity to some kind of semantic equivalence. Therefore we can determine many such relationships. To start with the most strongest relationship, we may define:

$A_1 r_0 A_2 \Leftrightarrow A_1$ and A_2 are equivalent on the text level.

This equivalence is a very restrictive concept, because it is satisfied only by identical texts (considering also white spaces and line breaks).

When we slightly weaken the relation r_0, we can define another relation as follows:

$A_1 r_1 A_2 \Leftrightarrow A_1$ and A_2 are syntactically equivalent.

The syntactic equivalence means that both texts are identical with respect to the relation r_0, but there is no control of repeated white spaces and line breaks.

Going further in this direction, we define another relation:

$A_1 r_2 A_2 \Leftrightarrow A_1$ and A_2 are "weakly" syntactically equivalent.

The "weak" syntactic equivalence means that the conditions of r_1 are fulfilled, with the exception of consistent changes of the names of constants, variables and adding or removing labels and other keywords connected to references (e.g. "then") if such an operation does not destroy the structure of references (e.g. it is unacceptable to remove a label which occurs later in some inference). But even a change of the order of references in some inference may cause the lack of equivalence with respect to relationship r_2.

Yet another relation:

$A_1 r_3 A_2 \Leftrightarrow A_1$ and A_2 are "weakly" syntactically equivalent excluding justifications is similar to the relation r_2, but is weaker than r_2 because it permits some alterations to the justification of reasoning steps. The structure of proofs remains the same, but there is a possibility of changing the organization of inferences, sometimes also changing the tree of references. Of course it is unacceptable to change the organization of inferences to such an extent so that a correct text is not accepted by the VERIFIER anymore.

Another relationship looks this:

$A_1 r_4 A_2 \Leftrightarrow$ the abstracts of the articles A_1 and A_2 are "weakly" syntactically equivalent.

A MIZAR abstract is the MIZAR text with all proofs and justifications removed [8], [11], [10]. Checking the equivalence of MIZAR abstracts is a specific form of syntactic equivalence, because it permits for full liberty in justifying proved facts if the weak syntactical equivalence is preserved.

$A_1 r_5 A_2 \Leftrightarrow A_1$ and A_2 are semantically equivalent.

The relation r_5 stands for the semantic equivalence of justified contents. The way that facts are justified is not important, and neither is their order nor the structure of their notations. Every theorem justified in the first text must have a semantic equivalent in the second text, and vice versa. Here the semantic equivalence of theorems is understood as the mutual justifiability of the theorems by one another.

$A_1 r_6 A_2 \Leftrightarrow A_1$ and A_2 are "weakly" semantically equivalent.

The relation r_6 holds when every theorem from one MIZAR text can be justified by some theorems taken from the other text, and vice versa.

There are, of course, other levels of equivalence, e.g. the tautological equivalence of theorems. However for my considerations it was enough to focus mostly on the 7 defined (mostly syntactic) ones.

All the relations defined above are of course equivalence relations, and they form a sequence as below:

$$r_0 \subseteq r_1 \subseteq r_2 \subseteq \cdots \subseteq r_6$$

Since the relations are contained in one another, if texts are equivalent with respect to r_i, then for every j such that $i \leq j \leq 6$ the relation r_j holds too. However if two text are not equivalent with respect to r_i, they are not equivalent with respect to r_j ($0 \leq j \leq i$) either. Therefore to determine which of the seven relations occur between two texts and which do not, it is enough to find a number i ($0 \leq i < 6$) such that the relation $r_i + 1$ is occurs between these texts, but the relation r_i does not. In such a situation we say that both texts are equivalent on the $i + 1$ level. In the extreme cases both texts may be equivalent with respect to the relationship r_0 (it means that all seven relations hold and texts are equivalent on the zero level) or may be not equivalent with respect to the relationship r_6, what means that any of the relations does not hold at all.

3   Examples of Equivalent Texts

Below I show fragments of MIZAR texts equivalent with respect to all of the seven relations to better represent the essence of the relations described above. They show the modifications that preserve the relation with the original text. The fragment of the MIZAR text presented below is taken from the article called XBOOLE_1 (from the MML database) and it shows an elementary theorem from set theory. For simplification of further descriptions, the fragment below will be called THO:

```
theorem
  for X,Y,Z be set holds X c= Y implies X \/ Z c= Y \/ Z
proof
  let X,Y,Z be set;
  assume A1: X c= Y;
  let x be set; assume x in X \/ Z;
  then x in X or x in Z by XBOOLE_0:def 2;
  then x in Y or x in Z by A1,TARSKI:def 3;
  hence thesis by XBOOLE_0:def 2;
end;
```

Beginning with the strongest relation, the text that should be in the relation r_0 with TH0 has to look exactly like TH0.

The relation r_1 requires the syntactic equivalence, so for example the text which is in the relation r_1 with TH0 may look as follows:

```
theorem
 for X , Y , Z be set holds
  X c= Y implies X\/Z c= Y\/Z
proof
 let X , Y , Z be set;
 assume A1: X c= Y;
 let x be set;
 assume x in X\/Z;
 then x in X or x in Z by XBOOLE_0:def 2;
 then x in Y or x in Z by A1,TARSKI:def 3;
 hence thesis by XBOOLE_0:def 2;
end;
```

According to the definition, differences may concern only white spaces between consecutive words, and the line breaks. It is obvious that the text above is not in the relationship r_0 with TH0.

The text which is equivalent with TH0 with respect to the relation r_2 (but is not equivalent with respect to r_1) may look as follows:

```
theorem
 for A , B , C be set holds
  A c= B implies A\/C c= B\/C
proof
 let A , B , C be set;
 assume et1: A c= B;
 let a be set;
 assume et2: a in A\/C;
 a in A or a in C by et2,XBOOLE_0:def 2;
 then et3: a in B or a in C by et1,TARSKI:def 3;
 thus thesis by et3,XBOOLE_0:def 2;
end;
```

As you can see the names of variables changed (A, B, C instead of X, Y, Z) as well as the references to local statements. All local references in the text THO were realized with the aid of the keywords "then" and "hence". In this case, however, we have references through labels. It is obviously in accordance with the definition of r_2 relationship.

The text below is not in the relation r_2 with the text THO, but it is so with respect to the next relation – r_3:

```
theorem
 for A , B , C be set holds
  A c= B implies A\/C c= B\/C
proof
 let A , B , C be set;
 assume et1: A c= B;
 let a be set;
 assume et2: a in A\/C;
 a in A or a in C by XBOOLE_0:def 2,et2;
 then et3: a in B or a in C by TARSKI:def 3,et2,et1;
 thus thesis by et3,XBOOLE_0:def 2;
end;
```

It is the "weak" syntactic equivalence excluding justifications. The differences in justifications lie in the change of the order of references in two cases, and in one case (in the sentence labeled et3) in the adding of the redundant reference labeled et2. It causes the change of the reference tree (there appears the extra edge linked the sentences et2 and et3; this edge is not necessary to justify sentence et3), but the proof is accepted by the MIZAR VERIFIER all the time. As you can see the conditions given in the definition of the relation r_3 are fulfilled. It is possible to imagine (although the example above does not show that situation) that instead of some – local or external – references there can appear another one, to a completely different sentence (for example to some other theorem from the MML database). But the inference with that reference is sufficient to justify current sentence. In that situation the tree of references is changed more significantly than previously. Instead of an edge, there appears another one which can lead to the theorem from other article from the MML database. However it does not affect the r_3 relation between given text and the text THO.

The text below does not show the relation r_3. It does show the relationship r_4, which requires "weak" syntactic equivalence between the abstracts of the MIZAR texts:

```
theorem
 for A , B , C be set holds
  A c= B implies A\/C c= B\/C
proof
 let A , B , C be set;
 assume that
  et1: A c= B and
  et2: not A\/C c= B\/C;
```

```
consider a be set such that
 et3: a in A\/C and
 et4: not a in B\/C by et2,TARSKI:def 3;
 a in A or a in C by et3,XBOOLE_0:def 2;
 then a in B or a in C by et1,TARSKI: def 3;
 then a in B\/C by XBOOLE_0:def 2;
 hence contradiction by et4;
end;
```

After removing all proofs from both above texts, they are equivalent with regard to the relation r_2, therefore the whole texts are equivalent with regard to the relation r_4. The equivalence r_3 does not occur, because there was a completely different kind of proof used (proof by contradiction). In the tree of references, not only edges are changed, but also the points. However this relation allows to change proofs freely, i.e. the form, the length and also the concepts used.

In the definition of r_5 the word "syntactic" does not occur anymore. In its place there occurs the word "semantic" (as we mentioned before, the semantic equivalence of theorems is understood here as the mutual justifiability of the theorems by one another). In the case of the relation r_4 the changes of the form of proofs constitute significant weakening of the relation. However considering the texts' semantics instead of their syntax gives a completely different (much weaker) meaning to the essence of the texts equivalence. The text below is in the relation r_5 (not in the relation r_4) with the text THO:

```
reserve A,B,C for set;

theorem
 not (A\/C c= B\/C) implies not (A c= B)
proof
 assume that
  et1: not A\/C c= B\/C and
  et2: A c= B;
 consider a be set such that
  et3: a in A\/C and
  et4: not a in B\/C by et1,TARSKI:def 3;
 a in A or a in C by et3,XBOOLE_0:def 2;
 then a in B or a in C by et2,TARSKI: def 3;
 then a in B\/C by XBOOLE_0:def 2;
 hence contradiction by et4;
end;
```

Here we can see the change in the statement of the theorem and its proof, the type of variables A, B, C is reserved before the beginning of the theorem block. However the information carried by this theorem is the same as the information carried by THO. It means that every reference to the above theorem may be replaced with a reference to THO (and vice versa) without the loss of the correctness of the proved sentences. In more complex contexts, it is also possible that the order of proved theorems in some article can be changed. It is only required that every correctly checked theorem should have its equivalent in the second text, and vice versa. Please note that it does not necessarily entail the same number of theorems in both texts, e.g. when there are, say, two equivalent theorems in the first text, and there is just one of them in the second text.

There is a problem with illustrating the relation r_6 using the same example, because it is rather indistinguishable from the relation r_5 with reference to THO since the text THO consists of one theorem only. It means that a lot of texts that are equivalent with respect to r_6, are also equivalent with respect to r_5. However we can consider a text which consists of n theorems, where all theorems have the form of equivalences, and the second text which includes $2 \cdot n$ theorems (every equivalence is changed into two implications). Obviously, the relation r_5 does not hold in this case, but r_6 does. This is one of the simplest cases. There are situations in which there are two semantically equivalent texts in which the first text has n theorems, the second text has m $(n \neq m)$, and any two theorems from those texts are not semantically equivalent (and they are not semantically included in one another).

4   Article Transformations Corresponding to the Equivalence Relations

Having described the above definitions of equivalence relations, we can look at the auxiliary software used for the management of the MML database. Through the analysis of the equivalence between the original and the resulting text we can determine which level of equivalence connect both texts and which relations hold there. The level of equivalence indicates how high is the level of alteration of the modifying software to the original text. When this level is closer to zero, the alteration of that application to the original text gets smaller. The equivalence on the zero level means that the application does not change the original text at all.

Below I will describe the level of alteration for several existing auxiliary applications which are used to modify the Text-Proper part of MIZAR articles from the MML database. Most auxiliary MIZAR applications only determine the parameters which characterize specific transformations – other software is used to actually modify the text of articles on the basis of previously determined parameters. For example, the application called CHKLAB writes to a file the coordinates of unused labels, and then the application called EDTFILE removes those labels from the original text (using the file made by CHKLAB as input). For simplicity, we may assume that it is the auxiliary application that itself modifies the original text (for example the application CHKLAB removes unused labels). That simplification makes much clearer the considerations of equivalences between the original and the resulting text. It certainly should not cause ambiguity.

5 Auxiliary Software Used for Managing the MML Database as Transformations Corresponding to the Equivalence Relations

The CHKLAB application mentioned above checks the use of labels and removes labels which are not used. The text after removing redundant labels does not have to be syntactically equivalent to the original text (it is not in the relation r_1 with the original text). However they are in the relation r_2 (weak syntactic equivalence). A similar situation takes place in the case of application called RENTHLAB. It is the application which "normalizes" the names of labels by changing them by means of a fixed pattern (theorems receive names Th1, Th2... local labels A1, A2... etc.). The change of label names is carried out consistently, new names appear in the place the original labels occurred, as well as in the places where there are references leadings to these labels. All the changes concern only the names of labels, so (similarly as in the case above) the level of alteration of that application to the original text is 2.

The situation in the case of applications like RELPREM or SORTREF is different. The first application removes redundant references from every inference. The relation r_2 is not fulfilled in this case, because even the removing of one reference causes the change of the references structure. It is not significant in the case of the relation r_3 which allows differences in justifications. The situation with the application called SORTREF is similar. It does not change the number and the kind of references, but it can change their order. Its purpose is the alphabetical sorting of references in every justification. As the result, we have the lack of equivalence on the second level. In this case the level of alteration to the original text is 3.

The level of alteration to the original text showed by applications like INACC, RELINFER, RELITERS or TRIVDEMO is even higher. The first program is designed to enable the elimination of redundant sentences. In this case redundant means that a sentence is not necessary for the proof skeleton and there are not any references that point to that sentence. After removing these sentences the proof becomes significantly different – the relation r_3 does not hold, now we can think only about equivalence of abstracts of the MIZAR texts (the relation r_4 is fulfilled). The application called RELINFER removes proof steps which are unnecessarily stated as individual steps (all reference lists referring to the removed sentence are extended by all references used in the removed statement). The application called RELITERS works in a similar way to RELINFER, but it analyzes only consecutive steps of the iterative equality. Both applications can change proofs essentially, but both do not destroy the image of abstracts (it is the fourth level of alteration to the original text). A similar situation takes place in the case of the application called TRIVDEMO. It is aimed at finding situations where reasoning in a given proof-end block can be replaced with a simple justification. Such a significant alteration to the form of justification (whole proof-end block containing many steps may be removed) does not violate the form of the abstract of the article.

The table below presents the levels of alteration to the original text of the applications discussed above:

	r_0	r_1	r_2	r_3	r_4	r_5	r_6
CHKLAB	-	-	+	+	+	+	+
INACC	-	-	-	-	+	+	+
RELINFER	-	-	-	-	+	+	+
RELITERS	-	-	-	-	+	+	+
RELPREM	-	-	-	+	+	+	+
RENTHLAB	-	-	+	+	+	+	+
SORTREF	-	-	-	+	+	+	+
TRIVDEMO	-	-	-	-	+	+	+

It is obvious that any useful application does not return text which is equivalent to the original text on the zero level. An example of an application with the first level of alteration will be described in the next section (as the new auxiliary software). Basic tools used for the development of the MML database do not change the essence of abstracts, all the more they do not change the semantics of proved sentences, so in the table above all applications have the level of alteration to the original text not bigger than four.

6   New Auxiliary Applications as the Transformations Corresponding to the Equivalence Relations

As it was mentioned above, the research on data-mining and the robustness of the MIZAR system [2] have given rise to new auxiliary applications. Some of them may be used to improve the existing texts (for example REMEQTH). Others may prepare text for carrying out additional experiments or for using them with yet other applications (e.g. DELINKER).

The routine use of applications introducing changes to formalized text depends on their level of alteration to the original text. The higher the level, the more different the resulting text is from the original text and the decision whether to apply that application (e.g. on the whole MML database) seems for the maintainers more difficult to make.

Below I will describe the level of alteration to the original text of new auxiliary applications which were created during the testing of the robustness of the MIZAR system.

Two simple applications: UNHEREBY which changes all occurrences of the word "hereby" into "thus now", and TOHEREBY which realizes the opposite operation by changing all phrases "thus now" (but only the ones that are not separated by a label) into "hereby", are applications which do not change the tree of references. But they change the content of proofs outside justifications, therefore the text which is the result of their work is not equivalent to the original text with respect to the relation r_4. The comparison of abstracts of both texts (before and after running these applications) does not show differences, therefore both applications have the fourth level of alteration to the original text.

The problem with shortened form of external references [2] is solved by two applications: SEPREF and UNSEPREF. SEPREF changes all short forms into full forms, and UNSEPREF realizes the opposite operation – changes full into shortened forms. Both applications can change the syntactic image of justified inferences, but they do not have any influence on the shape of the references tree. Therefore they fulfill the conditions given in the definition of the r_2 relation, which means that they alter the original text on the second level.

Another pair of applications was created to solve the problem of changing references realized through the keywords "then" and "hence" into the references realized through labels (and vice versa). The application called LINKER finds all references by a label into the directly preceding sentences and changes them into references by "then" (or "hence" in the case when the thesis is affirmed). This operation concerns only the organization of the given inference, but it can lead to the change in the order of references in the given inference. The reference realized by "then" or "hence" is treated as the first reference in the given inference. If the label which pointed to the directly preceding statement before running the application LINKER was not in the first place in the inference, then after running LINKER the order of references will be changed. Just this situation is the cause of classifying LINKER as an application which alters the original text on the third level. The second application called DELINKER removes references by the words "then" and "hence" and changes them into references by a label. DELINKER alters the original text even on the higher level than LINKER. It is possible that in a justification there exists a reference by the keyword "then" into the block of local reasoning, which begins with the keyword "hereby". It is not possible to label that block in the current version of the MIZAR system. The solution to this problem is to change the occurrences of keyword "hereby" into the phrase "thus now" (it is the task carried out by the application called UNHEREBY), and set the label between "thus" and "now". All these operations must be executed by DELINKER and it is the reason why it is considered among applications which alter the original text on the fourth level.

The following applications: FORMATER and LINE80 do not change the text syntactically. FORMATER reorganizes the text according to specific rules, without changing its syntax. There are two things that can be changed. There are: the number of white spaces between consecutive words and the place of breaking the lines. This application is not used to edit the official version of the MML database, but it is useful for carrying out various experiments. Thanks to FORMATER it is possible to determine precisely how many changes to the text size are introduced by a certain experiment (by invoking that application before and after the experiment). It is possible that after executing FORMATER the length of some lines will be more than 80 characters. The MIZAR system in the standard mode restricts the line length to 80 characters. This problem is solved by the application called LINE80 which breaks all lines which are too long. Because of their very editorial character, the alteration to the original text of both these applications is on the first level.

There is also an application which has somewhat larger level of alteration to the original text. It is called EVERYLAB and it changes names of all labels and adds labels everywhere where it is possible. EVERYLAB is used for some experiments. It does not violate the tree of references, so it satisfies the conditions described in the definition of r_2.

Some applications were created solely with the aim to perform certain experiments on the MML database. There is for example an application called PERMREF. It is the application which establishes the order of references at random. PERMREF is used in the experiment of testing permutability of references in the MIZAR system [2]. Similarly to the PERMREF there is an application called MONOTONE which adds one random reference from among all references available at that moment of parsing the document. That process allows for carrying out the experiment of testing monotonicity of references in the MIZAR system. Thanks to these experiments it was possible to find out and to correct several situations in which the MIZAR system was not robust (it was not permutable or monotone) [2]. Another application called CHKLPREM removes redundant references (it starts checking from the first element of the list of references). It is used in the experiment which adds redundant references at the end of the list of references, and next "cleans" this list with the help of the CHKLPREM application. That experiment shows some situations in which after adding one reference, more than one reference can be removed from the list instead. All these three applications described above change only the organization of inferences, but they may influence the tree of references. Therefore their alteration to the original text is on the third level.

Another application which is designed for testing the robustness of the MIZAR system is REMEQTH. It finds out and removes all repetitions of the semantically equivalent sentences. It is possible that there are justified two or more semantically equivalent sentences in one article or in one proof (unfortunately it is quite a frequent situation in the MML database). The application REMEQTH removes all of them, except the first one of course. It is possible that this operation will cause differences in abstract of both texts, and it classifies the REMEQTH as application with alteration to the original text on the fifth level (but it is still true that every statement justified in one of these texts has a semantic equivalent in the second one).

The last application that I wish to discussed is DELOCAL, which moves every justified sentence onto as high as possible level of the proof. Every sentence is pulled out of its original block and moves onto the higher level when it does not depend on any variable and any label defined in the current block. This process finally allows for finding a larger number of semantically equivalent sentences while using the REMEQTH utility. Some sentences are located in different sub-blocks and therefore they are not comparable. It is possible that after delocalization (the output of DELOCAL) such sentences may be located

on a common level and it will make it possible to remove all repetitions. The
structure of proofs may be changed during this process, but it cannot change
the abstracts of processed texts. Therefore the application DELOCAL alters the
original text on the fourth level.

The table below presents the levels of alteration to the original text of the
applications discussed in this section:

	r_0	r_1	r_2	r_3	r_4	r_5	r_6
CHKLPREM	-	-	-	+	+	+	+
DELINKER	-	-	-	-	+	+	+
DELOCAL	-	-	-	-	+	+	+
EVERYLAB	-	-	+	+	+	+	+
FORMATER	-	+	+	+	+	+	+
LINE80	-	+	+	+	+	+	+
LINKER	-	-	-	+	+	+	+
MONOTONE	-	-	-	+	+	+	+
PERMREF	-	-	-	+	+	+	+
REMEQTH	-	-	-	-	-	+	+
SEPREF	-	-	+	+	+	+	+
TOHEREBY	-	-	-	-	+	+	+
UNHEREBY	-	-	-	-	+	+	+
UNSEPREF	-	-	+	+	+	+	+

7   The Final Remarks

Relations described in this paper and various levels of alteration to the original text determined for some auxiliary applications used for modifying the
MML database elements determine a classification of these applications. The
higher the level of alteration, the more serious modifications of the text, and
then the greater difference in comparison to the original text. This factor shows
how useful the auxiliary applications are. But on the other hand, the application which appears to be more useful, has a smaller chance of being practically
used in the revision process or in modification of the whole MML database. It
is caused by the respect for the original content and form created by the authors of articles. This principle is not obeyed only if the Library Committee of
MML decides that there are really serious reasons for such actions. Therefore
the characterization of the level of alteration to the original text for auxiliary
applications which take part in a planned modification of the MML database
will allow to make a more well-thought-out decision: "to do that modification
or not?". It is also crucial because the number of auxiliary applications con-
stantly rises, certainly it is motivated by a constant growth of the interest
in research on the data-mining and the robustness of systems for formalized
mathematics.

Acknowledgments

I thank all those who helped me with writing this paper for their valuable hints and advice. In particular I want to thank A. Naumowicz and A. Trybulec.

References

1. Ewa Bonarska: An Introduction to PC MIZAR, Fondation Ph. le Hodey, Brussels, 1990
2. Robert Milewski: Robustness of Systems for Formalizing Mathematics – Testing Monotonicity and Permutability of References in MIZAR, Mechanized Mathematics and Its Applications, 4, pp. 51–58, 2005
3. Michał Muzalewski: An Outline of PC MIZAR, Fondation Philippe le Hodey, Brussels, 1993
4. Piotr Rudnicki: An Overview of the MIZAR Project, Proceedings of the 1992 Workshop on Types for Proofs and Programs, Chalmers University of Technology, Bastad, 1992
5. Piotr Rudnicki, Andrzej Trybulec: On Equivalents of Well-foundedness. An experiment in MIZAR Journal of Automated Reasoning, 23, pp. 197–234, Kluwer Academic Publishers, 1999
6. Andrzej Trybulec: Some Features of the MIZAR Language, ESPRIT Workshop, Torino, 1993
7. Freek Wiedijk: Comparing Mathematical Provers, Proceedings of the Second International Conference on Mathematical Knowledge Management, LNCS 2594, Springer-Verlag, pp. 188–202, 2003
8. Freek Wiedijk: MIZAR: An Impression, http://mizar.org/project/mizarintro.ps
9. MIZAR Home Page, http://mizar.org/
10. MIZAR Lecture Notes, 4th Edition, Shinshu University, Nagano, 2001 http://markun.cs.shinshu-u.ac.jp/kiso/projects2/proofchecker/mizar/Mizar4/index-e.html
11. MIZAR Users Guide in Japanese, Shinshu University, Nagano 1994, English translation: http://markun.cs.shinshu-u.ac.jp/kiso/projects2/proofchecker/mizar/Mizar-E/Miz-etit.htm

Translating a Fragment of Weak Type Theory into Type Theory with Open Terms

G.I. Jojgov

Technische Universiteit Eindhoven,
The Netherlands
G.I.Jojgov@tue.nl

1 Introduction

One of the main application areas of interactive proof assistants is the formalization of mathematical texts. This formalization not only allows mathematical texts to be handled electronically, but also to be checked for correctness. Due to the level of detail required in the formalization, formalized texts eliminate ambiguities that may be present in an informally presented mathematical texts.

The process of formalization begins with such an informal text given either on paper or kept in the author's mind. Then this text needs to be written down in a computer-processable formal language and fed into a theorem prover. In this theorem prover all the details of the proofs in the text need to be fully worked out. Once this is done, we have a completely formalized version of the original mathematical text. The main *problem* when formalizing is how to manage the complexity of the details needed for a completely formal proof while at the same time ensuring maximum reliability of the whole formalization process.

This paper is a part of an ongoing effort to study the whole process of formalization starting from the informal document and ending with the full formalization. Our approach to formalization is to use several intermediate steps designed to reduce the possibility of introducing errors during formalization and at the same time to explore the opportunities for computer assistance.

The contribution of this paper is the description of a set of formal rules that allow us to automatically make one of these steps - namely the step from our vernacular language Weak Type Theory (WTT) to the language of a theorem prover (Type Theory). We also present an implementation of these rules and report some early results from experiments with it.

1.1 The Formalization Path

We refer to the stages one goes through while formalizing a piece of mathematical text as the *formalization path*. It is clear that this path in real situations may pass through some of the stages several times due to finding and correcting errors made in earlier stages. But since the iterative nature of the formalization path is not the focus of attention in this article, we assume the simplest case when the path goes through each stage only once.

M. Kohlhase (Ed.): MKM 2005, LNAI 3863, pp. 389–403, 2006.

At the first stage we start with the informally presented mathematical text and write it down in a formal language called Weak Type Theory (see section 1.2). This language is designed to express a rich mathematical base while imposing minimal correctness and well-formedness criteria on the texts written in it. The advantage is that we have the text in a formal (i.e. machine-processable) form while remaining close to the original text because we didn't have to satisfy strict correctness criteria. Being close to the original is very important because it helps the author find any errors in this crucial (with respect to reliability) first stage. This first formalization stage is interesting subject of study by itself (see [6]).

The second stage on the path that we consider in this paper is to translate a text written in Weak Type Theory into Type Theory with open terms. The Type Theory with open terms is an extension of the target theorem prover's type theory with meta-variables that represent unknown terms. The open terms are needed to represent the parts of the proofs that are implicit in the informal text and which need to be provided to get a full formalization.

This second stage in effect gives semantics to the expressions in WTT by translating them into type theory. At this stage one needs to make many "formalization choices", i.e. choices that may otherwise be irrelevant to the proof (for example whether to use unary or binary representation of the naturals) but need to be made in order to get a full formalization.

The contribution of this paper is to provide a set of rules that allow automatic translation of a fragment of WTT into a type theory with open terms.

The last stage of the formalization path is to provide instantiations for all meta-variables (open places) in the incomplete formalization produced by the second stage. This process is exactly in the application domain of (interactive) theorem provers – we have a text in the language of the prover and we need to construct terms of known types in given contexts. Therefore a conventional theorem prover based on type theory is an excellent tool to conduct this last phase of the formalization path.

1.2   Formalizing Mathematical Texts Using Weak Type Theory

Weak Type Theory (abbreviated WTT) is a formal language proposed by Nederpelt [8] (its meta-theory was worked out by Kamareddine and Nederpelt in [4]) based on principles derived from the Automath Project of De Bruijn [9]. In WTT a mathematical text is structured in *books* consisting of *lines* and each line can be a *definition* or a *statement*. As suggested by its name, WTT has a very weak typing system. It is based not on a particular logic, but rather on the linguistic structure of the text. The weak types are set for sets, noun for nouns, adj for adjectives, term for terms and stat for statements (propositions).

An example of a mathematical text written in WTT (using the flag notation for contexts) is given in Figure 1. In WTT one forms terms from variables, constants, binders, applications of adjectives to nouns and from the sorts SET and STAT. Constants are either atomic (provided with their weak types in a preface of the book) or defined in a definition line. The scope of the defined

$$\boxed{Positive := Adj_{x:\text{nat}}0 < x}$$

$$\boxed{p, q:\text{nat}}$$

$$divides(p, q) := \exists_{k:\text{nat}}(kp = q)$$

$$\boxed{q:\text{nat}}$$

$$divisor(q) := Noun_{p:\text{nat}}divides(p, q)$$

$$proper(q) := Adj_{p:divisor(q)}p \neq q$$

$$perfect := Adj_{n:\text{nat}}(\textstyle\sum_{p:(proper(n)\ divisor(n))} p) = n$$

$$prime := Adj_{n:\text{nat}}(1 < n) \wedge \forall_{p:divisor(n)}(p = 1 \vee p = n)$$

$$\forall_{k:(Positive\ \text{nat})}((2^k - 1)\ \text{is}\ prime) \Rightarrow (2^{k-1}(2^k - 1)\ \text{is}\ perfect)$$

Fig. 1. An example of a mathematical text written in WTT

constants starts after the line in which they are defined and extends to the end
of the book (in practical applications one would like of course to have the freedom
to introduce constants locally, but here we do not consider this option). Below
we present the grammar of the fragment of WTT that we consider in this paper.

$$
\begin{array}{lll}
\text{term} & \mathcal{T} ::= x \mid c[\vec{\mathcal{P}}] \mid \lambda_{\mathcal{Z}}\mathcal{T} \\
\text{adjective} & \mathcal{A} ::= Adj_{\mathcal{Z}}(\mathcal{S}_p) \mid c[\vec{\mathcal{P}}] \\
\text{noun} & \mathcal{N} ::= c[\vec{\mathcal{P}}] \mid Noun_{\mathcal{Z}}(\mathcal{S}_p) \mid \mathcal{A}\mathcal{N} \\
\text{set} & \mathcal{S}_s ::= x \mid c[\vec{\mathcal{P}}] \mid Set_{\mathcal{Z}}(\mathcal{S}_p) \\
\text{statement} & \mathcal{S}_p ::= x \mid c[\vec{\mathcal{P}}] \mid \mathcal{S}_p{\rightarrow}\mathcal{S}_p \mid \forall_{\mathcal{Z}}(\mathcal{S}_p) \mid \exists_{\mathcal{Z}}(\mathcal{S}_p) \mid \mathcal{T}\ \text{is}\ \mathcal{A}
\end{array}
$$

$$
\begin{array}{lll}
\text{argument} & \mathcal{P} ::= \mathcal{S}_s \mid \mathcal{S}_p \mid \mathcal{T} \\
\text{declaration} & \mathcal{Z} ::= x{:}\mathsf{SET} \mid x{:}\mathsf{STAT} \mid x{:}\mathcal{N} \mid x{:}\mathcal{S}_s \\
\text{context} & \Gamma ::= \emptyset \mid \Gamma, \mathcal{Z} \mid \Gamma, \mathcal{S}_p \\
\text{definition} & \mathcal{D} ::= c[\vec{x}] := (\mathcal{T}/\mathcal{S}_p/\mathcal{S}_s/\mathcal{A}/\mathcal{N}) \\
\text{line} & l ::= \Gamma \triangleright \mathcal{D} \mid \Gamma \triangleright \mathcal{S}_p \\
\text{book} & \mathcal{B} ::= \emptyset \mid \mathcal{B} \circ l
\end{array}
$$

The binders have the general form $\mathcal{B}_{\mathcal{Z}}M$ where $\mathcal{B}$ is the name of the binder,
$\mathcal{Z}$ is a declaration and M is a term, adjective, noun, set or a statement. WTT
introduces a number of binders like the usual $\forall_{x:A}(Px)$, $\exists_{x:A}(Px)$, $\lambda_{x:A}(Px)$,
but also binders to create adjectives $Adj_{x:A}(Px)$, nouns $Noun_{x:A}(Px)$ and sets
$Set_{x:A}(Px)$, etc. For an extensive presentation of WTT and more examples we
refer the reader to [4]. Unless explicitly stated otherwise, the word 'term' in the
context of WTT refers to any expression of the categories $\mathcal{T}$, $\mathcal{A}$, $\mathcal{N}$, $\mathcal{S}_s$ or $\mathcal{S}_p$. If
we want to say that t is in $\mathcal{T}$ we will write that t has weak type $\mathtt{term}$.

1.3 Type Theory with Open Terms

As we mentioned in Section 1.1 our goal is to translate (a fragment of) WTT
into a type theory with open terms. In order to fix the setting, we will assume

that our target type theory is λHOL which is a type-theoretic presentation of higher-order logic. The introduction of open terms to λHOL is done by allowing meta-variables in the syntax of the terms. Each meta-variable is declared by a name m, a list of its parameters $x_1{:}A_1 \ldots x_n{:}A_n$ and a type A and the declaration is denoted by $m[x_1{:}A_1 \ldots x_n{:}A_n]{:}A$. Instances of meta-variables are formed by providing actual arguments $M_1 \ldots M_n$ for the formal parameters $x_1 \ldots x_n$ that match the declared types: $m[M_1 \ldots M_n]$.

The intuition is that a meta-variable $m[x_1{:}A_1 \ldots x_n{:}A_n]{:}A$ represents an unknown term of type A in a context $x_1{:}A_1 \ldots x_n{:}A_n$. When a solution for the meta-variable is found, its declaration can be converted to a definition ([7]): $m[x_1{:}A_1 \ldots x_n{:}A_n]{:}=M{:}A$.

The term language of λHOL is given by the following

$$T ::= x \mid \mathsf{Prop} \mid \mathsf{Type} \mid \mathsf{Kind} \mid m[\vec{T}] \mid TT \mid \lambda x{:}T.T \mid \Pi x{:}T.T$$

Due to space restrictions, we refer the reader to [2, 3] for more examples, the typing rules and the properties of the extension with open terms. Here we will only mention that all standard properties of the typing system extend without great problem to the case of open terms.

2   The Translation

In this section we discuss the translation of a fragment of WTT into λHOL with open terms. As we already mentioned this means giving semantics of the different elements (terms, declarations, lines, etc.) of WTT by providing meaning for them in the type theory. We need to state here however that such a translation can never be complete in the sense that every weakly well-typed term is translated into a strongly well-typed term. This is natural because the weak typing of WTT cannot reject all (strong) typing errors like for example the application of a non-function to an argument.

2.1   Semantic Model

In a translation we need to specify how each WTT element is interpreted in Type Theory. This can be done in different ways and choosing the right one is not straightforward since it may depend on external factors like for example libraries of the theorem prover, etc.

The model chosen in this article is as follows:

A **book** is translated into a context containing declarations of the global symbols from the preface of the book, followed by declarations of meta-variables and definitions. These are added consecutively to the context by translating the lines in the book.

A **definition line** $\Gamma \vdash c := M$ is translated into a definition $c[\Delta]{:}=N{:}A$ where Δ is the translation of the context Γ and N is the translation of M and A is its type. A **statement line** $\Gamma \vdash \phi$ is translated into a meta-variable declaration

$m[\Delta]{:}\psi$ where Δ is the translation of Γ and ψ is the translation of ϕ. During the translation of the contexts and terms in a line we may need to introduce new meta-variables that are also added to the context.

The translation of a WTT context is not reduced to translating the type of the declared variables, or at least this is not straightforward. Instead, we declare by mutual induction the translation of WTT **terms** and **declarations**. The reason for the separation is that λHOL does not support comprehension on the level of types. This means that expressions like *Positive nat* or $Set_{x:nat}0 < x$ cannot be translated into λHOL types. Fortunately, declarations of variables of those types can be translated. For example $x{:}Positive\ nat$ is translated into $x{:}nat, h{:}(Positive\ x)$ and $x{:}(Set_{y:nat}0 < y)$ is translated into $x{:}nat, h{:}(0 < x)$.

A WTT context contains variable declarations of the form $x{:}A$ and assumptions of the form ϕ where ϕ is a statement (i.e. has a weak type stat). As the examples above show the translation of a declaration may result in more than one λHOL-declaration.

Furthermore, the translation of WTT-terms may require the introduction of extra meta-variables. An example when we need to introduce a meta-variable during the translation is given in Figure 2. In general, meta-variables appear

---

$\boxed{m, n : \mathsf{nat}}$

$\quad \boxed{m \neq 0}$

$\quad divides(m, n) := \exists_{k:\mathsf{nat}}(n = km)$

$\quad positive := Adj_{n:\mathsf{nat}}(0 < n)$

$\quad \boxed{y : positive\ \mathsf{nat}}$

$\quad divides(y, 2y)$

Fig. 2. The translation of the last line requires the introduction of a meta-variable for the proof-obligation in the context of *divides*

---

because some of the declarations in contexts are split into multiple declaration in type theory during the translation. Most of the meta-variables will turn out to be standing for unknown proof objects, but this is not always the case.

The rules for the translation of declarations and terms define by mutual induction two relations:

$$\Gamma_1 \vdash x{:}A \longrightarrow \Gamma_2 \vdash \vec{y}{:}\vec{B}$$

and

$$\Gamma_1 \vdash M \longrightarrow \Gamma_2 \vdash N$$

where the first judgment should be read as: The translation of the WTT-declaration $x{:}A$ in the λHOL-context Γ_1 results in the λHOL-declarations $\vec{y}{:}\vec{B}$ in λHOL-context Γ_2. The second judgement similarly states that the translation of the WTT-term M (in Γ_1) is the λHOL-term N (in Γ_2).

The following section describes the rules defining the translation of declarations and terms.

2.2 Rules for the Translation of Declarations and Terms

The *Set* binder. The *Set* binder creates a set by restricting a given set with a predicate. A declaration $x{:}(Set_{z:N}\phi)$ intuitively means that x belongs to the set of all those z of type N for which ϕ holds. The rule below translates declarations of variables with *Set* as a main constructor in their type.

$$\frac{\begin{array}{c}\Gamma_1 \vdash x{:}N \longrightarrow \Gamma_2 \vdash x{:}A, \vec{y}{:}\vec{B} \\ \Gamma_2, x{:}A, \vec{y}{:}\vec{B} \vdash \phi[z/x] \longrightarrow \Gamma_3, x{:}A, \vec{y}{:}\vec{B} \vdash C\end{array}}{\Gamma_1 \vdash x{:}(Set_{z:N}\phi) \longrightarrow \Gamma_3 \vdash x{:}A, \vec{y}{:}\vec{B}, h{:}C}$$

As usual, this rule says that we can conclude

$$\Gamma_1 \vdash x{:}(Set_{z:N}\phi) \longrightarrow \Gamma_3 \vdash x{:}A, \vec{y}{:}\vec{B}, h{:}C$$

if we have already inferred that

$$\Gamma_1 \vdash x{:}N \longrightarrow \Gamma_2 \vdash x{:}A, \vec{y}{:}\vec{B}$$

and

$$\Gamma_2, x{:}A, \vec{y}{:}\vec{B} \vdash \phi[z/x] \longrightarrow \Gamma_3, x{:}A, \vec{y}{:}\vec{B} \vdash C$$

To summarize, a declaration $x{:}(Set_{z:N}\phi)$ is translated into a series of declarations $x{:}A, \vec{y}{:}\vec{B}, h{:}C$ where A will turn out to be a type representing a set and the rest are assumptions on x.

The *Noun* binder. The *Noun* binder creates nouns from existing ones using comprehension (restricting predicate) very much like the *Set* binder.

$$\frac{\begin{array}{c}\Gamma_1 \vdash x{:}N \longrightarrow \Gamma_2 \vdash x{:}A, \vec{y}{:}\vec{B} \\ \Gamma_2, x{:}A, \vec{y}{:}\vec{B} \vdash \phi[x/z] \longrightarrow \Gamma_3, x{:}A, \vec{y}{:}\vec{B} \vdash C\end{array}}{\Gamma_1 \vdash x{:}(Noun_{z:N}\phi) \longrightarrow \Gamma_3 \vdash x{:}A, \vec{y}{:}\vec{B}, h{:}C}$$

If x is to be of type $Noun_{z:N}\phi$, it needs to be of type N (which gives rise to the first premise of the rule) and then it needs to satisfy the predicate $\phi[x/z]$ (the second premise).

Function declaration. The $\rightarrow$ constant of WTT can be used to introduce (non-dependent) function types. It has two arguments A and B of weak type set and the type of $A{\rightarrow}B$ is also set. Therefore we are allowed to form function types that use sets constructed by comprehension. As we have already seen, those do not have direct translations, but can only be translated when used in declarations.

The rule for declarations of functional variables is the following:

$$\frac{\begin{array}{c} \Gamma_1 \vdash x{:}M \longrightarrow \Gamma_2 \vdash \vec{x}{:}\vec{A} \\ \Gamma_2, \vec{x}{:}\vec{A} \vdash y{:}N \longrightarrow \Gamma_3, \vec{x}{:}\vec{A} \vdash \vec{y}{:}\vec{B} \end{array}}{\Gamma_1 \vdash f{:}M{\to}N \longrightarrow \Gamma_3 \vdash f_i{:}\Pi\vec{x}{:}\vec{A}.B_i[(f_j\vec{x})/y_j]_{j<i}}$$

Example: Let positive be an adjective on the natural numbers. Suppose we want to declare a function $f{:}(\text{nat}{\to}\text{positive nat})$. Since $y{:}\text{positive nat}$ maps to $y{:}\text{nat}, h{:}(\text{positive } y)$, we have the following diagram:

$$\frac{\begin{array}{c} \vdash x{:}\text{nat} \longrightarrow \vdash x{:}\text{nat} \\ \vdash x{:}\text{nat} \vdash y{:}\text{positive nat} \longrightarrow x{:}\text{nat} \vdash y{:}\text{nat}, h{:}(\text{positive } y) \end{array}}{\vdash f{:}\text{nat}{\to}\text{positive nat} \longrightarrow \vdash \begin{array}{l} f{:}\Pi x{:}\text{nat.nat} \\ f_1{:}\Pi x{:}\text{nat.}(\text{positive } (f\ x)) \end{array}}$$

Other declarations. Except for the case of an application of an adjective to a noun (see below), the rest of the term constructors in declarations we can give compositional translation. This means that if A translates to B, then the declaration $x{:}A$ translates to the declaration $x{:}B$. This is captured by the following rule:

$$\frac{\Gamma_1 \vdash A \longrightarrow \Gamma_2 \vdash B}{\Gamma_1 \vdash x{:}A \longrightarrow \Gamma_2 \vdash x{:}B}$$

Hence, in the rest of this section we will only describe rules that define the translations of terms (i.e. the relation $\Gamma_1 \vdash M \longrightarrow \Gamma_2 \vdash N$).

The $\forall$ binder. The rule for the $\forall$ binder is the following:

$$\frac{\begin{array}{c} \Gamma_1 \vdash x{:}N \longrightarrow \Gamma_2 \vdash \vec{y}{:}\vec{B} \\ \Gamma_2, \vec{y}{:}\vec{B} \vdash A \longrightarrow \Gamma_3, \vec{y}{:}\vec{B} \vdash \phi \end{array}}{\Gamma_1 \vdash \forall x{:}N.A \longrightarrow \Gamma_3 \vdash \Pi\vec{y}{:}\vec{B}.\phi}$$

This means that to translate $\forall x{:}N.A$, we first translate the declaration $x{:}N$ into a series of declarations $\vec{y}{:}\vec{B}$. Then in a context extended with those declarations we translate the body A. The resulting term ϕ is the body of the Π abstraction to which the $\forall$ binder is translated.

Variables and sorts. The rules for translating variables and sorts are straightforward. The sort STAT in WTT is mapped to the Prop in the type theory and SET is mapped into Type.

$$\Gamma \vdash \text{STAT} \longrightarrow \Gamma \vdash \text{Prop} \qquad \Gamma \vdash \text{SET} \longrightarrow \Gamma \vdash \text{Type} \qquad \Gamma \vdash x \longrightarrow \Gamma \vdash x$$

The fact that STAT and SET are not weakly typable ensures that their only use will be in declarations since we assume that the book we translate is weakly well-typed.

Constants. As we already mentioned, due to the fact that in λHOL we cannot represent all sets directly as terms we allow set/noun comprehension only in variable declarations. However in WTT one may introduce a definition that gives a name for a set/noun and the use it later in the book. It is clear that such definitions cannot be directly represented by terms and for that reason we expand any occurrences of such definitions that occur later in the book.

Hence we may either assume that the book we are translating does not contain such definitions or we can introduce a rule like the one below to handle defined constant sets/nouns.

$$\frac{\Gamma_1 \vdash M[M_1/x_1 \ldots M_n/x_n] \longrightarrow \Gamma_2 \vdash N}{\Gamma_1 \vdash c[M_1 \ldots M_n] \longrightarrow \Gamma_2 \vdash N}$$

where the constant c is defined to be the set/noun M in context containing the variables $x_1 \ldots x_n$. The disadvantage of this rule is that it is only sound for sets and nouns that are defined in a context that contains no assumptions (see the treatment of assumptions in the other cases below).

Let c be an atomic constant (i.e. non-defined constant from the preface) or a defined constant that is not a set/noun. Let Δ be the context of length k describing the parameters of c and let $x_1{:}A_1, \Delta_1, \ldots, x_k{:}A_k, \Delta_k$ be a sequence of declarations obtained by translating Δ. Each group $x_i{:}A_i, \Delta_i$ corresponds to the ith element of Δ. Then the declaration of c in λHOL will be

$$c[x_1{:}A_1, \Delta_1, \ldots, x_k{:}A_k, \Delta_k]{:}A$$

for some A (if c is defined, A is found by inferring the type of the defining term, in case c is in the preface, it has to be provided externally). The rule below shows how constant instances are translated:

$$\Gamma_0 \vdash M_1 \longrightarrow \Gamma_1 \vdash N_1$$

$$\ldots$$

$$\Gamma_{n-1} \vdash M_n \longrightarrow \Gamma_n \vdash N_n$$

$$\overline{\Gamma_0 \vdash c[M_1 \ldots M_n] \longrightarrow \Gamma_{n+1} \vdash c[t_1, \vec{s}_1[\vec{y}], \ldots, t_k, \vec{s}_k[\vec{y}]]}$$

Note that $n \leq k$ because in WTT we do not provide arguments for the assumptions in the context. Here t_i is either a term N_j when the ith component of Γ is a declaration or an instance of a fresh meta-variable $r_j[\vec{y}]$ if it was an assumption. The terms $\vec{s}_i[\vec{y}]$ are also fresh meta-variables introduced to stand for the unknown arguments. The context of each of these meta-variables is the current context and their types are computed from the corresponding A_i or Δ_i by substituting the previous arguments in them. More formally, the context Γ_{n+1} is computed in the following way: assume that $\Gamma_n \equiv \Gamma^M, \Gamma^V$ where Γ^M declares only meta-variables and Γ^V declares the variables $\vec{y}$. Then Γ_{n+1} is Γ^M followed by declarations of meta-variables r_j or s_j standing for the elements of $x_1{:}A_1, \Delta_1, \ldots, x_k{:}A_k, \Delta_k$ that do not have a corresponding N_j. After that we concatenate the context with the declarations of the variables Γ^V.

To illustrate this rule we show the following example: Let *divides* be a predicate defined in context $\Gamma \equiv x{:}\mathsf{nat}, 0 < x, y{:}\mathsf{positive\ nat}$ where positive is an adjective on the naturals. Then the translation of Γ yields the context x:nat, h_1:$0 < x$, y:nat, h_2:(positive y). To show how this relates to the rule above, we present the instantiation of the variables from the rule:

$$
\begin{array}{llll}
x\text{:nat} & x_1 = x & A_1 \equiv \text{nat} & \Delta_1 \equiv \emptyset \\
h_1\text{:}0 < x & x_2 \equiv h_1 & A_2 \equiv 0 < x & \Delta_2 \equiv \emptyset \\
y\text{:nat}, h_2\text{:(positive } y) & x_3 \equiv y & A_3 \equiv \text{nat} & \Delta_3 \equiv h_2\text{:(positive } y)
\end{array}
$$

Suppose we need to translate the occurrence $divides[2, 4]$. Then $M_1 = 2$, $M_2 = 4$. According to the rule, $t_1 = 2$, $t_2 = r_1$, $t_3 = 4$ where r_1 is a fresh meta-variable of type $(0 < x)[2/x]$ which after the substitution is $0 < 2$. Since Δ_1 and Δ_2 are empty, there are no meta-variables $\vec{s}_1$ and $\vec{s}_2$. However Δ_3 is non-empty and we introduce a fresh meta-variable s_3^1 for its single component. The type of this meta-variable is computed by substituting 2 for x and 4 for y in Δ_3. Hence it is (positive y)$[2/x][4/y]$ which of course is (positive 4).

To sum up, the constant occurrence $divides[2, 4]$ will be translated into $divides[2, r_1[], 4, s_3^1[]]$ where $r_1[]$:$(0 < 2)$ and $s_3^1[]$:(positive 4) are fresh meta-variables standing for the proof obligations $0 < 2$ and (positive 4) as expected by the declaration of the arguments of $divides$ in Γ. The $[]$ in $r_1[]$ denote the fact that (in this particular case) the local context in which the proof of $0 < 2$ needs to be found is empty.

Adjectives. Adjectives can be created using the Adj binder. We choose to model adjectives as predicates (i.e. as terms of type $A_1 \to \ldots \to A_n \to \text{Prop}$).

$$
\frac{\begin{array}{c} \Gamma_1 \vdash x\text{:}N \longrightarrow \Gamma_2 \vdash \vec{y}\text{:}\vec{B} \\ \Gamma_2, \vec{y}\text{:}\vec{B} \vdash A \longrightarrow \Gamma_3, \vec{y}\text{:}\vec{B} \vdash \phi \end{array}}{\Gamma_1 \vdash Adj_{x\text{:}N}.A \longrightarrow \Gamma_3 \vdash \lambda\vec{y}\text{:}\vec{B}.\phi}
$$

Adjectives can be applied to nouns to produce other nouns. We cannot translate directly such applications of adjectives to nouns, but this is possible if they occur in a declaration:

$$
\frac{\begin{array}{c} \Gamma_1 \vdash x\text{:}N \longrightarrow \Gamma_2 \vdash \vec{x}\text{:}\vec{B} \\ \Gamma_2, x\text{:}\vec{B} \vdash A \longrightarrow \Gamma_3, \vec{x}\text{:}\vec{B} \vdash E \\ \Gamma_3, \vec{x}\text{:}\vec{B} \vdash E : \Pi z\text{:}P.\Pi\vec{y}\text{:}\vec{C}.\text{Prop} \end{array}}{\Gamma_1, x\text{:}AN \longrightarrow \Gamma_4, \vec{x}\text{:}\vec{B}, h\text{:}(E\,x\,\vec{m})}
$$

In this rule $\vec{m}$ are fresh meta-variables standing for the arguments y whose declarations are added to Γ_3 in order to obtain Γ_4. Here we also use type inference in λHOL in order to obtain the type of E. As an example of this rule consider the declaration

$$
x\text{:}(Adj_{y\text{:prime nat}}0 < y)\text{nat}
$$

First, the declaration y:prime nat is translated into y:nat, h:(prime y). Since $0 < y$ is mapped to itself, the adjective $Adj_{y\text{:prime nat}}0 < y$ is mapped into

$$
\lambda y\text{:nat}.\lambda h\text{:(prime } y).0 < y
$$

which is of type $\Pi y\text{:nat}.\Pi h\text{:(prime } y).\text{Prop}$

Since nat is a variable, it is translated to itself and the whole application $x{:}(Adj_{y{:}\mathsf{prime\,nat}}0 < y)\mathsf{nat}$ is hence mapped (after β-reduction) to

$$x{:}\mathsf{nat}, h{:}(0 < x)$$

with an extra proof obligation $(\mathsf{prime}\,x)$ added to the context.

There are two more term constructions in WTT that involve adjectives. These are $(t\,\mathsf{is}\,A)$ saying that the term t has the property A and $(t\,\mathsf{is}\,N)$ saying that t is a term of type N where N is a noun. Examples of such terms are $(1{+}3)\,\mathsf{is\,positive}$ or $(2 + 2)\,\mathsf{is\,positive\,nat}$.

We didn't include is as a term constructor in WTT since we can model it with two constants that take a term and an adjective or a noun to produce a statement. Their semantics is however a bit different from the one of constants and for that reason we give the rules for is here:

$$\frac{\begin{array}{c} \Gamma_1 \vdash t \longrightarrow \Gamma_2 \vdash s \\ \Gamma_2 \vdash A \longrightarrow \Gamma_3 \vdash E \\ \Gamma_3 \vdash E : \Pi x{:}U.\Pi\vec{y}{:}\vec{B}.\mathsf{Prop} \end{array}}{\Gamma_1 \vdash t\,\mathsf{is}\,A \longrightarrow \Gamma_4 \vdash (E\,s\,\vec{m})}$$

$$\frac{\begin{array}{c} \Gamma_1 \vdash t \longrightarrow \Gamma_2 \vdash s \\ \Gamma_2 \vdash x{:}N \longrightarrow \Gamma_3 \vdash x{:}U, \vec{y}{:}\vec{B} \end{array}}{\Gamma_1 \vdash t\,\mathsf{is}\,N \longrightarrow \Gamma_4 \vdash (\lambda\vec{x}{:}U.\lambda\vec{y}{:}\vec{B}.\mathsf{true})(s)(\vec{m})}$$

where in both rules Γ_4 is obtained from Γ_3 by adding the declarations of the metavariables $\vec{m}$ of types $\vec{B}[s/x]$ with local contexts containing the variables in Γ_3. The first diagram shows the rule for $(t\,\mathsf{is}\,A)$ where t is a term and A is an adjective. After translating t we translate A to a term E whose type should be of the form $\Pi x{:}U.\Pi\vec{y}{:}\vec{B}.\mathsf{Prop}$. We introduce meta-variables for each of the arguments $\vec{y}$ with appropriate types and the result of the translation is an application of the predicate E to s and the meta-variables representing the assumptions on s required by the adjective A.

The case $(t\,\mathsf{is}\,N)$ is handled in a very similar way. In this case after translating t to s we translate a declaration $x{:}N$ (for a fresh x) to a list of declarations $x{:}U, \vec{y}{:}\vec{A}$. Then we introduce meta-variables $\vec{m}$ of types $\vec{A}[s/x]$. As a result the translation of $(t\,\mathsf{is}\,N)$ is a redex that β-equals true. This "trick" is necessary to ensure that the translation of t has the right type U.

3   Example

To illustrate the rules for the translation we will show how they apply to the WTT text shown on Figure 1. The first two lines are rather straightforward, we need not do much. The WTT line

$$Positive := Adj_{x{:}\mathsf{nat}}0 < x$$

is mapped into

$$Positive := (\lambda x{:}\mathsf{nat}.0 < x) : (\Pi x{:}\mathsf{nat}.\mathsf{Prop})$$

Since the declaration $x{:}\mathsf{nat}$ in the binder *Adj* contains only the "simple" type nat, all we have to do is apply the rule for the *Adj* binder and translate it to a λ-binder.

The following line: $divides(p,q) := \exists_{k{:}\mathsf{nat}}(kp = q)$ is also easy to translate, we only map the WTT-binder $\exists$ into its type-theoretical version:

$$divides(p,q) := \exists_{k{:}\mathsf{nat}}(kp = q) : \mathsf{Prop}$$

The next line is a definition of a the noun *divisor*. As we mentioned earlier, nouns can only be translated in declaration positions. We cannot translate defined complex nouns and instead we unfold the definition in the places where it occurs in the book. Therefore, the next line from the example is

$$proper(q) := Adj_{p{:}Noun_{p{:}\mathsf{nat}}divides(p,q)}p \neq q$$

According to the rules for translating adjectives, we first have to translate the declaration $p{:}Noun_{p{:}\mathsf{nat}}divides(p,q)$ in context $q{:}\mathsf{nat}$.

$$\frac{\vdash p{:}\mathsf{nat} \longrightarrow \vdash p{:}\mathsf{nat} \qquad p{:}\mathsf{nat} \vdash divides(p,q) \longrightarrow p{:}\mathsf{nat} \vdash divides(p,q)}{\vdash p{:}Noun_{p{:}\mathsf{nat}}divides(p,q) \longrightarrow \vdash p{:}\mathsf{nat}, h{:}divides(p,q)}$$

$$\frac{\vdash p{:}Noun_{p{:}\mathsf{nat}}divides(p,q) \longrightarrow \vdash p{:}\mathsf{nat}, h{:}divides(p,q) \qquad p{:}\mathsf{nat}, h{:}divides(p,q) \vdash p \neq q \longrightarrow p{:}\mathsf{nat}, h{:}divides(p,q) \vdash p \neq q}{\vdash Adj_{p{:}Noun_{p{:}\mathsf{nat}}divides(p,q)}p \neq q \longrightarrow \vdash \lambda p{:}\mathsf{nat}\lambda h{:}divides(p,q)p \neq q}$$

Here we have skipped the application of the rule for constants that does the trivial translation of $divides(p,q)$ into $divides(p,q)$. The rules above show the translation of the whole adjective once we have computed the translation of the declaration.

The translation of the next line is the most complicated one in the example. We see on Figure 3 that the line

$$perfect := Adj_{n{:}\mathsf{nat}}(\textstyle\sum_{p{:}(proper(n)\ divisor(n))} p) = n$$

generates four output lines. This has to do with the way we have chosen to formalize the $\sum$ binder. (sum f ϕ M D L) computes the sum of those $f(i)$ for which the decidable predicate ϕ holds. $?M$ is a natural number that is an upper bound on those numbers for which ϕ holds ($?L$) and $?D$ is a proof that ϕ is decidable. See Section 5.2 for more details.

The last two lines are also straightforward and we will not go into further detail, except to say that the last line, which is a statement, is translated into a declaration of a meta-variable standing for the unknown proof of that statement.

$Positive := \lambda x{:}\text{nat}.0 < x : \Pi x{:}\text{nat}.\text{Prop}$

$\boxed{p, q{:}\text{nat}}$

$divides(p, q) := \exists_{k{:}\text{nat}}(kp = q) : \text{Prop}$

$\boxed{q{:}\text{nat}}$

$proper(q) := \lambda p{:}\text{nat}.\lambda h{:}divides(p, q).p \neq q$

$?M : \text{nat}$

$?D : \Pi p{:}\text{nat}.(divides(p, n) \wedge (proper(n)\ p)) \vee \neg(divides(p, n) \wedge (proper(n)\ p))$

$?L : \Pi p{:}\text{nat}.(divides(p, n) \wedge (proper(n)\ p)) \Rightarrow p \leq ?M$

$perfect := \lambda n{:}\text{nat}.(\text{sum}\ \lambda p.p\ \lambda p.divides(p, n) \wedge (proper(n)\ p)\ ?M\ ?D\ ?L) = n$

$prime := \lambda n{:}\text{nat}.(1 < n) \wedge \Pi p{:}\text{nat}.\Pi h{:}divides(p, n).(p = 1 \vee p = n)$

$?Stm_1 : \Pi k{:}\text{nat}.\Pi h{:}(Positive\ k)(prime\ (2^k - 1)) \Rightarrow (perfect\ 2^{k-1}(2^k - 1))$

Fig. 3. The translation of the WTT text from Figure 1. Some types in the definitions are omitted for readability. See Section 5.2 for explanation of the translation of $\sum$.

4   Properties

As we already mentioned in Section 2, we cannot hope that every well-typed WTT-book is translatable into a well-typed context of type theory. We could strengthen the rules above to produce only typable translations and rejecting books that do not translate to a typable context, but this will ammount to adding a typechecker for type theory to the translation rules. This approach may well be more efficient in a practical implementation, but here we separate the translation from the typechecking.

There are however properties that hold for the translation:

Property 1 (Context property).
Let $\Gamma_1 \vdash M \longrightarrow \Gamma_2 \vdash N$ or $\Gamma_1 \vdash x{:}A \longrightarrow \Gamma_2 \vdash \vec{y}{:}\vec{B}$. Then

1. (Preservation of variables) $(x{:}A) \in \Gamma_1$ if and only if $(x{:}A) \in \Gamma_2$.
2. (Weakening with meta-variables) if $(m[\Delta]{:}A) \in \Gamma_1$ then $(m[\Delta]{:}A) \in \Gamma_2$.

One can show (1) by induction on the generation of the relation $\longrightarrow$. (2) holds since none of the rules removes or changes meta-variable declarations in the context. A rule may only add new ones.

5   Implementation and Experiments

The rules describing the translation in this article have been implemented in OCAML. The implemented program can read a WTT book from file and type-check it, after which it performs the translation into a context containing meta-variables and declarations in λHOL. Such a context can then be exported into a Coq [1] script file. The prototype also includes a type inference algorithm for λHOL with open terms as such an algorithm is used in the translation rules.

5.1 Export to Coq

The export of a context to a Coq script is rather trivial. Every variable is translated in a variable declaration, every definition is mapped to a Coq definition and every meta-variable is mapped into a lemma without proof. The local variables in the definitions and the meta-variables are abstracted away, so the meta-variables $m[x_1:A_1, x_2:A_2]:B$ and $m[x_1:A_1]:\Pi x_2:A_2.B$ will be mapped to lemmata stating the same proposition.

For the sake of using the Coq library, some variables with special names are mapped into the corresponding Coq notion. Examples are the type of the natural numbers nat and operations and relations on it (plus, mult, le, etc.).

5.2 Introducing a $\sum$ Binder

Early experiments with the implementation point to some problems with this way of exporting of contexts to Coq however. One such experiment was to add a new binder $\sum$ for summation so that one is able to write down statements like:

$$\left(\sum_{p:divisor(2^{k-1}(2^k-1))} p\right) = \left(\sum_{t:divisor(2^{k-1})} \left(\sum_{s:divisor(2^k-1)} .ts\right)\right)$$

where $p{:}divisor(n)$ means that p is a divisor of n (belongs to the type of all divisors of n).

To be able to handle the general case $\sum_{x:N} .t$ where N is any term of weak type set, we define in Coq (using fixpoints) a function with the following type:

```
sum:
  forall f:nat->nat,
  forall phi:nat->Prop,
  forall M:nat,
  forall dec:(forall n:nat, {phi n}+{~phi n}),
  forall lim:(forall n:nat, (phi n)-> n<= M),
  nat
```

The idea is that (sum f ϕ M D L) represents $\sum_{\{x:nat|\phi(x)\}} f(x)$ where ϕ is a decidable predicate (D is a proof of that) and M is an upper bound on the values for which ϕ holds (L is a proof of that).

Apart from the question of whether this is a good way to formalize the $\sum$ binder, we observed the following. Every time the $\sum$ binder is encountered we introduce new meta-variables for the bound and the two proof obligations. This implies that if the same term occurs twice as a subterm in a formula, it will be translated to applications of sum with different meta-variables as arguments. The meta-variables however are exported into lemmata in Coq and this leads to non-unifiable terms in Coq. Since in Coq one cannot work directly with meta-variables, this implies that to resolve this problem we need to employ unification during the translation process. Another open path to explore in this direction is to see whether the mechanism of implicit arguments that is present in Coq can be utilized in some way to resolve the problem without having to build unification into the translation procedure.

6   Conclusions, Related and Further Work

By its nature, the translation does not produce terms that are always strongly typable. Integrating a type-checker may make an implementation of the rules more efficient, but the rules will become much more complicated.

As the reader may have noticed, we show only one property of the translation given in Property 1. There is a conceptual problem here that lies behind this apparently lack of some other 'expected' properties. Normally, when translating from one system to another, one formulates a correctness lemma for the translation which usually states that the semantics is preserved in some form. Here however we are in the case when we actually give semantics to our fragment of WTT by the presented translation.

As a future work, we need to test the implementation on more and bigger practical examples in order to evaluate and if necessary refine the translation presented in this paper. As noted by one of the referees, it might also be interesting to look at the possibilities of exporting the translated terms to Isabelle [11, 15] since there is better support for explicit manipulation of meta-variables.

Many different aspects of the problem of the transition between formal and informal presentations of mathematical texts has been studied in many different contexts. For example, Wiedijk [16] proposes to use formal proof sketches which are machine checkable presentations of math texts in which (some) proofs are omitted and Thery [14] proposes colourings as way to present the structure of proofs. Kamareddine, Maarek and Wells [5, 6] also study the process of formalizing mathematical texts and propose sophisticated extensions of WTT suitable for practical applications and show how the initial step from informal to a formal language can be facilitated by a computer. Another aspect of the same problem is which language is a good vernacular for expressing mathematics on the computer. Here we use WTT, but there are other approaches as for example the declarative proof styles of Mizar [13] and Isar [10]. We are aware of the work on translations between different formal systems in the Grammatical Framework (GF) of Ranta [12]. The main conceptual difference of what we do here is that GF performs semantics-preserving translations, while in our case the translation actually gives the semantics of the WTT text.

References

1. The Coq Development Team. *The Coq Proof Assistant Reference Manual – Version V8*, 2004. http://coq.inria.fr/.
2. Herman Geuvers and Gueorgui Jojgov. Open Proofs and Open Terms: a Basis for Interactive Logic. In J. Bradfield, editor, *Proceedings of CSL'02*, number 2471 in LNCS, pages 537–552. Springer, 2002.

3. G. I. Jojgov. *Incomplete Proofs and Terms and Their Use in Interactive Theorem Proving.* PhD thesis, Eindhoven University of Technology, 2004.
4. F. Kamareddine and R. Nederpelt. A refinement of de Bruijn's formal language of mathematics. *Journal of Logic, Language and Information,* 2001.
5. Fairouz Kamareddine, Manuel Maarek, and J.B. Wells. Mathlang: Experience-driven development of a new mathematical language. In *MKM Symposium 2003.* Elsevier, 2003.
6. Fairouz Kamareddine, Manuel Maarek, and J.B. Wells. Flexible encoding of mathematics on the computer. In *Proceedings of MKM 2004,* LNCS 3119, 2004.
7. Conor McBride. *Dependently Typed Functional Programs and their Proofs.* PhD thesis, University of Edinburgh, 1999.
8. R. Nederpelt. Weak Type Theory: A formal language for mathematics. Technical report, Eindhoven University of Technology, May 2002.
9. R.P. Nederpelt, J.H. Geuvers, and R.C. de Vrijer, editors. *Selected Papers on Automath,* volume 133 of *Studies in Logic and Foundations of Mathematics.* North Holland, 1994.
10. Tobias Nipkow. Structured Proofs in Isar/HOL. In *Proceedings of TYPES 2002,* LNCS 2646, 2003.
11. L. Paulson. Introduction to Isabelle. Technical report, University of Cambridge, 1993.
12. A. Ranta. Grammatical Framework: A Type-Theoretical Grammar Formalism. *Journal of Functional Programming,* 2003.
13. P. Rudnicki. An overview of the Mizar project. In *Proceedings of the 1992 Workshop on Types for Proofs and Programs,* 1992. http://www.mizar.org.
14. Laurent Thery. Colouring Proofs: A Lightweight Approach to Adding Formal Structure to Proofs. In *Proceedings of User Interfaces for Theorem Provers UITP'03,* ENTCS 103, 2003.
15. Markus Wenzel Tobias Nipkow, Lawrence Paulson. *Isabelle/HOL. A Proof Assistant for Higher-Order Logic.* LNCS 2283. Springer Verlag, 2002.
16. F. Wiedijk. Formal Proof Sketches. In *Proceedings of TYPES 2003,* LNCS 3085, 2003.

Author Index

Lecture Notes in Artificial Intelligence (LNAI)

Vol. 3620: H. Muñoz-Ávila, F. Ricci (Eds.), Case-Based Reasoning Research and Development. XV, 654 pages. 2005.

Vol. 3614: L. Wang, Y. Jin (Eds.), Fuzzy Systems and Knowledge Discovery, Part II. XLI, 1314 pages. 2005.

Vol. 3613: L. Wang, Y. Jin (Eds.), Fuzzy Systems and Knowledge Discovery, Part I. XLI, 1334 pages. 2005.

Vol. 3607: J.-D. Zucker, L. Saitta (Eds.), Abstraction, Reformulation and Approximation. XII, 376 pages. 2005.

Vol. 3601: G. Moro, S. Bergamaschi, K. Aberer (Eds.), Agents and Peer-to-Peer Computing. XII, 245 pages. 2005.

Vol. 3596: F. Dau, M.-L. Mugnier, G. Stumme (Eds.), Conceptual Structures: Common Semantics for Sharing Knowledge. XI, 467 pages. 2005.

Vol. 3593: V. Mařík, R. W. Brennan, M. Pěchouček (Eds.), Holonic and Multi-Agent Systems for Manufacturing. XI, 269 pages. 2005.

Vol. 3587: P. Perner, A. Imiya (Eds.), Machine Learning and Data Mining in Pattern Recognition. XVII, 695 pages. 2005.

Vol. 3584: X. Li, S. Wang, Z.Y. Dong (Eds.), Advanced Data Mining and Applications. XIX, 835 pages. 2005.

Vol. 3581: S. Miksch, J. Hunter, E.T. Keravnou (Eds.), Artificial Intelligence in Medicine. XVII, 547 pages. 2005.

Vol. 3577: R. Falcone, S. Barber, J. Sabater-Mir, M.P. Singh (Eds.), Trusting Agents for Trusting Electronic Societies. VIII, 235 pages. 2005.

Vol. 3575: S. Wermter, G. Palm, M. Elshaw (Eds.), Biomimetic Neural Learning for Intelligent Robots. IX, 383 pages. 2005.

Vol. 3571: L. Godo (Ed.), Symbolic and Quantitative Approaches to Reasoning with Uncertainty. XVI, 1028 pages. 2005.

Vol. 3559: P. Auer, R. Meir (Eds.), Learning Theory. XI, 692 pages. 2005.

Vol. 3558: V. Torra, Y. Narukawa, S. Miyamoto (Eds.), Modeling Decisions for Artificial Intelligence. XII, 470 pages. 2005.

Vol. 3554: A.K. Dey, B. Kokinov, D.B. Leake, R. Turner (Eds.), Modeling and Using Context. XIV, 572 pages. 2005.

Vol. 3550: T. Eymann, F. Klügl, W. Lamersdorf, M. Klusch, M.N. Huhns (Eds.), Multiagent System Technologies. XI, 246 pages. 2005.

Vol. 3539: K. Morik, J.-F. Boulicaut, A. Siebes (Eds.), Local Pattern Detection. XI, 233 pages. 2005.

Vol. 3538: L. Ardissono, P. Brna, A. Mitrović (Eds.), User Modeling 2005. XVI, 533 pages. 2005.

Vol. 3533: M. Ali, F. Esposito (Eds.), Innovations in Applied Artificial Intelligence. XX, 858 pages. 2005.

Vol. 3528: P.S. Szczepaniak, J. Kacprzyk, A. Niewiadomski (Eds.), Advances in Web Intelligence. XVII, 513 pages. 2005.

Vol. 3518: T.-B. Ho, D. Cheung, H. Liu (Eds.), Advances in Knowledge Discovery and Data Mining. XXI, 864 pages. 2005.

Vol. 3508: P. Bresciani, P. Giorgini, B. Henderson-Sellers, G. Low, M. Winikoff (Eds.), Agent-Oriented Information Systems II. X, 227 pages. 2005.

Vol. 3505: V. Gorodetsky, J. Liu, V.A. Skormin (Eds.), Autonomous Intelligent Systems: Agents and Data Mining. XIII, 303 pages. 2005.

Vol. 3501: B. Kégl, G. Lapalme (Eds.), Advances in Artificial Intelligence. XV, 458 pages. 2005.

Vol. 3492: P. Blache, E.P. Stabler, J.V. Busquets, R. Moot (Eds.), Logical Aspects of Computational Linguistics. X, 363 pages. 2005.

Vol. 3490: L. Bolc, Z. Michalewicz, T. Nishida (Eds.), Intelligent Media Technology for Communicative Intelligence. X, 259 pages. 2005.

Vol. 3488: M.-S. Hacid, N.V. Murray, Z.W. Raś, S. Tsumoto (Eds.), Foundations of Intelligent Systems. XIII, 700 pages. 2005.

Vol. 3487: J.A. Leite, P. Torroni (Eds.), Computational Logic in Multi-Agent Systems. XII, 281 pages. 2005.

Vol. 3476: J.A. Leite, A. Omicini, P. Torroni, P. Yolum (Eds.), Declarative Agent Languages and Technologies II. XII, 289 pages. 2005.

Vol. 3464: S.A. Brueckner, G.D.M. Serugendo, A. Karageorgos, R. Nagpal (Eds.), Engineering Self-Organising Systems. XIII, 299 pages. 2005.

Vol. 3452: F. Baader, A. Voronkov (Eds.), Logic for Programming, Artificial Intelligence, and Reasoning. XI, 562 pages. 2005.

Vol. 3451: M.-P. Gleizes, A. Omicini, F. Zambonelli (Eds.), Engineering Societies in the Agents World V. XIII, 349 pages. 2005.

Vol. 3446: T. Ishida, L. Gasser, H. Nakashima (Eds.), Massively Multi-Agent Systems I. XI, 349 pages. 2005.

Vol. 3445: G. Chollet, A. Esposito, M. Faúndez-Zanuy, M. Marinaro (Eds.), Nonlinear Speech Modeling and Applications. XIII, 433 pages. 2005.

Vol. 3438: H. Christiansen, P.R. Skadhauge, J. Villadsen (Eds.), Constraint Solving and Language Processing. VIII, 205 pages. 2005.

Vol. 3430: S. Tsumoto, T. Yamaguchi, M. Numao, H. Motoda (Eds.), Active Mining. XII, 349 pages. 2005.

Vol. 3419: B.V. Faltings, A. Petcu, F. Fages, F. Rossi (Eds.), Recent Advances in Constraints. X, 217 pages. 2005.

Vol. 3416: M.H. Böhlen, J. Gamper, W. Polasek, M.A. Wimmer (Eds.), E-Government: Towards Electronic Democracy. XIII, 311 pages. 2005.

Vol. 3415: P. Davidsson, B. Logan, K. Takadama (Eds.), Multi-Agent and Multi-Agent-Based Simulation. X, 265 pages. 2005.

Vol. 3413: K. Fischer, M. Florian, T. Malsch (Eds.), Socionics. X, 315 pages. 2005.

Vol. 3403: B. Ganter, R. Godin (Eds.), Formal Concept Analysis. XI, 419 pages. 2005.

Vol. 3398: D.-K. Baik (Ed.), Systems Modeling and Simulation: Theory and Applications. XIV, 733 pages. 2005.

Vol. 3397: T.G. Kim (Ed.), Artificial Intelligence and Simulation. XV, 711 pages. 2005.